Mechanical Properties and Performance of Engineering Ceramics and Composites IV

Mechanical Properties and Performance of Engineering Ceramics and Composites IV

A Collection of Papers Presented at the 33rd International Conference on Advanced Ceramics and Composites
January 18–23, 2009
Daytona Beach, Florida

Edited by
Dileep Singh
Waltraud M. Kriven

Volume Editors
Dileep Singh
Jonathan Salem

A John Wiley & Sons, Inc., Publication

Published by John Wiley & Sons, Inc., Hoboken, New Jersey.
Published simultaneously in Canada.

Library of Congress Cataloging-in-Publication Data is available.

ISBN 978-0-470-45752-8

Printed in the United States of America.

10 9 8 7 6 5 4 3 2 1

Contents

Preface

This volume is a compilation of papers presented in the Mechanical Behavior and Performance of Ceramics & Composites and Geopolymers and Other Inorganic Polymers symposia during the 33rd International Conference & Exposition on Advanced Ceramics and Composites held January 18-23, 2009, at Daytona Beach, Florida.

The Mechanical Behavior and Performance of Ceramics & Composites symposium addressed the cutting-edge topics on mechanical properties and reliability of ceramics and composites and their correlations to processing, microstructure, and environmental effects. The symposium included over 100 presentations representing 10 countries. Symposium topics included:

- Ceramics and composites for engine applications
- Design and life prediction methodologies
- Environmental effects on mechanical properties
- Mechanical behavior of porous ceramics
- Reliability of small scale systems
- Ultra high temperature ceramics
- Ternary compounds
- Mechanics & characterization of nanomaterials and devices
- Novel test methods and equipment
- Processing - microstructure - mechanical properties correlations
- Ceramics & composites joining and testing
- NDE of ceramic components

This meeting also marked the seventh meeting on geopolymers within an ACerS sponsored conference. This years Focused Session on Geopolymers included 22 speakers. While the session provided a lively forum for discussion, the resulting 7 papers are published in this issue. This relatively large, international gathering of geopolymer researchers was directly due to the generous financial support for speakers, provided by the US Air Force Office of Scientific Research through Dr. Joan Fuller, Program Director of Ceramic and Non-Metallic Materials, Directorate of Aerospace and Materials Science.

To organize these symposia and publish a volume requires significant time and effort from many individuals. Sincere thanks to the symposium organizers, invited speakers, session chairs, presenters, manuscript reviewers, and conference attendees for their enthusiastic participation and contributions. Finally, credit also goes to the dedicated, tireless and courteous staff at The American Ceramic Society in making this symposium a success.

DILEEP SINGH
Argonne National Laboratory

WALTRAUD M. KRIVEN
University of Illinois at Urbana-Champaign

Introduction

The theme of international participation continued at the 33rd International Conference on Advanced Ceramics and Composites (ICACC), with over 1000 attendees from 39 countries. China has become a more significant participant in the program with 15 contributed papers and the presentation of the 2009 Engineering Ceramic Division's Bridge Building Award lecture. The 2009 meeting was organized in conjunction with the Electronics Division and the Nuclear and Environmental Technology Division.

Energy related themes were a mainstay, with symposia on nuclear energy, solid oxide fuel cells, materials for thermal-to-electric energy conversion, and thermal barrier coatings participating along with the traditional themes of armor, mechanical properties, and porous ceramics. Newer themes included nano-structured materials, advanced manufacturing, and bioceramics. Once again the conference included topics ranging from ceramic nanomaterials to structural reliability of ceramic components, demonstrating the linkage between materials science developments at the atomic level and macro-level structural applications. Symposium on Nanostructured Materials and Nanocomposites was held in honor of Prof. Koichi Niihara and recognized the significant contributions made by him. The conference was organized into the following symposia and focused sessions:

Symposium 1	Mechanical Behavior and Performance of Ceramics and Composites
Symposium 2	Advanced Ceramic Coatings for Structural, Environmental, and Functional Applications
Symposium 3	6th International Symposium on Solid Oxide Fuel Cells (SOFC): Materials, Science, and Technology
Symposium 4	Armor Ceramics
Symposium 5	Next Generation Bioceramics
Symposium 6	Key Materials and Technologies for Efficient Direct Thermal-to-Electrical Conversion
Symposium 7	3rd International Symposium on Nanostructured Materials and Nanocomposites: In Honor of Professor Koichi Niihara
Symposium 8	3rd International symposium on Advanced Processing & Manufacturing Technologies (APMT) for Structural & Multifunctional Materials and Systems

Symposium 9	Porous Ceramics: Novel Developments and Applications
Symposium 10	International Symposium on Silicon Carbide and Carbon-Based Materials for Fusion and Advanced Nuclear Energy Applications
Symposium 11	Symposium on Advanced Dielectrics, Piezoelectric, Ferroelectric, and Multiferroic Materials
Focused Session 1	Geopolymers and other Inorganic Polymers
Focused Session 2	Materials for Solid State Lighting
Focused Session 3	Advanced Sensor Technology for High-Temperature Applications
Focused Session 4	Processing and Properties of Nuclear Fuels and Wastes

The conference proceedings compiles peer reviewed papers from the above symposia and focused sessions into 9 issues of the 2009 Ceramic Engineering & Science Proceedings (CESP); Volume 30, Issues 2-10, 2009 as outlined below:

- Mechanical Properties and Performance of Engineering Ceramics and Composites IV, CESP Volume 30, Issue 2 (includes papers from Symp. 1 and FS 1)
- Advanced Ceramic Coatings and Interfaces IV Volume 30, Issue 3 (includes papers from Symp. 2)
- Advances in Solid Oxide Fuel Cells V, CESP Volume 30, Issue 4 (includes papers from Symp. 3)
- Advances in Ceramic Armor V, CESP Volume 30, Issue 5 (includes papers from Symp. 4)
- Advances in Bioceramics and Porous Ceramics II, CESP Volume 30, Issue 6 (includes papers from Symp. 5 and Symp. 9)
- Nanostructured Materials and Nanotechnology III, CESP Volume 30, Issue 7 (includes papers from Symp. 7)
- Advanced Processing and Manufacturing Technologies for Structural and Multifunctional Materials III, CESP Volume 30, Issue 8 (includes papers from Symp. 8)
- Advances in Electronic Ceramics II, CESP Volume 30, Issue 9 (includes papers from Symp. 11, Symp. 6, FS 2 and FS 3)
- Ceramics in Nuclear Applications, CESP Volume 30, Issue 10 (includes papers from Symp. 10 and FS 4)

The organization of the Daytona Beach meeting and the publication of these proceedings were possible thanks to the professional staff of The American Ceramic Society (ACerS) and the tireless dedication of the many members of the ACerS Engineering Ceramics, Nuclear & Environmental Technology and Electronics Divisions. We would especially like to express our sincere thanks to the symposia organizers, session chairs, presenters and conference attendees, for their efforts and enthusiastic participation in the vibrant and cutting-edge conference.

DILEEP SINGH and JONATHAN SALEM
Volume Editors

Mechanical Properties and Performance

R & D OF ADVANCED CERAMICS ACTIVITIES IN CHINA AND SHANGHAI INSTITUTE OF CERAMICS CHINESE ACADEMY OF SCIENCES (SICCAS)

Dongliang Jiang
The State Key Lab. of High Performance Ceramics and Superfine Microstructure
Shanghai Institute of Ceramics, Chinese Academy of Sciences
1295 Ding-Xi Rd, Shanghai 200050, China

ABSTRACT

There has been a long term and sustained research and development effort in advanced ceramics in China over last few decades. In this presentation, an overview of recent research and development in the area of advanced ceramics and ceramic matrix composites will be presented. Some key results from national projects funded by Ministry of Science and Technology and National Nature Scientific Foundation of China will be provided. In addition, recent results of R & D activities in Shanghai Institute of Ceramics (SICCAS) especially in ceramic matrix composites, transparent ceramics, bio-ceramics, and mesoporous materials will also be presented.

CMC development is one of the most active research areas in SICCAS including the preparation and characterization of C_f/SiC and SiC_f/SiC composites. Various techniques such as chemical vapour infiltration (CVI), hot pressing, and liquid/vapour silicon infiltration (L/VSI) were used to fabricate high performance CMCs, depending on the different application requirements. A modified chemical vapor infiltration (CVI) process, which had been defined as temperature-pulsing CVI was successfully used to deposit SiC matrix into carbon fiber preforms. The final properties of CMCs were strongly dependent on the processing conditions and the nano-SiC particle size. Nano-SiC particulate phases impregnated into the fiber bundles show the reduced interaction between fibers and matrix. Active fillers in polymer impregnation and pyrolysis (PIP) aid to form new phases in the matrix to lower the volume shrinkage and provide improved performances for the composites. In the area of transparent ceramics, Al_2O_3, Nd-YAG, Ce-YAG, La-Y_2O_3, Si_3N_4 and AlON systems are being studied extensively. Among them, the in line transmittance of transparent Al_2O_3 has reached 70%; the average output power of 1at% Nd-YAG laser transparent ceramic with size $22\times39\times4.5mm^3$, at 500Hz is around 100W; The transmittance of Ce-YAG scintillation transparent ceramic at 550nm was 80%. The development of high quality transparent ceramics is mainly associated with the understanding of the fundamental principles underlying the advanced processing technologies and nanopowder synthesis and dispersion technology. An overview will be presented on the research and development efforts in biomaterials especially in hard implant materials such as teeth, knee, bone repair, and biocompatible coatings. In addition, some examples of recent development in mesoporous materials developed for drug delivery will also be provided. Finally, the colloidal processing processes for developing high performance ceramics were also referred.

1. BRIEF OVERVIEW OF THE RECENT RESEARCH ACTIVITY IN CHINA

Most of China's research program is funded by the Ministry of Science and Technology, including the Basic Sciences (973 Program) and High Technology Research and Development Programs (863 program). In accordance with the government's guiding principles for the national development of science and technology, the National Natural Scientific Foundation of China also provide financially supports for basic research and applied basic research, the funding is usually smaller except for the key and major ones. Some basic and application programs can also be funded by other Department, local government and companies.

In the field of materials science especially in advanced ceramics, the already funded 973 programs include nano-materials & technology; information functional ceramics, thermo-electric materials, bio-ceramics, new energy materials, C/C composites etc. In consideration of the environmental, energy and human health issues, the funded 863 program include solid oxide fuel cell (SOFC), bio-ceramics, advanced ceramics, composites etc. In recent years, the NNSFC, Chinese Academy of Sciences and local government of Shanghai put a lot of money for research on transparent ceramics, ultrahigh temperature ceramics (UHTC) and composites etc.

2. MAIN RESEARCH ACTIVITIES IN SICCAS

In this part, the major research activities of advanced ceramics in SICCAS including the CMC, transparent ceramics, and biomaterials will be addressed. The low cost, eco-friendly processing technologies for developing high performance ceramics were also referred.

1. CMC processing

Continuous fiber reinforced ceramic matrix composites, such as C_f/SiC, SiC_f/SiC composites have been widely recognized as the most promising candidates for brake disks, heat exchangers, advanced aero-engines, fusion power reactors and space usage for their outstanding characteristics including high toughness, low density, thermal and chemical stability, radiation tolerance and so on[1,2]. Especially, the C_f/SiC composites, for their relatively lower cost, larger-scale production and better thermal stability at elevated temperature, have been extensively investigated. Conventionally, continuous fiber reinforced SiC composites can be prepared by chemical vapor infiltration (CVI)[3,4], hot pressing (HP)[5,6], reaction sintering (RS)

[7], and polymer impregnation and pyrolysis (PIP)[8] etc techniques.

1.1 Temperature pulsing chemical vapor infiltration [9,10,11,12]

A novel route of temperature pulsing chemical vapor infiltration (T-pulsing CVI) was developed to prepare interfacial coatings. In the T-pulsing CVI process, thermal pulses were created by combining induction heating with water cooling, in which the power from the frequency generator was modulated by an artificial intelligence industrial controller.

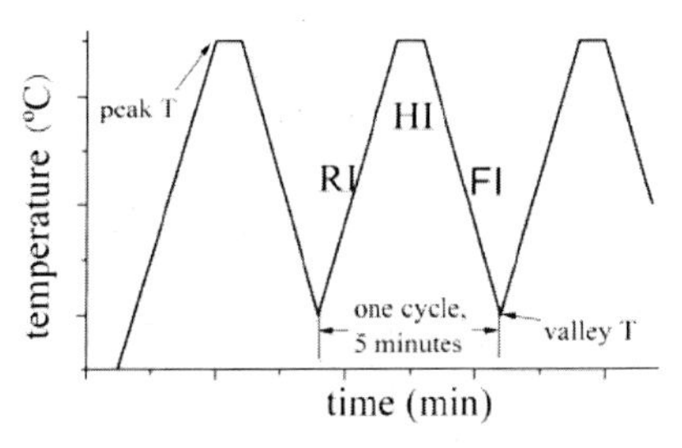

Fig.1 The temperature pulse rate curve

Fig.1 shows the temperature–time set curve from the artificial intelligence industrial controller. One cycle contained in the curve includes three steps: (i) linear temperature rising (RI), (ii) temperature holding (HI), and (iii) linear temperature falling (FI). Methane and MTS were employed as precursors of PyC and SiC, respectively. The hydrogen to MTS mole ratio was maintained at 10, and the flux of CH_4 was 10 $cm^3 \cdot min^{-1}$. Argon was used as dilute gas.

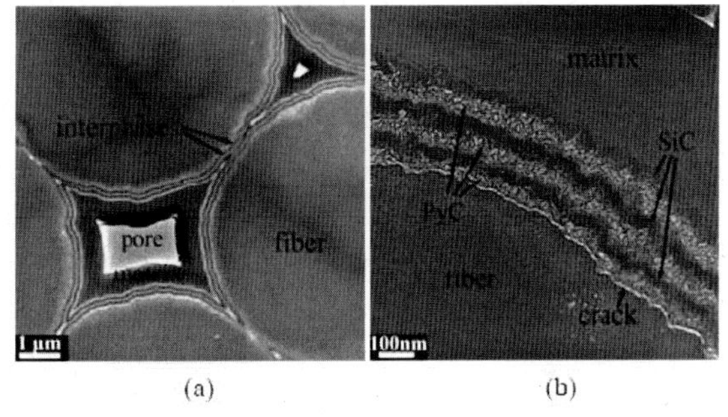

Fig.2 TEM micrographs: (a) inter-fiber configuration; (b) multilayered interfacial coatings

Typical microstructures of the interfacial layers fabricated by T-pulsing CVI were characterized by transmission electron microscopy (TEM), Fig.2. Fig.2a shows the inter-fiber configuration of the composites, in which multilayered interphases can be observed clearly. Fig.2b shows the TEM micrograph of the interfacial region, indicating that the homogeneous nano-scale PyC and SiC alternating layers were formed. The temperature-pulsing CVI can be successfully applied to design and precisely adjust the composition and thickness of the interfacial coatings in CMCs.

As shown in Fig.3, the interfacial debonding can be clearly observed. This fracture behavior will be more effective for improving the mechanical properties of the composites through the stress transfer and crack deflection mechanism. Load-displacement curve from bending test of the as prepared 3D C_f/SiC composite shows the typically linear and nonlinear parts.

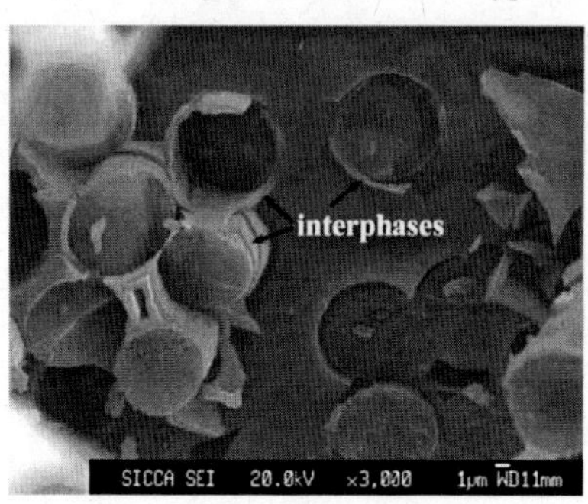

Fig.3 Fractography of the composite showing the interfacial debonding

The flexural strength is about 474.0±36.9MPa, indicating that T-pulsing CVI may provide another way to design and fabricate CMCs with well controlled interfacial layers and improved properties.

1.2 C_f/SiC composite prepared by HP [6,13,14,15]

High performance C_f/SiC composites were prepared through hot pressing (HP) by using SiC nanopowder as the matrix. The strength and fracture toughness of C_f/SiC composites would increase with the decrease in SiC particle size and the increase in the modulus of carbon fiber at a proper sintering temperature and pressure. Fig.4 shows the fracture behaviors of the composite with SiC particle size as 50nm, the highest bending strength (500.1MPa) and fracture toughness (16.9MPa·$m^{1/2}$) were obtained at 1850°C under 20MPa, using M40JB fiber as the reinforcement. Fiber pull-out was clearly observed (Fig.4a). The strong interaction between the SiC matrix and the fiber might lead to the potential damage of the fiber during fabrication, as shown in Fig. 4b.

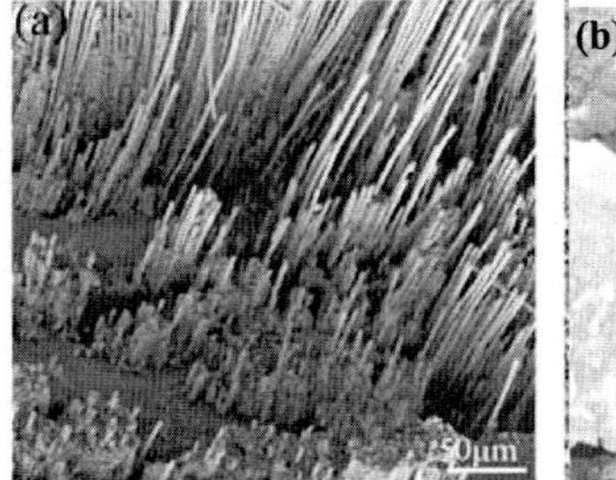

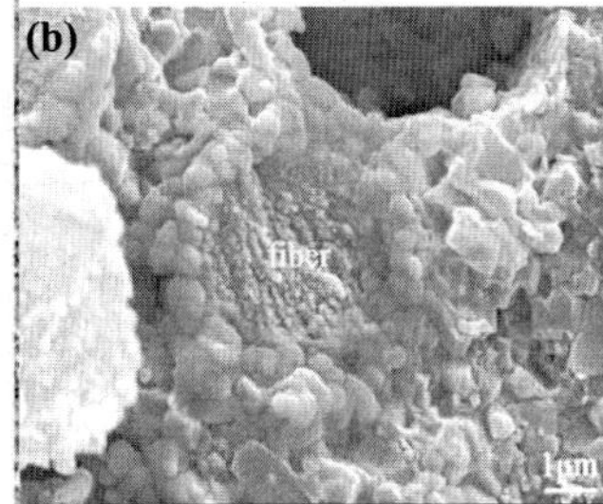

Fig.4 Fracture behaviors of the composite with different SiC particle inclusion: (a) SiC particle size is 50nm, (b) Fracture surface showing the strong fiber/matrix interaction when use sub-micrometer SiC particle (500nm)

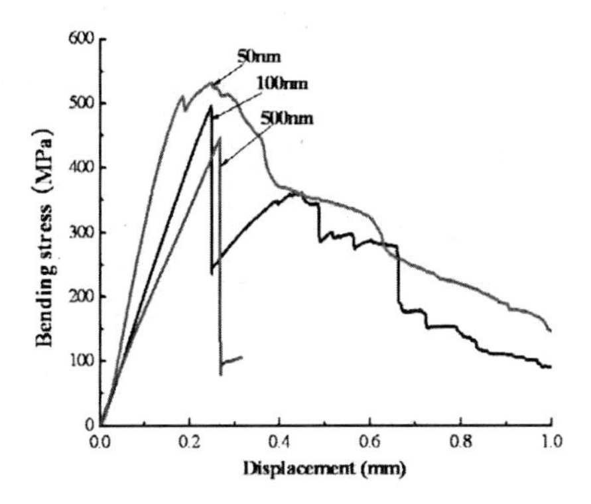

Fig.5 Stress-displacement curves of C/SiC composites using different SiC particle for matrix formation

Fig.5 indicates that using nano-SiC as the matrix, the prepared C_f/SiC composite demonstrates the typically non-catastrophic fracture behavior. With the increase in the particle size of SiC, the fracture toughness decreases correspondingly. As using 500nm SiC powder, the fracture toughness is down to 7.5 MPa·$m^{1/2}$, similar to that of brittle ceramics.

1.3 Vapor silicon infiltration (VSI) process[16,17]

The bulk density and the open porosity of C_f/SiC composites prepared by VSI are shown in Fig.6 as a function of infiltration temperature. The sintered density reached 2.25 g/cm^3 with the porosity as 6% at 1700°C. The fracture surface (Fig. 7 a) showed the dense matrix after VSI at 1700°C. An obvious fiber pull-out (Fig. 7 b) was observed after fracture although the silicon content is quite high (14.5 vol %).

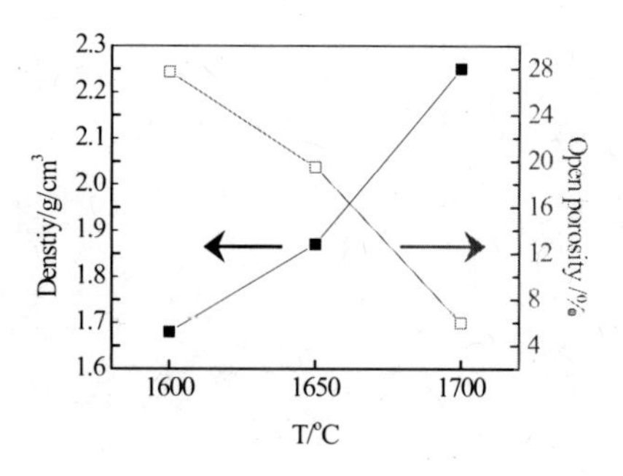

Fig.6 Density and porosity of the VSI C_f/SiC

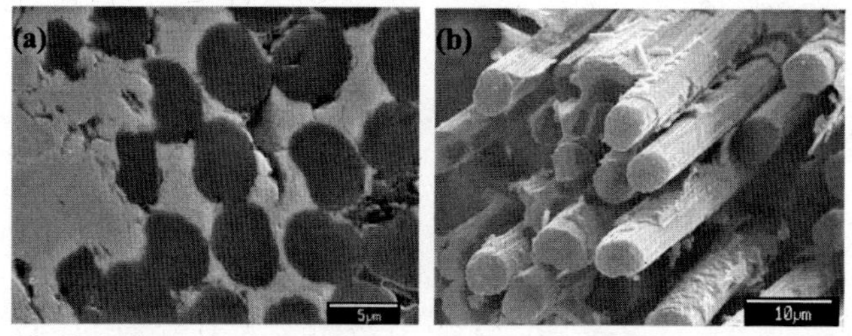

Fig.7 Cross section (a) and fracture surface of the VSI composites (b)

Stress-displacement curves (Fig.8) showed the non-catastrophic fracture behavior of VSI composites. It shows that the work of fracture (WOF) decreased after the composite was densified. WOF decreased from 12.13 to 9.54 kJ·m^{-2} with the increase in density, when temperature increased from 1650°C to 1700°C. However, the strength kept almost the same (239.5±35.6MPa at 1650°C and 238.9±41.2MPa at 1700°C).

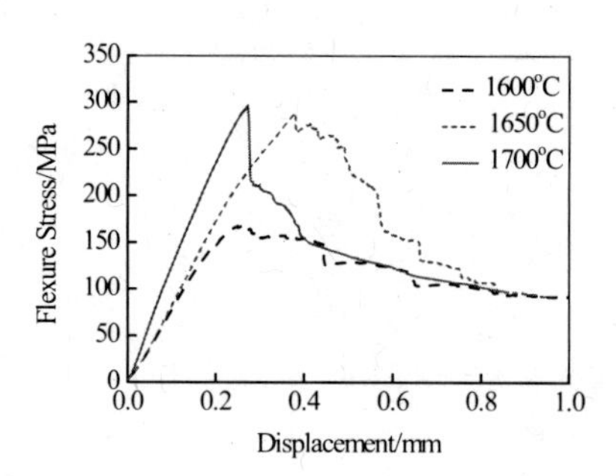

Fig.8 Stress-displacement curves of the VSI composites

1.4 Active filler enhanced PIP process (R-PIP)[18,19,20,21,22]

The incorporation of active fillers to the matrix during polymer impregnation can enhance the PIP process by reducing the volume shrinkage of the matrix. Usually, active filler will react with the reactive atmosphere and the decomposed species of the preceramic polymer. These reactions usually lead to the volume expansion and could be used to compensate the polymer shrinkage during the pyrolysis process. Aluminum (Al) is the well recognized active filler. Fig.9 shows the variation of density as the PIP cycles, using the 2.5D preforms. It is clear that the bulk density increases quickly for the initial several cycles with Al-loading. Stress/displacement curves of the composites are shown in Fig.10. With Al loading, the composite shows higher flexural strength due to the strong bonding caused by reactions between the matrix and fibers.

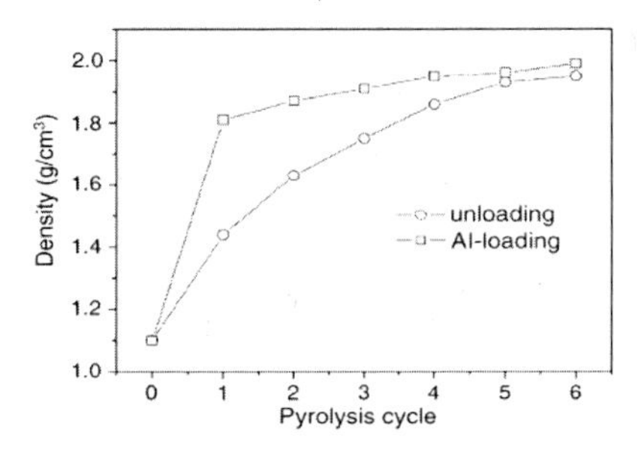

Fig.9 Relationship between the density and the pyrolysis cycle of the composites with and without active filler loading

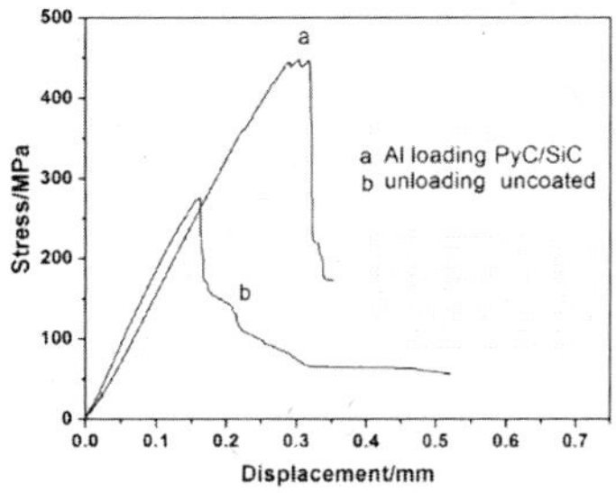

Fig.10 Stress/displacement curves of the 2.5D composites with and without active filler loading

Another role for active filler is to increase the anti-oxidation behavior of CMC composites. The lifetime of CMCs which are partly made of carbon is strongly depended on the efficiency of the anti-oxidation systems used to reduce the oxygen permeability. Extensive studies have been undertaken to improve the oxidation resistance of carbon fiber reinforced CMCs, and most are related to the application of boron-bearing species.

Dong et al proposed a facile route to fabricate carbon fiber reinforced ceramic matrix composites (C_f/SiC-BN) by an active-filler-controlled polymer pyrolysis (AFCOP) process. In the proposed process, boron was introduced into the carbon fibers as active filler to form some boron-bearing species by in-situ reactions during the subsequent heat-treatment process. The composites were prepared by PIP using PCS as the polymer precursor. XRD patterns of the obtained composites confirmed the presence of H-BN. With the presence of BN, the oxidation of the composites was greatly improved. The weight losses of C_f/SiC and C_f/SiC-BN after being oxidized at 800°C for 10h were ~36% and ~16% respectively and most of the carbon fibers in

C_f/SiC-BN composites were well retained while those in C_f/SiC composites were oxidized, Fig.11.

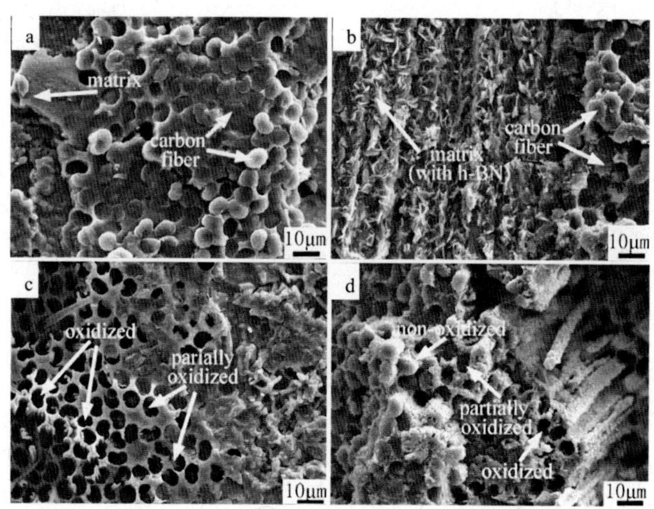

Fig.11. SEM observation on the fracture surface of the composites: (a) without filler before oxidation, (b) with B filler before oxidation, (c) without filler after oxidization at 800°C for 10h, (d) with B filler after oxidization at 800°C for 10h.

The above mentioned techniques (T-pulsing CVI, HP, L/VSI and R-PIP) can be used successfully for microstructure design and properties optimization. Deposition of multilayered coating and matrix can be controlled easily by T-pulsing CVI. Combined with nano-SiC infiltration, HP shows its superiority on fabrication of high performance C_f/SiC composites. L/VSI is the facile and economic way to prepare CMCs, and microstructure control was proved to be effective for improving the final properties of the composite. R-PIP can improve the efficiency of regular PIP process, and the flexural strength is shown to be highly dependent on the surface coating properties of the fibers. However, precise microstructure control and process optimization is still an important issue for the future study.

2. TRANSPARENT CERAMICS

Compared with single crystals, transparent ceramics offer several advantages, for example, lower cost, easy fabrication of materials of large size, and especially those with high melting points. Currently, transparent ceramics have found a wide range of application in lighting, solid state lasers, scintillators, windows and battle field etc.

Nanotechnology and vacuum or atmosphere sintering approaches make the fabrication of highly transparent ceramic materials possible. To prepare transparent ceramics, a strict control of the characteristics of initial powder was required. Nano particles with narrow size distribution, mono-dispersion and high chemical purity is recommended for the subsequent densification process. The nanopowder is commonly prepared by sol-gel technology, co-precipitation, thermal decomposition of salts, solid-phase synthesis, combustion, freezing, and hydrothermal methods etc. The sample can be fully densified by hot pressing, hot isostatic pressing or pressureless sintering in vacuum or atmosphere. To reduce the grain growth during the densification process, usually a long holding time at relatively low temperature are recommended.

2.1 Transparent ceramics for lighting -Al_2O_3

Since the first translucent polycrystalline alumina was developed by Coble in the 1960s, continued efforts are made to improve the optical qualities of alumina ceramics because they are a potential alternative to corundum single crystal. It was believed that the optical qualities of Al_2O_3 ceramics could be improved by increasing the purity and density and by controlling the microstructures. However, the needed degree of optical transparency has not yet been achieved after half a century's effort. In fact, α-Al_2O_3 (corundum) with hexagonal crystal structure is optically uniaxial and birefringent. Therefore, it is impossible to prepare transparent alumina ceramics with randomly oriented grains due to the existence of the numerous interference of optical axis.

In a recent study, Wang et.al.[23] proposed a simple route to prepare transparent polycrystalline alumina ceramics with well controlled orientation of individual grains. They used slip casting process to prepare the green bodies and simultaneously applied a strong magnetic field to the slurries. It was found that the grains have been successfully orientated in the alumina ceramics after sintering, Fig.12. The in-line optical transmission is much higher than that of the randomly orientated Al_2O_3 because the optical axes of the individual grains are parallel to each other. Samples that were slip cast in a strong magnetic field exhibited higher optical transparency than those prepared without using a magnetic field, Fig.13. To prepare transparent alumina, Wang's group also investigated the gelcasting process based on epoxy system [24,25].

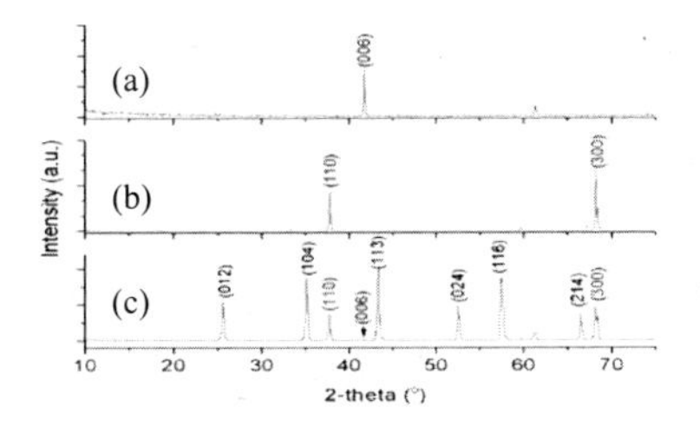

Fig.12. X-ray diffraction patterns of alumina slices cut, (a) perpendicular to and (b) parallel to the magnetic field, and (c) slip cast without a magnetic field.

Fig.13. Polycrystalline alumina ceramics from slip casting (a) with and (b) without a magnetic field. The samples are placed 8 mm above the paper.

2.2 Transparent ceramic for laser (Nd:YAG)

In 1995, Ikesue et al.[26,27] first demonstrated the possibility of fabricating transparent Nd:YAG ceramics of sufficient quality for solid-state lasers with reasonable efficiency. Since then, Nd:YAG laser ceramics have attracted considerable attention because the optical quality has been improved greatly and highly efficient laser oscillations could be obtained with comparable efficiency to YAG single crystals.[28] Compared with YAG single crystal laser materials, YAG ceramics have several prominent advantages: (1) easy fabrication; (2) scalability in size; (3) high doping concentration; (4) better homogeneity of the doping ions; (5) ease of achieving composite structure and so on. Ceramic technology also makes it easier to incorporate several dopant ions into the YAG material compared with single crystals grown from a melt, thus the lattice deformation leading to optical birefringence and wave front distortion can be circumvented.

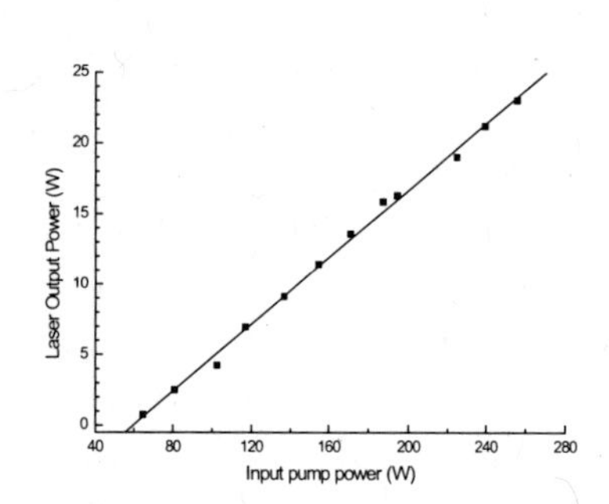

Fig.14 Laser power input-output curve with pump frequency of 1000Hz, obtained 23W continuum laser output

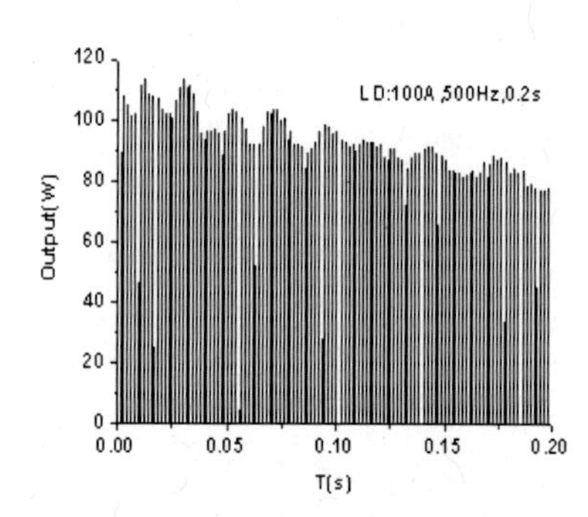

Fig.15 Nd:YAG transparent ceramics heat capacity, average output power reached 100W

In SICCAS, Pan et al have successfully developed the transparent Nd: YAG ceramics for laser [29,30,31,32,33,34]. A continuum laser output of 23W has been reported, Fig.14. For Nd:YAG transparent ceramics heat capacity, the average output power is now 100W, Fig.15. They also tried a laminar-structured of YAG/1.0 at. % Nd: YAG/YAG transparent ceramic to improve the laser performance [35]. Other laser ceramics (such as Yb:YAG [36] , YAG:Tb[37], YAG:Eu [38]etc.) in SICCAS were also well studied.

2.3 Transparent ceramics for scintillator

The development of scintillator and phosphor materials started at the end of 19th century. The $CaWO_4$ and ZnS-based phosphor powders were introduced in practice very soon after. In the late forties the first single crystal scintillators NaI:Tl and CsI:Tl appeared and has been widely used till today. The latest class of materials scintillation optical ceramics came into use relatively recent when appropriate technology was developed to obtain competitive material properties mainly for materials where single crystals cannot be prepared or their production is extremely expensive[39].

An interest in new scintillator materials is pushed by increasing number of medical, industrial or scientific applications, requiring higher material performance. The main transparent ceramic systems include $(Y,Gd)_2O_3$:Eu,Pr(YGO), Gd_2O_2S:Pr,Ce,F(GOS), and $Gd_3Ga_5O_{12}$:Cr,Ce (GGG). In recent years, new systems (Ce doped $BaHfO_3$ and Eu^{3+} or Tb^{3+}-doped Lu_2O_3) were also under study.

The studied transparent ceramics system in SICCAS including Eu^{3+}, Pr^{3+} doped Lu_2WO_6[40,41], Ce doped YSAG[42], $LuBa_3B_9O_{18}$[43], LuAG:Ce[44,45], $BaBPO_5$[46,47], Eu:Gd_2O_3-HfO_2[48], Ce:$SrHfO_3$[49] and $La_2Hf_2O_7$:Ti[50] etc. A series of transparent ceramics were successfully prepared with well improved properties. Recently, Shi et al [51] prepared Ce:YSAG with near complete absorption for lights at wavelengths below 480nm. The transparent ceramics show remarkably enhanced light emission intensity and much suppressed concentration quenching effect than those of Ce:YAG ceramics at the same Ce doping levels. Liu et al reported the LuAG:Ce transparent ceramics fabricated by the solid-state reaction method. The relative density reaches 99.5%. The transmittance in the visible light region reaches 70% [52]. Shi et al reported the Eu doped Lu_2O_3 from a novel co-precipitation process. The optical in-line transmittance in the visible wavelength region reached 80% [53]. The Ce doped LuAG and the Eu doped Lu_2O_3 ceramics are shown in Fig 16.

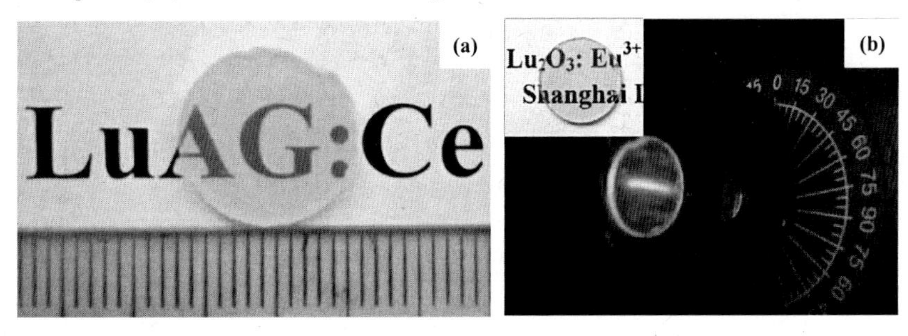

Fig.16 Optical images of (a) Ce doped LuAG and (b) the Eu doped Lu_2O_3 transparent ceramics

2.4 Transparent ceramics for lens

The first transparent ceramic lens for the digital camera was developed successfully by CASIO. The material (LUMICERA) has the same light transmitting qualities as optical glass commonly used in today's conventional camera lenses, however, it has two very important properties: high refractive index (2.08) and superior strength. This will make it possible to create zoom lenses for cameras with greatly reduced profiles.

In recent years, the development of transparent ceramics lens is also in active. A complex system $Ba(MgTa)O_3$, with cubic lattice structure, has been studied. The material showed excellent properties with high refractive index (2.05-2.10) at 580nm and the light transmission rate can reach 76%. The material preparation process is still under refining for further improving the performance.

In addition to the above mentioned transparent ceramic systems, there are other systems studied in SICCAS including the infrared ceramics for windows, domes and armors such as SiAlON, AlON, Y_2O_3[54,55] etc.

Transparent ceramics with excellent physical, mechanical and other properties have been a key material for commercial (high-tech industry) and defense applications. The present results showed that the development of high quality transparent ceramics is mainly associated with the understanding of the fundamental principles underlying the advanced processing technologies.

3. BIOMATERIALS

Having existed for around half a century, the development of biomaterials is not a new area of science. But the demand for implants and prostheses made from biomaterials is constantly growing. The materials need to be very robust and active for promoting tissue regeneration. Scientists have developed bone implants from metal alloys and ceramics. The most promising substances are made of Ti, Al_2O_3 or ZrO_2. Because they do not interact with biological processes in the body, they are not usually rejected. In addition, interaction between biomaterials and natural tissues is an important subject for biomaterial science. Such information is essential to aid the design of new biocompatible biomaterials.

The research activities of biomaterials in SICCAS cover a wide range including bioactive materials and tissue engineering scaffolds, nano-biomaterials for controlled drug release, bio-labeling and diagnostic, biomaterials surface engineering for medical implants, optical fiber materials for medical device etc. In this paper, two research areas will be introduced.

3.1 Biomaterials surface coating

The research of biomaterials surface coating started from 1959 using Plasma spraying technique on the surface of Ti alloys for bone replacement. With almost 50 years of scientific research and technical buildup, the technologies for biological coating have been well developed and found orthopedic applications [56,57,58]. In order to improve the cytocompatibility and antibacterial properties, different techniques have been developed including loading drugs, nano silver powders, surface treatment etc[59,60,61,62]. In recent years, novel bioactive coatings have also been developed, including wollastonite, dicalcium silicate, nano-TiO_2 and nano-ZrO_2

coatings, which possess excellent bioactivity and biocompatibility as well as high bonding strength to Ti-6Al-4V substrate[63,64,65]. In order to improve the bioactivity, biocompatibility and antibacterial property of titanium and its alloys, Si-, Ag-, Ca-, P- and Na-ion was also incorporated into the coating through ion implantation technique[66,67]. Fig.17 showed the plasma spraying system and the coated parts for hip replacement etc. applications.

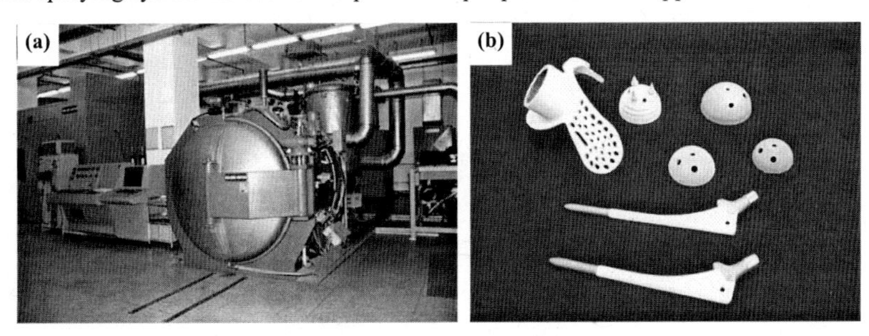

Fig.17 Optical images of (a) Vacuum-atmospheric plasma spraying system and (b) biocompatible coatings

3.2 Nano-biomaterials for controlled drug release

Amorphous mesoporous silica materials with a non-toxic nature, tunable pore diameter, and very high specific surface area with abundant Si-OH bonds on the pore surface are promising candidates for use as carriers in drug delivery systems[68]. Conventional mesoporous silica materials (such as MCM-41 and SBA-15) exhibit sustained-release properties, but their drug storage capacity is relatively low, and also the irregular bulk morphology is not perfect for drug delivery. In this report, two advances in SICCAS will be addressed.

Stimuli-responsive controlled drug release. Recently, Shi's group has successfully synthesized HMS spheres with a 3D pore-network shell [69,70]. The drug storage capacity of this system is three times higher than that of the reported MCM-41 system [71]. Subsequently, a novel stimuli-responsive controlled drug-release system was developed by using PAH/PSS multilayers as a coating to cap the mesopore openings of drug-loaded hollow mesoporous silica spheres. In this way, the drug release rate from the new system can be well-controlled by changing the pH value (or the salt concentration) of the release medium. This system, which combines the advantages of both high drug storage capacity and the property of stimuli-responsive controlled release, has potential applications in drug delivery [72], Fig.18.

Drug targeted delivery. For drug delivery system, the efficiency of drugs to the targeted delivery has always a hot research area and has attracted considerable attention in recent years. Shi's group reported a new kind of uniform magnetic nanocomposite sphere (MCMS) with a magnetic core/mesoporous silica shell structure.[73,74] Based on this advance, a novel kind of rattle-type hollow magnetic mesoporous sphere (HMMS) with Fe_3O_4 particles encapsulated in the cores of mesoporous silica microspheres were successfully fabricated by Sol-gel reactions on hematite particles[75,76]. The ring-shaped cavity between the magnetic core and the mesoporous silica shell has been generated by a simple hydrothermal treatment and H_2 reduction process. Using IBU as a model drug, hollow magnetic mesoporous spheres can realize a significantly higher storage capacity of drug because of the rattle-type structure than the corresponding magnetic/mesporous spheres without the cavities. The release experiments identify a sustained-release behavior of IBU from the present drug storage system and the release process follows a Fick's law. This hollow magnetic mesoporous spheres are a promising candidate material as drug targeted delivery carriers. Fig.19 shows the possible mechanism for the formation and the relevant microstructure of mesospheres.

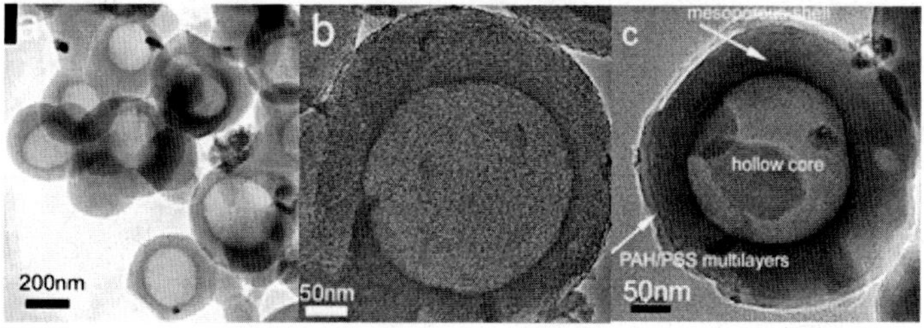

Fig.18 TEM images (a, b) of hollow mesoporous silica sphere (HMS) with a 3D pore network, core-shell structure, (c) with polyelectrolyte multilayer coating as a stimuli-responsive controlled drug-delivery system.

In addition, for the drug delivery system, there are also extensive works in SICCAS using mesoporous bioactive glass [77], porous HA[78,79], and Fe_2O_3 hollow spheres[80,81] etc composites.

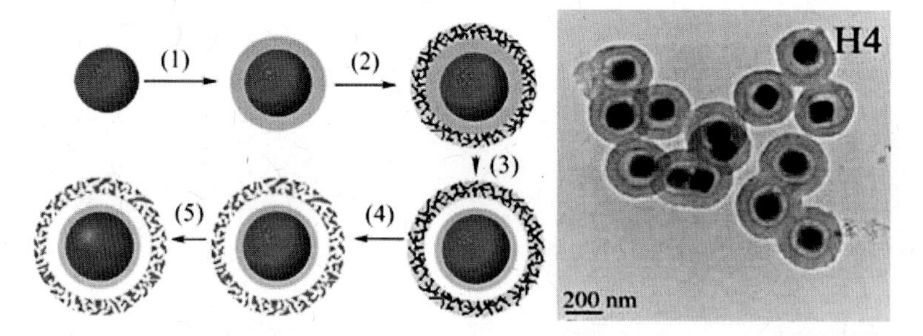

Fig.19 Schematic illustration of the synthesis and the TEM image of MFeCMS nanospheres.

From dental repairs to controlled drug release or total organ/joint replacement, almost every human being on earth will be exposed to one biomaterial or another during their lifetime. Artificial organs are necessary to support part or all of their essential functions thereby improving quality of life. In biomaterial science, interaction between biomaterials and natural tissues is still an important subject. The informations are essential to aid the design of new biocompatible materials.

4. ADVANCED PROCESSING

Colloidal processing, which has the potential for eliminating detrimental heterogeneities and avoiding their reintroduction during the successive processing steps, has long been reported as the most potential cost effective and eco-friendly route to prepare ceramic composites with improved stability at low cost. In colloidal processing, particle dispersion is often the limiting factor, affecting both the rheology and the homogeneity of suspensions and the properties of samples before and after sintering. There are several colloidal processing techniques, including tape casting, gel casting, direct coagulation casting etc. In this report, the study on aqueous tape casting and gelcasting process in SICCAS will be addressed briefly.

4.1 Tape casting

Tape casting is a prominent process to produce thin and flat green sheets for the fabrication of ceramic substrates and multilayered structures. Organic solvents are commonly used as the liquid vehicle, but the volatility and toxicity of these solvents has lead to increasing interest in research on the aqueous tape casting process.
A lot of ceramic tapes have been prepared in SICCAS, including silicon carbide [82], silicon nitride[83], titanium carbide[84], titanium nitride[85], hydroxyapatite[86], alumina[87], zirconia[88].
The design of laminated structure was guided by finite element method (FEM) to optimize the mechanical and other properties. A series of laminated composites with well improved properties were developed, including SiC/C[89], SiC/TiC[90], Al_2O_3/TiC[91],Al_2O_3/HA[92], Al_2O_3/Ni[93], TiN/Al_2O_3[94] etc. Fig. 20 showed the FEM simulation of a symmetrical layered SiC/S10T ceramic with gradual thermal residual stress distribution. The initial results were shown in Table 1. Compared with pure SiC and 90%SiC-10%TiC (S10T) ceramics fabricated by the same process, the designed composites showed excellent mechanical properties. The tested strength was close to the theoretical value. The strengthening and toughening mechanisms of the ceramic were ascribed to surface compressive residual stress [90].

Table 1 Comparison of tested strength of SiC/S10T layered ceramic with theoretical

Calculated residual stress (MPa)	Tested residual stress (MPa)	Theoretical strength(MPa)	Tested strength (MPa)
-126	-129	840	834

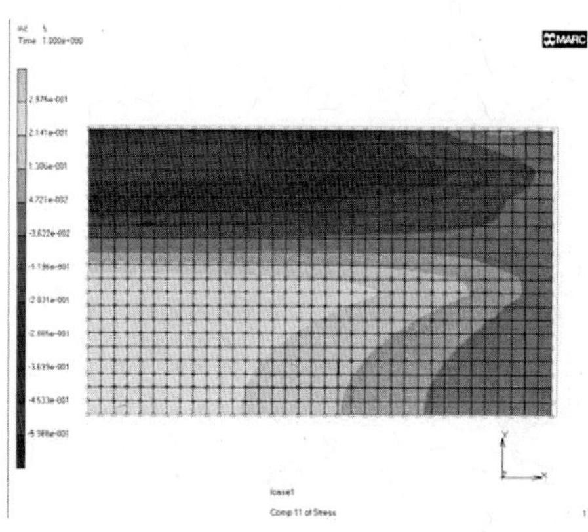

Fig.20 FEM analysis of SiC/ S10T system

4.2 Gelcasting

Gelcasting is an attractive highly versatile fabrication process to prepare ceramic green body with high-quality and complex shape [95]. The process is based on the polymerization of organic monomers and solidification of slurry to green body. In gelcasting, slurry made from ceramic powder and a water-based monomer solution is prepared and poured into a mold, followed by the polymerization *in-situ* to immobilize the particles in a gelled part. The samples were removed from the mold when it was still wet, then dried and fired. In recent years, gelcasting has been widely studied to produce ceramic materials. Jiang's group studied the gelcasting of SiC[96], TiC[97], Al_2O_3, B_4C[98], HA[99], HA/ZrO_2, ZrB_2[100] etc. For SiC system, after gelcasting and drying, the relative density and the bending strength of the green body was 57% and 40MPa, respectively. After sintering at 2200°C, SiC samples can be densified to a relative density of 98%. The mechanical properties of the obtained pieces are satisfying, with the flexural strength, toughness and the hardness as 539±89MPa, 3.46±0.31 $MPa{\cdot}s^{1/2}$ and 26.2±0.97 GPa, respectively. Further study on this system is still under way.

Colloidal processing is proved to be a facile, environmental conscious route to obtain ceramic composites with large size, complex shape and high quality at low cost. However, the in depth research on the surface chemistry, rheology, polymer science etc. is still needed for understanding its principal mechanism especially for nano powder processing.

ACKNOWLEDGEMENTS

This work was supported by the Ministry of Science and Technology, Chinese Academy of sciences, the National Natural Science Foundation of China (No. 50772128), Shanghai Science and Technology Committee (No. 07DJ14001, No.07pj14094), and the State Key Laboratory of High Performance Ceramics and Superfine Microstructures.

REFERENCES

[1] L.F. Cheng, Y.D. Xu, L.T. Zhang, and R. Gao, "Effect of Glass Sealing on the Oxidation Behavior of Three Dimensional C/SiC Composites in Air," *Carbon*, **39,** 1127-1133(2001).

[2] X.B. He and H. Yang, "Preparation of SiC Fiber-Reinforced SiC Composites," *J. Mater. Process. Technol.*, **159,** 135-138(2005).

[3] T.M. Besmann, B.W. Sheldon, R.A. Lowden, and D.P. Stinton, "Vapor-Phase Fabrication and Properties of Continuous-Filament Ceramic Composites," *Science*, **253,** 1104-1109(1991).

[4] R. Naslain, J. Lamon, R. Pailler, X. Bourrat, A. Guette, and F. Langlais, "Micro/ Minicomposites: a Useful Approach to the Design and Development of Nonoxide CMCs," *Composites: Part A*, **30,** 537-547(1999).

[5] S.M. Dong, Y. Katoh, and A. Kohyama, "Processing, Optimization and Mechanical Evaluation of Hot Press 2D Tyranno-SA/SiC Composites," *J. Euro. Ceram. Soc.*, **23,** 1223-1231 (2003).

[6] S.M. Dong, Y. Katoh, and A. Kohyama, "Preparation of SiC/SiC Composites by Hot Press, Using Tyranno-SA Fiber as Reinforcement," *J. Am. Ceram. Soc.*, **86[1]** 26-32(2003).

[7] T. Taguchi, N. Igawa, R. Yamada, and S. Jitsukawa, "Effect of Thick SiC Interphase Layers on Microstructure, Mechanical and Thermal Properties of Reaction-Bonded SiC/SiC Composites," *J. Phys. Chem. Solids*, **66,** 576-580 (2005).

[8] S.M. Dong, Y. Katoh, A. Kohyama, S.T. Schwab, and L.L. Snead, "Microstructural Evolution and Mechanical Performances of SiC/SiC Composites by Polymer Impregnation/Microwave Pyrolysis (PIMP) Process," *Ceram. Int.*, **28,** 899-905 (2002).

[9] M. Yuan, Z.R. Huang, and S.M. Dong et al., "Carbon/Silicon Carbide Composites with Interphases Processed by Temperature-Pulsing Chemical Vapor Infiltration Technique," *J. Inorg. Mater.*, **22[2]** 305-310 (2007).

[10] Q. Zhou, S.M. Dong, and X.Y. Zhang, et al., "Carbon Fiber Surface Coating by Forced Pressure-Pulsed CVI," *J. Inorg. Mater.*, **21[6]** 1378-1384(2006).

[11] Y.S. Ding, S.M. Dong, and Q. Zhou, et al., "Preparation of C/SiC Composites by Hot Pressing, Using Different C Fiber Content as Reinforcement," *J.Am.Ceram.Soc.*, **89[4]** 1447-1449(2006).

[12] M. Yuan, Z.R. Huang, and S.M. Dong, et al., "Microstructure of Multi-Layered Interphases Processed by Temperature-Pulsing Chemical Vapor Infiltration," *Phys. Status Solidi. A*, **203[8]** R58-R60(2006).

[13] Y.S. Ding, S.M. Dong, and L. Gao, et al., "Effect of Sintering Temperature on Microstructure and Properties of C-f/SiC Composites," *J. Inorg. Mater.*, **23[6]** 1151-1154 (2008).

[14] X.H. Qin, B.L. Xiao, and S.M. Dong, et al., "SiCf/SiC Composites Reinforced by Randomly Oriented Chopped Fibers Prepared by Semi-Solid Mechanical Stirring Method and Hot Pressing," *J. Mater. Sci.*, **42[10]** 3488-3494(2007).

[15] S.M. Dong, Y.S. Ding, and D.L. Jiang, et al., "Effects of Preparation Conditions on the Microstructure and Properties of Hot Pressed SiC/SiC Composites," *J.Inorg.Mater.*, **20[4]** 883-888(2005).

[16] Q. Zhou, S.M. Dong, and YS Ding, et al., "Effect of Interphase on Mechanical Pproperties of C-f/SiC Composites Fabricated by Vapor Silicon Infiltration," *J. Inorg. Mater.*, **22,**1142-1146 (2007).

[17] Q. Zhou, S.M. Dong, and X.Y. Zhang, et al., "Fabrication of C-f/SiC Composites by Vapor Silicon Infiltration," *J.Am.Ceram.Soc.*, **89[7]** 2338-2340 (2006).

[18] Z. Wang, S.M. Dong, and XY Zhang, et al., "Fabrication and Properties of C-f/SiC-ZrC Composites," *J.Am.Ceram.Soc.*, **91[10]** 3434-3436(2008)

[19] Y.Z. Zhu, Z.R. Huang, and S.M. Dong, et al., "Manufacturing 2D Carbon-Fiber-Reinforced SiC Matrix Composites by Slurry Infiltration and PIP Process," *Ceram. Inter.*, **34[5]**1201-1205(2008).

[20] Y.Z. Zhu, Z.R. Huang, and S.M. Dong, et al., "Pressureless Preparation and Properties of C/SiC Composites," *J. Inorg. Mater.*, **22 [4]** 685-689(2007).

[21] Y.Z. Zhu, Z.R. Huang, and S.M. Dong, et al., "Polymer-Filler Drived Ceramics with Low Shrinkage Using PCS/SiC/Al Mixture," *Phys. Status Solidi-R*, **1[2]** R59-R61(2007).

[22] Y.Z. Zhu, Z.R. Huang, and S.M. Dong, et al., "Fabricating 2.5D SiCf/SiC Composite Using Polycarbosilane/SiC/Al Mixture for Matrix Derivation," *J.Am.Ceram.Soc.*, **90[3]** 969-972 (2007).

[23] X.J. Mao, S.W. Wang, S.Z. Shimai, and J.K. Guo, "Transparent Polycrystalline Alumina Ceramics with Orientated Optical

Axes," *J. Am. Ceram. Soc.*, **91[10]** 3431-3433(2008).

[24] X.J. Mao, S.Z. Shimai, M.J. Dong, and S.W. Wang, "Gelcasting of Alumina Using Epoxy Resin as a Gelling Agent," *J. Am. Ceram. Soc.*, **90[3]** 986-988 (2007).

[25] X.J. Mao, S.Z. Shimai, and M.J. Dong, et al., "Investigation of New Epoxy Resins for the Gel Casting of Ceramics," *J.Am.Ceram.Soc.*, **91[4]** 1354-1356 (2008).

[26] A. Ikesue, I. Furusato, and K. Kamata, ''Fabrication of Polycrystallin, Transparent YAG Ceramics by a Solid-State Reaction Method,'' *J. Am. Ceram. Soc.*, **78[1]** 225-228(1995).

[27] A. Ikesue, T. Kinoshita, K. Kamata, and K. Yoshida, ''Fabrication and Optical Properties of High Performance Polycrystalline Nd: YAG Ceramics for Solid-State Lasers,'' *J. Am. Ceram. Soc.*, **78[4]** 1033-1040 (1995).

[28] H. Yagi, T. Yanagitani, K. Takaichi, K. Ueda, and A. A. Kaminskii, ''Characterizations and Laser Performances of Highly Transparent Nd^{3+}: $Y_3Al_5O_{12}$ Laser Ceramics,'' *Opt. Mater.*, **29[10]**1258-1262 (2007).

[29] J. Li, Y.S. Wu, Y.B. Pan, et al., "Fabrication of Cr^{4+}, Nd^{3+}: YAG Transparent Ceramics for Self-Q-switched Laser," *J .Non-Cryst. Solids*, **352[23-25]** 2404-2407(2006).

[30] Y.S. Wu, J. Li, and Y.B. Pan, et al., "Diode-Pumped Passively q-Switched Nd : YAG Ceramic Laser with a Cr^{4+}: YAG Crystal-Satiable Absorber," *J.Am.Ceram.Soc.*, **90[5]** 1629-1631 (2007).

[31] J. Li, Y.B. Pan, and F.G. Qiu, et al., "Synthesis of Nanosized Nd:YAG Powders via Gel Combustion," *Ceram.Inter.*, **33[6]** 1047-1052(2007).

[32] J. Li, Y.B. Pan, and F.G. Qiu, et al., "Nanostructured Nd:YAG Powders via Gel Combustion: The Influence of Citrate-to-Nitrate Ratio," *Ceram.Inter.*, **34[1]**141-149(2008).

[33] J. Li, Y.S. Wu, and Y.B. Pan, et al., "Densification and Microstructure Evolution of Cr^{4+}, Nd^{3+}: YAG Transparent Ceramics for Self-Q-Switched Laser," *Ceram.Inter.*, **34[7]** 1675-1679(2008).

[34] J. Li, Y.S. Wu, and Y.B. Pan, et al., "Fabrication, Microstructure and Properties of Highly Transparent Nd : YAG Laser Ceramics," *Opt. Mater.*, **31[1]** 6-17(2008).

[35] J. Li, Y.S. Wu, and Y.B. Pan, et al., "Laminar-Structured YAG/Nd : YAG/YAG Transparent Ceramics for Solid-State Lasers," *Int. J. Appl. Ceram. Tec.*, **5[4]** 360-364(2008).

[36] Y.S. Wu, J.Li, and Y.B. Pan, et al., "Diode-Pumped Yb:YAG Ceramic Laser," *J.Am.Ceram.Soc.*, **90,** 3334-3337(2007).

[37] J.J. Zhang, J.W. Ning, and X.J. Liu, et al., "Synthesis of Ultrafine YAG: Tb Phosphor by Nitrate-Citrate Sol-Gel Combustion Process," *Mater. Res. Bull.*, **38[7]** 1249-1256(2003).

[38] J.J. Zhang, J.W. Ning, and X.J. Liu, et al., "Low-Temperature Synthesis of Single-Phase Nanocrystalline YAG: Eu Phosphor," *J. Mater. Sci. Lett.*, **22[1]** 13-14(2003).

[39] M. Nikl, "Scintillation Detectors for X-Rays," *Meas. Sci. Technol.*, **17,** R37-R54(2006).

[40] H.L. Li, X.J. Liu, and L.P. Huang, "Fabrication of Transparent Cerium-Doped Lutetium Aluminum Garnet (LuAG:Ce) Ceramics by a Solid-State Reaction Method," *J.Am.Ceram.Soc.*, **88[11]** 3226-3228(2005).

[41] Q.W. Chen, Y. Shi, and L.Q. An, et al., "Fabrication and Photoluminescence Characteristics of Eu^{3+}-Doped Lu_2O_3 Transparent Ceramics," *J.Am.Ceram.Soc.*, **89[6]** 2038-2042(2006).

[42] T. Feng, J.L. Shi, and X.G. Jin, et al., "Effect of Sc Substitution for Al on the Ooptical Properties of Ttransparent Ce : YSAG Ceramics," *J.Am.Ceram.Soc.*, **91[7]** 2394-2397(2008).

[43] C.J. Duan, W.F. Li, and J.L. Yuan, et al., "Synthesis, Crystal Structure and X-Ray Excited Luminescent Properties of $LuBa_3B_9O_{18}$," *J. Alloy Compd.*, **458[1-2]** 536-541(2008).

[44] H.L. Li, X.J. Liu, and R.J. Xie, et al., "Cerium-Doped Lutetium Aluminum Garnet Phosphors and Optically Transparent Ceramics Prepared from Powder Pprecursors by a Urea Homogeneous pprecipitation Method," *Jpn. J. Appl. Phys.*, **47[3]** 1657-1661(2008).

[45] H.L. Li, X.J. Liu, and R.J. Xie, et al., "Fabrication of Transparent cCerium-Doped Lutetium Aaluminum Garnet Ceramics by Co-Precipitation Routes," *J.Am.Ceram.Soc.*, **89[7]** 2356-2358 (2006).

[46] C.J. Duan, X.Y. Wu, and W.F. Li, et al., "$Ba_3BP_3O_{12}$: Eu^{2+} - A Potential Scintillation Material," *Appl. Phys. Lett.*, **87[20]** 201917 (2005).

[47] C.J. Duan, J.L. Yuan, and J.T. Zhao, "Luminescence Properties of Efficient X-ray Phosphors of $YBa_3B_9O_{18}$, $LuBa_3(BO_3)_3$, α-$YBa_3(BO_3)_3$ and $LuBO_3$," *J. Solid State Chem.*, **178[12]** 3698-3702(2005).

[48] J.Y. Chen, Y. Shi, and T. Feng, et al., "Synthesis of Eu^{3+} Doped Gd_2O_3-HfO_2 Nanopowder for Radiation Detection," *J. Alloy Compd.*, **391[1-2]** 181-184(2005).

[49] Y.M. Ji, D.Y. Jiang, and L.S. Qin, et al., "Preparation and Luminescent Properties of Nanocrystals of Ce^{3+}-Activated $SrHfO_3$," *J. Cryst. Growth*, **280[1-2]** 93-98(2005).

[50] Y.M. Ji, D.Y. Jiang, and J.L. Shi, "$La_2Hf_2O_7$: Ti^{4+} Ceramic Scintillator for X-Ray Imaging," *J Mater. Res.*, **20[3]**

567-570(2005).

[51] T. Feng, J.L. Shi, and D.Y. Jiang, "Preparation of Transparent Ce:YSAG Ceramic and Its Optical Properties," *J. Euro. Ceram. Soc.*, **28,** 2539-2543(2008).

[52] H.L. Li, X.J. Liu, and L.P. Huang, "Fabrication of Transparent Cerium-Doped Lutetium Aluminum Garnet (LuAG:Ce) Ceramics by a Solid-State Reaction Method," *J. Am. Ceram. Soc.*, **88[11]** 3226-3228 (2005).

[53] Q. W. Chen, Y. Shi, L.Q. An, J.Y. Chen, and J.L. Shi, "Fabrication and Photoluminescence Characteristics of Eu^{3+}-Doped Lu_2O_3 Transparent Ceramics," *J. Am. Ceram. Soc.*, **89[6]** 2038-2042 (2006).

[54] Y.H. Huang, D.L. Jiang, and J.X. Zhang, et al., "Fabrication of Transparent Yttria Ceramics Through Gel-freezing Dry Method," *J.Inorg.Mater.*, **23[6]** 1135-1140(2008).

[55] Y.H. Huang, D.L. Jiang, J.X. Zhang, and Q.L. Lin, "Fabrication of Lanthanum doped yttria transparent ceramics," *Chinese Sci. Bull.*, Accepted.

[56] W.C. Xue, X.Y.Liu, X.B. Zheng, and C.X. Ding, "In Vivo Evaluation of Plasma-Sprayed Wollastonite Coating," *Biomaterials*, **26,** 3455-3460(2005).

[57] Y.T. Xie, X.Y. Liu, X.B.Zheng, C.X. Ding, and P.K. Chu, "Improved Stability of Plasma-Sprayed Dicalcium Silicate/Zirconia Composite Coating," *Thin Solid Films*, **515,** 1214-121(2006).

[58] J. Jiang, K.F. Huo, Z.W. Wu, S.P. Chen, S.H. Pu, Z.L. Yu, X.Y. Liu, and P.K. Chu, "Silicon Induced DNA Damage Pathway and its Modulation by Titanium Plasma Immersion Ion Implantation," *Biomaterials*, **29,** 544-550(2008).

[59] X. Y. Liu, P. K. Chu, and C. X. Ding, "Surface Modification of Titanium, Titanium Alloys, and Related Materials for Biomedical Applications," *Mat. Sci. Eng: Reports*, **47,** 49-121(2004).

[60] X. Y. Liu, X. B. Zhao, R. K. Y. Fu, J. P. Y. Ho, C. X. Ding, and P. K. Chu, "Plasma-Treated Nanostructured TiO_2 Surface Supporting Biomimetic Growth of Apatite," *Biomaterials*, **26/31,** 6143- 6150(2005).

[61] Y. T. Xie, X. Y. Liu, A. P. Huang, C. X. Ding, and P. K. Chu, "Improvement of Surface Bioactivity on Titanium by Water and Hydrogen Plasma Immersion Ion Implantation," *Biomaterials*, **26/31,** 6129-6135(2005).

[62] W.C. Xue, X.Y. Liu, X.B. Zheng, and C.X. Ding, "In Vivo Evaluation of Plasma Sprayed Titanium Coating After Alkali Modification," *Biomaterials*, **26,** 3029-3037(2005).

[63] X.Y. Liu, A.P.Huang, C.X. Ding, and P. K. Chu, "Bioactivity and Cytocompatibility of Zirconia (ZrO_2) Films Fabricated by Cathodic Arc Deposition," *Biomaterials*, **27[21]** 3904-3911(2006).

[64] X. Y. Liu, X. B. Zhao, C. X. Ding, and P. K. Chu, "Light-Induced Bioactive TiO_2 Surface", *Appl. Phys. Lett.*, **88,** 013905 (2006).

[65] W.F. Li, X.Y. Liu, A.P. Huang, and P. K Chu, "Structure and Properties of Zirconia (ZrO_2) Films Fabricated by Plasma-Assisted Cathodic Arc Deposition," *J. Phys. D: Appl. Phys.*, **40,** 2293-2299(2007).

[66] X. Y. Liu, R. Poon, L. H. Li, P. K. Chu, and C. X. Ding, "Structure and Properties of Ca-Ion Implanted Titanium by Plasma Immersion," *Surf. Coat Tech.*, **191,** 43-48(2005).

[67] Y. T. Xie, X. Y. Liu, P. K. Chu, and C. X. Ding, "Nucleation and Growth of Calcium-Phosphate on Ca-Implanted Titanium Surface," *Surf. Sci.*, **600,** 651-656(2006).

[68] N. K. Mal, M. Fujiwara, Y. Tanaka, "Photocontrolled Reversible Release of Guest Molecules from Coumarin-Modified Mesoporous Silica," *Nature*, **421,** 350-353(2003).

[69] Y.S. Li, J.L. Shi, Z.L. Hua, H.R. Chen, M.L. Ruan, and D.S. Yan, "Hollow Spheres of Mesoporous Aluminosilicate with a Three-Dimensional Pore Network and Extraordinarily High Hydrothermal Stability," *Nano Lett.*, **3**, 609-612(2003).

[70] Y.S. Li, J.L. Shi, H.R. Chen, Z.L. Hua, L.X. Zhang, M.L. Ruan, J. Yan, D.S. Yan, "One-step Synthesis of Hydrothermally Stable Cubic Mesoporous Aluminosilicates with a Novel Particle Structure," *Microporous Mesoporous Mater.*, **60,** 51-56(2003).

[71] Y.F. Zhu, J.L. Shi, Y.S. Li, H.R. Chen, W.H. Shen, and X.P. Dong, "Hollow Mesoporous Spheres with Cubic Pore Network as a Potential Carrier for Drug Storage and Its in Vitro Release Kinetics," *J. Mater. Res.*, **20**, 54-61(2005).

[72] Y.F. Zhu, J.L. Shi, W.H. Shen, X.P. Dong, J.W. Feng, M.L. Ruan, and Y.S. Li, "Stimuli-Responsive Controlled Drug Release from a Hollow Mesoporous Silica Sphere/Polyelectrolyte Multilayer Core-Shell Structure," *Angew. Chem. Int. Ed.*, **44,** 5083-5087(2005).

[73] W.R.Zhao, J.L. Gu, L.X. Zhang, H.R. Chen and J.L. Shi, "Fabrication of Uniform Magnetic Nanocomposite Spheres with a Magnetic Core/Mesoporous Silica Shell Structure," *J. Am. Chem. Soc.*, **127**, 8916-8917(2005).

[74] W.R. Zhao, J.L. Shi, H.R. Chen, and L.X. Zhang, "Particle Size, Uniformity, and Mesostructure Control of Magnetic Core/Mesoporous Silica Shell Nanocomposite Spheres," *J. Mater. Res.*, **21**, 3080-3089(2006).

[75] W.R. Zhao, J.L. Gu, L.X. Zhang, H.R. Chen and J.L. Shi, "Fabrication of Uniform Magnetic Nanocomposite Spheres with a Magnetic Core/Mesoporous Silica Shell Structure," *J.Am.Chem.Soc.*, **127[25]** 8916-8917(2005).

[76] W.H. Zhang, X.B. Lu, J.H. Xu, Z.L. Hua, L.X. Zhang, J.L. Shi, and D.S. Yan, "Synthesis and Characterization of Biofunctionized Ordered Mesoporous Materials," *Adv.Funct.Mater.*, **14**, 544-552(2004).

[77] X. Li, X.P. Wang, H.R. Chen, P. Jiang, X.P. Dong, and J.L. Shi, "Hierarchically Porous Bbioactive Glass Scaffolds Synthesized with a PUF and P123 Contemplated Approach," *Chem. Mater.*, **19**, 322-4326(2007).

[78] J.X. Zhang, M. Fujiwara, Q. Xu, Y.C. Zhu, M. Iwasa and D.L. Jiang, "Synthesis of Mesoporous Calcium Phosphate Using Hybrid Templates," *Microporous and Mesoporous Materials*, **111[1-3]** 411-416(2008).

[79] M.-Y. Ma, Y.-J. Zhu, L. Li, and S.-W. Cao, "Nanostructured Porous Hollow Ellipsoidal Capsules of Hydroxyapatite and Calcium Silicate: Preparation and Application in Drug Delivery," *J. Mater. Chem.*, **18**, 2722-2727(2008).

[80] S.W. Cao, and Y.J. Zhu, "Surfactant-Free Preparation and Drug Release Property of Magnetic Hhollow Core/Shell Hierarchical Nnanostructures," *J. Phys.Chem. C*, **112[32]** 12149-12156(2008).

[81] S.-W. Cao, Y.-J. Zhu, M.-Y. Ma, L. Li, and L. Zhang, "Hierarchically Nanostructured Magnetic Hollow Spheres of Fe_3O_4 and γ-Fe_2O_3: Preparation and Potential Application in Drug Delivery," *J.. Phys. Chem. C*, **112**, 1851-1856 (2008).

[82] J.X. Zhang, D.L. Jiang, S.H. Tan, L.H. Gui and M.L. Ruan, "Aqueous Processing of SiC Green Sheets: I, Dispersant," *J. Mater. Res.*, **17[8]** 2012-2018(2002).

[83] J.X. Zhang, F. Ye, D.L. Jiang and M. Iwasa, "Dispersion of Si_3N_4 Powders in Aqueous Media," *Colloids and Surfaces, A: Physiochem. and Eng.Asp.*, **259**, 117-123(2005).

[84] J.X. Zhang, D.L. Jiang, S.H. Tan, L.H. Gui and M.L. Ruan, "Aqueous Processing of TiC Green Sheets," *J. Am. Ceram. Soc.*, **84 [11]** 2537-2541(2001).

[85] J.X Zhang, L.P. Duan, D.L Jiang, Q.L. Lin and M. Iwasa, "Dispersion of TiN in Aqueous Media," *J. Colloid Interf. Sci.*, **286[1]** 209-215(2005).

[86] T. Tian, D.L.Jiang, J.X. Zhang and Q.L. Lin, "Aqueous Tape Casting Process of Hydroxyapatite," *J.Euro. Ceram. Soc.*, **27[7]** 2671-2677(2007).

[87] J.X. Zhang, Q. Xu, H. Tanaka, M. Iwasa and D.L. Jiang, "Improvement of the Dispersion of Al_2O_3 Slurries Using EDTA-4Na," *J. Am. Ceram. Soc.*, **89[4]** 1440-1442 (2006).

[88] J.X. Zhang, F. Ye, J. Sun, D.L. Jiang and M. Iwasa, "Aqueous Processing of Fine ZrO_2 Particles," *Colloids and Surfaces, A: Physiochem. and Eng.Asp.*, **254**, 199-205 (2005).

[89] J.X. Zhang, D.L. Jiang, S.Y. Qin, and Z.R. Huang, "Fracture Behavior of Laminated SiC Composites," *Ceram. Inter.*, **30[5]** 697-703(2004).

[90] S.Y. Qin, D.L. Jiang, J.X. Zhang, and J.N. Qin, "Evaluation of Weak Interface Effect on Residual Stress of Layered SiC/TiC Composites by Finite Element and X-Ray Diffraction," *J. Mater. Res.*, **17[5]** 1118-1124(2002).

[91] Y.P. Zeng, D.L. Jiang, and T. Watanabe, "Fabrication and Properties of Tape-Cast Laminated and Functionally Gradient Alumina-Titanium Carbide Materials," *J.Am.Ceram.Soc.*, **83[12]** 2999-3003(2000).

[92] Y.P. Zeng, D.L. Jiang, and P. Greil, "Tape Csting of Aqueous Al_2O_3 Slurries," *J.Euro.Ceram. Soc.*, **20[11]** 1691-1697 (2000).

[93] K.H. Zuo, D.L. Jiang, and Q.L. Lin, "Al_2O_3/Ni Laminar Ceramics Shaped by Tape Casting and Electroless Plating," *J.Am.Ceram.Soc.*, **88[9]** 2659-2661(2005).

[94] J.-X. Zhang, Z.-R. Huang, D.-L. Jiang, S.H. Tan, Z.J. Shen, and M. Nygren, "Preparation of TiN/ Al_2O_3 Composites with Laminated Structures," *J.Am.Ceram.Soc.*, **85[5]** 1133-1138(2002).

[95] A. C. Young, O. O. Omatete, M. A. Janney, and P. A. Menchhofer, ''Gelcasting of Alumina,'' *J. Am. Ceram. Soc.*, **74[3]** 612-8 (1991).

[96] T. Zhang, Z.Q. Zhang, and J.X. Zhang, et al., "Preparation of SiC Ceramics by Aqueous Gelcasting and Pressureless Sintering," *Mat. Sci. Eng. A-Struct*, **443[1-2]** 257-261 (2007).

[97] D.L. Jiang, "Gelcasting of Carbide Ceramics," J. Ceram. Soc. Jpn., **116[1354]** 694-699(2008).

[98] J.X. Zhang, D.L. Jiang, and Q.L. Lin, "Aqueous Processing of Boron Carbide Powders," *J. Ceram. Soc. Jpn.*, **116[1354]** 681-684 (2008).

[99] B.Q. Chen, Z.Q. Zhang, and J.X. Zhang, et al., "Aqueous Gel-casting of Hydroxyapatite," *Mat. Sci. Eng. A-Struct.*, **435**, 198-203(2006).

[100] H. Zhang, Y.J. Yan, and Z.R. Huang, et al., "Preparation and Characterization of Stable ZrB_2-Based Ultra-High Temperature Ceramics Slurry by Aqueous Gelcasting," *High-Performance Ceram. V, PTS 1-2*, **368-372**, 1756-1757(2008).

FABRICATION OF SILICON NITRIDE - MULTI-WALLED NANOTUBE COMPOSITES BY DIRECT IN-SITU GROWTH OF NANOTUBES ON SILICON NITRIDE PARTICLES

Amit Datye, Kuang-Hsi Wu, S. Kulkarni,
Mechanical and Materials Engineering, Florida International University
Miami, Florida 33174

H. T. Lin
Oak Ridge National Laboratory
Oak Ridge, TN 37831-6068.

J.Schmidt
Fraunhofer Institute for Manufacturing Engineering and Applied Materials Research
Dresden, Germany

D.Hunn
Lockheed Martin Missiles and Fire Control
Dallas, Texas

Wenzhi Li, Latha Kumari
Department of Physics, Florida International University
Miami, Florida 33172

ABSTRACT

In this research, Silicon Nitride (Si_3N_4)-Carbon Nanotube (CNT) composites were fabricated by direct in-situ growth of CNTs on the Si_3N_4 mixtures using Chemical Vapor Depositon (CVD) followed by Spark Plasma Sintering (SPS). The SPS technique used to sinter these powders is characterized by high heating and cooling rates coupled with pressure which prevents grain coarsening and also allows for densification in a very short period of time compared to the conventional sintering methods. The CVD techniques for in – situ CNT growth ensures a more uniform dispersion in the matrix than traditional ex-situ CNT mixing methods.

The sintered samples were analyzed using Field Emission Scanning Electron Microscopy (FEGSEM), X Ray Diffraction (XRD), Raman Spectroscopy and High Resolution Transmission Electron Microscopy (HRTEM). FEGSEM analysis of the Si_3N_4-CNT powders show uniform distribution of multi-walled nanotubes (MWNTs) in the matrix without the formation of bundles seen with traditional ex-situ mixing of CNTs in ceramic compositions. FEGSEM analysis of the fractured surface shows a uniform distribution of CNTs in the ceramic matrix. The presence of CNTs in the matrix is confirmed by Raman Spectroscopy and HRTEM. The Si_3N_4-MWNT composite thus fabricated shows a more uniform distribution of CNTs in the matrix and excellent CNT retention after sintering at 1850°C. FEGSEM analysis shows a finer grain size due to the presence of CNTs at grain boundaries which inhibit the diffusion related grain growth.

INTRODUCTION

Carbon nanotubes (CNTs) owing to their extraordinary properties have been the subject of intensive research over the last few years. These properties include high tensile strength (60 GPa)[1; 2] very high youngs modulus 1- 5 TPa [2; 3], very high aspect ratio and high thermal conductivity[4]. Based on their intrinsic structure they could be either metallic or semiconductor.[5] CNTs therefore can be used as ideal reinforcing fibers in ceramic and polymer matrix composites.[6] It was found out that composites fabricated by reinforcing brittle ceramics with CNTs can enhance the mechanical properties of the ceramic. In recent years, ceramic materials with CNTs reinforcements have been intensively studied for their potential applications.[7-9] Traditional ex-situ mixing of CNTs with the ceramic matrix composite have been tried by Ma et al.[10] who synthesized a CNT–SiC composite, Siegel et al. who formed a Al_2O_3–CNT composite[11], by An et al.[6], who synthesized a SiCN–CNT composite and more recently Balazsi et al.[9; 15; 16], who made a Si_3N_4–CNT composite. While all the above researchers showed enhancement of certain properties they all fall below the theoretical predications of enhancement of properties due to CNT addition. The effective utilization of CNTs in ceramic composites depends strongly on its uniform dispersion, interface bonding and CNT retention after sintering. These problems have hindered the development of ceramic–CNT composites because not much improvement in the properties was witnessed in previous research.[9; 10; 13; 24; 25] In order to solve the problem of uniform dispersion Peigney et al. made a Fe/Al_2O_3–CNT composite[12-14] by using direct in-situ growth of CNTs on Al_2O_3 using Fe as a catalyst.

Out of a variety of ceramics Si_3N_4 has attracted much attention due to its melting temperature, lightness, high mechanical strength at elevated temperatures and excellent fracture toughness.[17-21] Furthermore it is also corrosion resistant and is impervious to attack by most molten metals. These properties have made Si_3N_4 an ideal ceramic for applications that involve operation at high temperatures such as nozzles, spouts, and missile nose cones, rotor blades for turbines, and parts for reciprocating engines. However, pure Si_3N_4 is hard to sinter as it has a low ratio of the solid vapor surfaces to that of grain boundary energy, which is the driving force for sintering in such ceramic materials.[22] However, it was found that by using sintering additives and pressure the grain boundary energy can be decreased and almost fully dense samples can be achieved after sintering. The sintering aids added create a Liquid Phase (LP) which aids in the densification of the Silicon Nitride (Si_3N_4). These sintering aids which are mostly metal oxides react with the SiO_2 on the surface of the Si_3N_4 and with the Si_3N_4 itself to form an oxy-nitride liquid.[23] This liquid wets the surrounding particles causing more nitride particles to dissolve in it. Supersaturation of this liquid causes beta silicon nitride to precipitate which promotes densification.[24] This mechanism is much more complicated due to the formation of intermediate compounds during sintering of Si_3N_4 which consume the liquid phase or alter the composition of the liquid.[24] Theoretical predictions suggest that by combining the properties of both the CNTs and Si_3N_4, a ceramic with exceptional properties can be obtained.

In this study we have shown that by utilizing unique Chemical Vapor Deposition technique to grow CNTs in-situ on Si_3N_4 based mixtures a uniform dispersion of CNTs in the ceramic matrix can be achieved. High CNT retention is achieved by employing the Spark Plasma Sintering technique which is characterized by high heating and cooling rates and a shorter hold time. Results show that these uniformly dispersed CNTs can be retained even after sintering the in-situ CNT grown mixtures at 1850°C.

EXPERIMENTAL DETAILS

Powder Processing

Silicon Nitride (Si_3N_4, Grade M11 higher purity – HC Starck – Average particle size = 0.5 microns), Alumina(Alpha – Al_2O_3 powder Nanocrystalline – Inframat Advanced Materials) and Yttria (Y_2O_3 Nanopowder 99.95% purity – Inframat Advanced Materials) and Magnesium Oxide were mixed in various compositions by weight in the ratios as shown in Table I below. The Alumina and Yttria powders from Inframat Advanced Materials had an average particle size of 150 nm. The mixtures were first mixed in a rolling mill using baffled bottles for 48 hours, followed by using Si_3N_4 balls in a plastic bottle for another 48 hours. The powder mixture samples were then analyzed using X-Ray Diffraction

(XRD) and Scanning Electron Microscopy (SEM). The base mixtures were then further mixed using a Turbula multidirectional shaker for 24 hours before SPS sintering or catalyst coating for in-situ growth.

Table I: Composition of Various Si_3N_4 Based Mixtures

Sample	Composition Weight %			
	Si_3N_4	Y_2O_3	Al_2O_3	MgO
1E	95	5	-	-
1G	95	-	5	-
1A	90	6	4	-
1F	95	-	-	5

In-Situ Growth of MWNT on Si_3N_4 Mixtures

Previous research has shown that cobalt, which was used as a catalyst in the CNT growth, gives the highest yield of nanotubes per weight percent catalyst[39] addition. CNTs can be grown by using a cobalt based catalyst precursor.[39,40] $Co(NO_3)·6H_2O$ (Cobalt(II) nitrate hexadydrate, 98+%, A.C.S. reagent, Sigma-Aldrich) and Si_3N_4 mixtures were mixed in ethanol, followed by sonication for 15 min. The mixture is ground into powder after overnight drying. For CNT synthesis, Si_3N_4 mixtures containing cobalt contents (Co/Si_3N_4 = 2% or 5 wt%) were placed in a tube furnace modified with stirrers for CNT growth at 750°C.

CNT Yield Characterization

The CVD grown powders were analyzed using a field emission-gun scanning electron microscope (FEGSEM - JEOL JSM-633OF) and X-ray diffraction (XRD). XRD was performed using a Siemens D5000 Diffraktometer using Cu Ka radiation operated at 40 kV and 40 mA and fitted with a graphite monochromator. The step size and scan rate were fixed at 0.01° and 1sec/step respectively. The yield of the in-situ growth of CNTs on Si_3N_4 mixtures was characterized using a TGA. The samples were heated in an oxidizing atmosphere with a heating rate of 20°C/min in the temperature range 50–800°C under 20ml/min ambient air flow rate and the weight change was recorded.

Spark Plasma Sintering (SPS)

Spark plasma sintering of the powders was carried out using an equipment from FCT Systeme GmbH (Rauenstein, Germany), which allows several modes of operation (constant and pulsed DC current, pulse and pause times between 1 and 255 ms, different atmospheres: vacuum, N_2, Ar, high pressure forces: up to 250 kN). The baseline data were obtained by the SPS sintering of the base powder mixtures at 40MPa in a 45 mm graphite die at 1750°C and 1850°C for 4 minutes at peak temperature. The in-situ CNT grown powders were placed in a 30 mm diameter graphite die and sintered to full density under vacuum in a SPS unit. The SPS conditions are as follows: Applied pressure 60 MPa and peak temperature 1850°C with a dwell time of 4 min for the Si_3N_4 based powder mixtures.

Microstructural Characterization

Microstructure of the sintered samples was analyzed using the FEGSEM. The density of the sintered samples was measured by the Archimedes method with distilled water as the immersion medium. The XRD analysis was carried out for the powders before and after sintering to determine the chemical composition of each nanocomposite. The fracture surface of the samples was also analyzed using the FEGSEM.

CNT Retention

The CNT composites were characterized using Raman Spectroscopy to verify the CNT retention in the composites after sintering. The SPS sintered samples were fractured and Raman spectroscopy was carried out on the fracture surface. The Raman spectroscopy measurements were conducted at room temperature with a Raman spectrometer in the back scattering mode. An argon ion laser tuned to 514 nm was used to collect the data. To avoid a heating effect, the laser power was operated at 3 mW after filter to excite the sample. Raman spectra were collected using a high throughput holographic imaging spectrograph with a volume transmission grating, holographic notch filter, and thermoelectrically cooled charge coupled device (CCD) detector (Spectra Physics) with a resolution of 4 cm^{-1}. A 15-min exposure was used for each spectral collection.

RESULTS AND DISCUSSION

FEGSEM, HRTEM, XRD and TGA Analysis of In-situ Grown CNTs on Si_3N_4 Mixtures

Figures 1 and 2 show the representative FEGSEM images of the Si_3N_4 mixture powders after the CNT growth process. Prior to the SEM studies, the powders were oxidized for 30 minutes at 300°C in air to remove the amorphous carbon. Figure 1 shows the FEGSEM image of Sample 1A with 2% Cobalt catalyst. It is evident from these photos that CNTs were uniformly distributed throughout the powders. Figure 2 shows the in-situ grown CNTs on Sample 1G, which was coated with 2% Cobalt catalyst. As can be seen in Figure 2, a mesh of CNT has covered some area of the powder. The extent of CNT growth depends on the composition of the powder.

Figure 1: In-situ CNTs after CVD on 1A with 2%Co catalyst

Figure 2: In-situ CNTs after CVD on mixture 1G with 2%Co catalyst

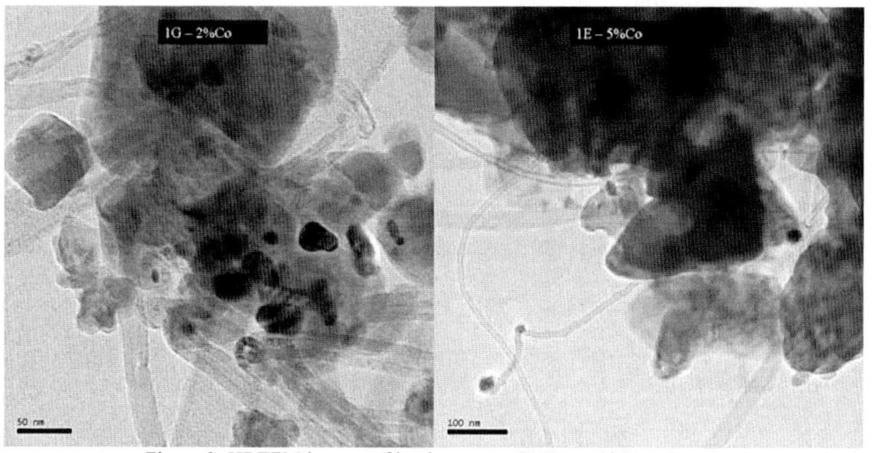

Figure 3: HRTEM images of in-situ grown CNTs on Si_3N_4 mixtures

The HRTEM studies in Figure 3, revealed that the CNTs produced in this study are MWNTs with an average diameter of 10-40nm and lengths from 1- 5 microns, evident in Figure 4. It is also noticed that the CNTs physically wrapped around the particles and there are no evident bundles in the mixture. In fact, the CNTs grown using this CVD process are all single CNTs.

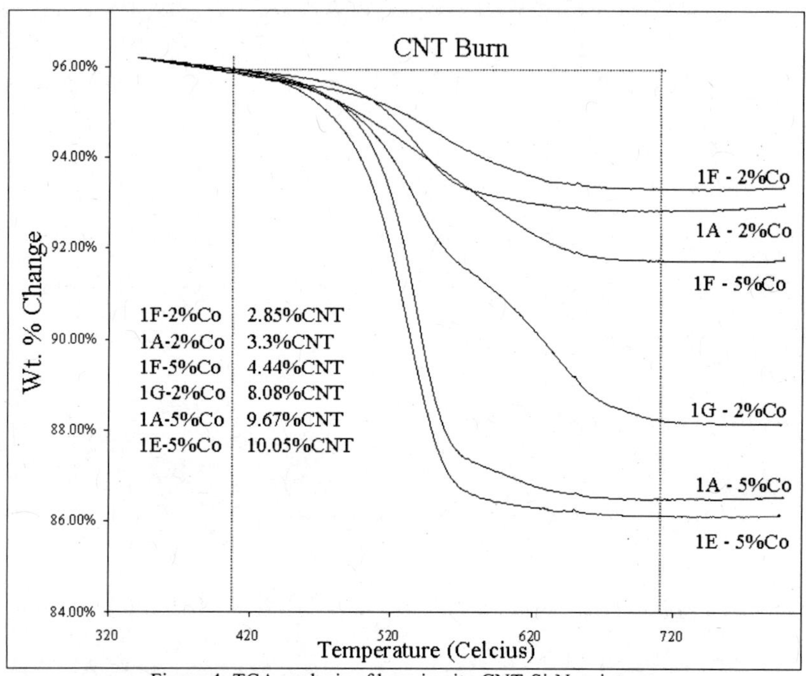

Figure 4: TGA analysis of base in-situ CNT-Si_3N_4 mixtures

Figure 4 shows the TGA analysis, which was conducted in dry air, of the mixtures after in-situ CNT growth. The TGA data show a sudden drop in the region ranging from 400°C to 600°C. Huang et al. showed that MWNTs start decomposing above 400°C in an oxidizing atmosphere with a rapid mass loss at 600°C, after which the curve flattens.[42] The weight percent loss of the sample from 400°C to 600°C can therefore be regarded as the yield of the CNTs in the CVD process. Amorphous carbon which deposited on the mixture during the CVD process was removed during the oxidation stage. The TGA data show expected trends - the CNT yield increases with an increasing catalyst content. The CNT yield for Sample 1F increases from ~2.85 to 4.44 wt% when the cobalt content increases from 2 to 5 wt%. An increase of cobalt catalyst from 2 to 5 wt% leads to a triple fold increase of CNT yield in Sample 1A, as illustrated in Figure 4. It has been well recognized experimentally by various researchers that adding more than 10 wt% CNT could lead to an adverse effect on the properties of the ceramic composite. Therefore a cobalt catalyst addition of greater than 5 wt% Co is not desirable.

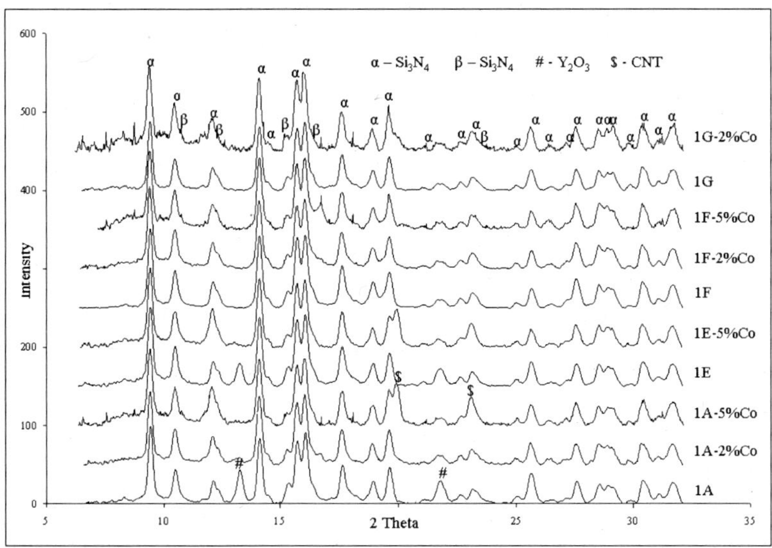

Figure 5: XRD on base Si_3N_4 mixtures and in-situ CNT- Si_3N_4 mixtures

Figure 5 shows the XRD data of the base powder mixtures before SPS sintering. These powders were heated in air at 300°C for 30 min to remove the amorphous carbon after the CNT growth. From this figure it can be seen that there is no reaction between any of the mixed powders. It is also noticed that the Si_3N_4 powder remains in its original α form.

Table I: Density Measurement of SPS Samples

Sample ID	Composition wt %						Relative Density
	Si_3N_4	Y_2O_3	Al_2O_3	MgO	Co	wt% CNT	
1E -5%Co	95	5	-	-	5	10.05	89.15%
1G -2%Co	95	-	5	-	2	8.01	88.96%
1A-2%Co	90	6	4	-	2	3.3	90.74%
1A-5%Co	90	6	4		5	9.67	89.16%
1F -2%Co	95	-	-	5	2	2.85	90.98%
1F -5%Co	95	-	-	5	5	4.44	89.50%
1E	95	5	-	-	-	-	96.10%
1G	95	-	5	-	-	-	98.67%

Density Measurement

The density measurements of the samples with and without CNTs after SPS are listed in Table II. It can be observed that there is a clear correlation between the amount of CNTs and the density, as expected. At 0% CNT the sample attained more than 95% of the theoretical density. Adding more

CNTs decreases the relative density. This decrease in density can be attributed to the hollow nature of the multi-walled nanotubes in the composite. The fact that almost all samples achieve near or above 90% of the theoretical density implies that the CNTs have been tightly integrated into the ceramic matrix without causing a large amount of porosity.

Microstructure of Fractured Surfaces of SPS Sintered Samples using FEGSEM
The FEGSEM images of the fracture surfaces of the base samples, as seen in Figures 6 and 7, show the presence of beta Si_3N_4 in the matrix. The fracture mode is mostly intergranular for the base sintered samples. Figures 8 -11 show the FEGSEM images of fracture surfaces of Samples 1A-2%Co, 1A-5 %Co, 1E-5 %Co and 1F-5 %Co, respectively. Again, through these micrographs, it is clear that the CNTs have been uniformly distributed in the ceramic matrix by means of the in-situ growth method. It can also be noticed that the fracture surface exhibits a mixed mode of intergranular and transgranular fracture, which is indicative of an enhancement in fracture toughness. CNTs distributed along the grain boundaries are clearly evident in Figures 8-11. Some CNT pull-outs are observed in most Si_3N_4-CNT samples, indicating that some CNTs were well-bonded in the matrix. In addition to more uniform CNTs distribution in the matrix, samples with higher CNT contents also appear to have smaller grain size. While the base Si_3N_4 mixture samples shows an average grain size of 2-5 μm, the grain size in Sample 1F with 4.44% CNT is ~1μm. This implies that, in addition to the strengthening effect, the CNT wrapping around the ceramic particles may serve as inhibitor for grain growth. Hence, addition of CNTs could minimize grain growth, leading to a finer grain ceramic-CNT nanocomposite with higher fracture toughness.

Figure 6: Fracture surface of sample 1E

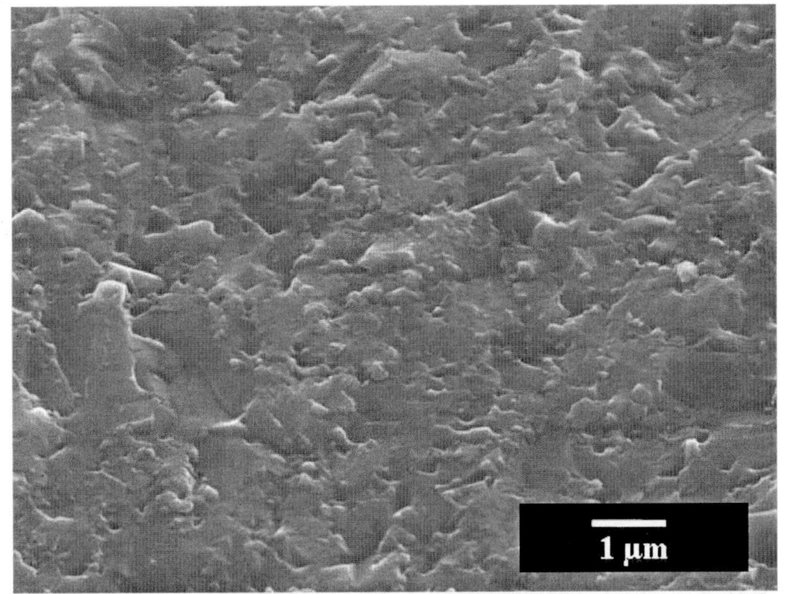

Figure 7: Fracture surface of sample 1G.

Figure 8: Fracture surface of sample 1A-2%Co.

Figure 9: Fracture surface of sample 1A-5%Co

Figure 10: Fracture surface of sample 1E-5%Co

Figure 11: Fracture surface of sample 1F-5%Co

In Figure 12, the XRD spectra of the base powders used in this study are plotted along with the results of Si_3N_4-CNT powders. From the graphs it can be seen that there is no unwanted reaction between the constituents however there is a significant amount of β phase formed after sintering in all samples when compared with the original powders.

Raman Spectroscopic Analysis

Figure 13 shows the Raman spectroscopy of the fracture surface of the SPS sintered CNT-grown samples. This figure clearly demonstrates that the CNTs are retained after the SPS sintering process. The peaks from the MWNTs seen at 1353 (D–band) and 1580 (G–band) in all the patterns are in good agreement with the MWNT modes reported in the literature.[43; 44] The other peaks seen in the pattern of 1A-2%Co sample were identified as the β Si_3N_4. These peaks are not seen in the other patterns as the laser spot on those cases was focused on the CNT rich area of the fracture surface of the sintered samples. The results confirm the important fact that the CNTs are retained after the SPS process.

HRTEM Studies

CNT retention is confirmed by the HRTEM studies on the samples sintered at 1850°C. As seen from Figure 14, the CNTs are present in all the sintered samples. The CNTs seem to be damaged due to the nature of the SPS process. However, the damage appears to be limited to the outer walls and the inner layers of CNTs remain to be intact. The MWNTs seem not to have undergone graphitization even at 1850°C.

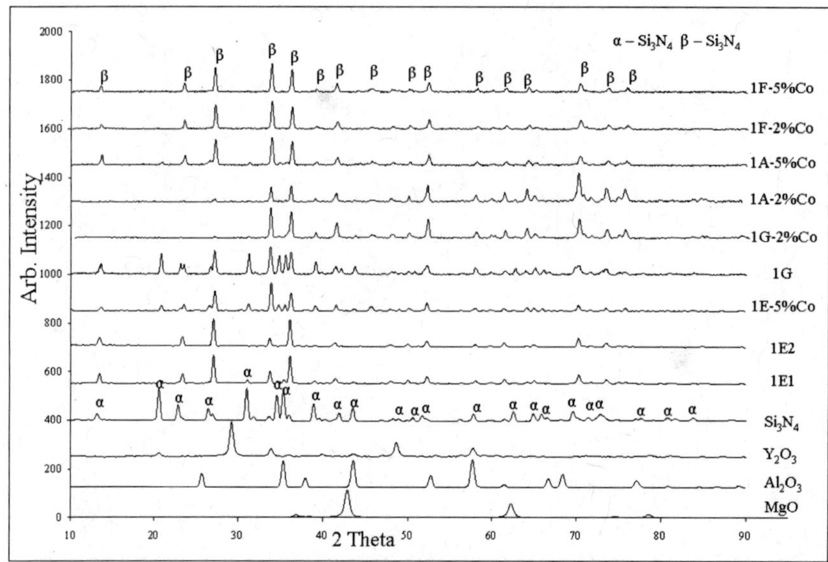

Figure 12: XRD patterns of samples after SPS

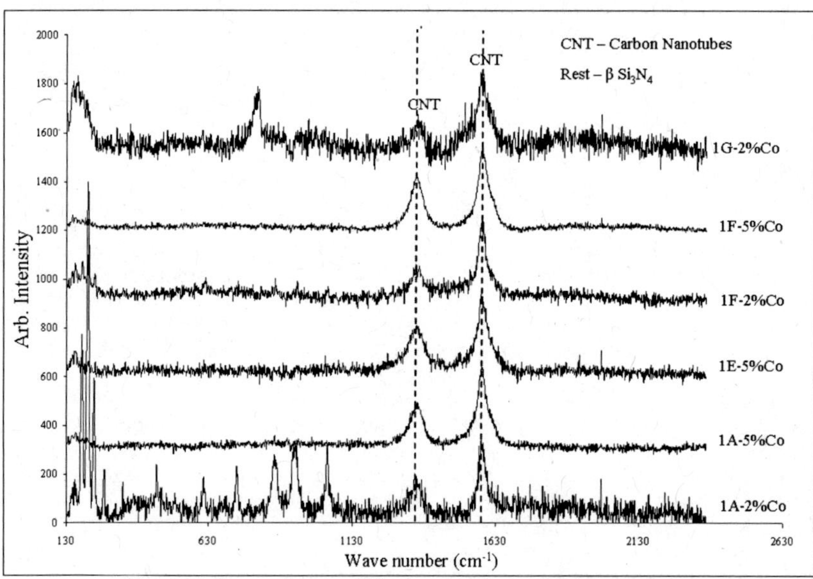

Figure 13: Raman Spectra of fractured surfaces of sintered samples

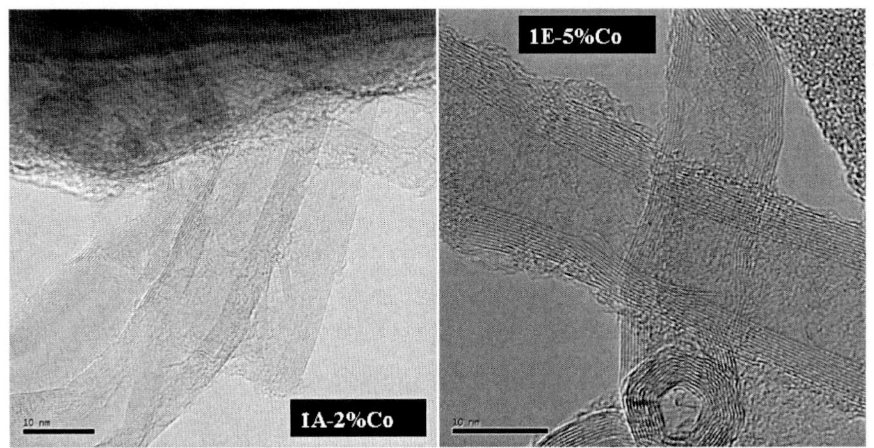

Figure 14: HRTEM of Sample 1A-2%Co and 1E–5%Co sintered at 1850C

CONCLUSION

In this research multi-walled carbon nanotubes have been directly grown in-situ on various Silicon Nitride – metal oxide mixtures. The grown MWNTs have been characterized using FEGSEM and the yield with different catalyst ratios has been analyzed using TGA. The Si_3N_4 powder mixtures with the in-situ grown nanotubes were then sintered to ~90% theoretical density using spark plasma sintering. Analysis of the fracture surfaces of the composite when sintered at 1850°C shows uniformly distributed CNTs in the entire ceramic matrix due to the direct growth of CNTs on the powder particles.

The study clearly demonstrated that a uniform distribution of CNTs by in – situ growth can be achieved. Further densification of the samples by the SPS technique results in a 90% theoretical density with a high degree of CNT retention. The method developed in this research is thus better suited for bulk manufacturing of the ceramic composites, with a uniform distribution of MWNTs in the matrix, than conventional sintering methods. Through this study, it is also observed that wrapping the CNTs around the ceramic particles, via this in-situ method, can minimize the grain growth of the ceramic powders and could lead to fine grain size composites. A smaller grain-size ceramic composite is highly desirable for the purpose of superior fracture toughness.

REFERENCES

1. M. F. Yu, O. Lourie, M. J. Dyer, K. Moloni, T. F. Kelly, and R. S. Ruoff, "Strength and breaking mechanism of multiwalled carbon nanotubes under tensile load," Science, 287 637–640.2000.
2. M. F. Yu, B. S. Files, S. Arepalli, and R. S. Ruoff, "Tensile loading of ropes of single wall carbon nanotubes and their mechanical properties," Physical Review Letters, 84[24] 5552-5555.2000.
3. B. I. Yakobson, C. J. Brabec, and J. Bernholc, "Nanomechanics of carbon tubes: Instabilities beyond linear response," Physical Review Letters, 76[14] 2511-2514.1996.
4. S. Berber, Y. K. Kwon, and D. Tomanek, "Unusually high thermal conductivity of carbon nanotubes," Physical Review Letters, 84[20] 4613-4616.2000.

5. N. Hamada, S. Sawada, and A. Oshiyama, "New One-Dimensional Conductors - Graphitic Microtubules," Physical Review Letters, 68[10] 1579-1581.1992.
6. L. N. An, W. X. Xu, S. Rajagopalan, C. M. Wang, H. Wang, Y. Fan, L. G. Zhang, D. P. Jiang, J. Kapat, L. Chow, B. H. Guo, J. Liang, and R. Vaidyanathan, "Carbon-nanotube-reinforced polymer-derived ceramic composites," Advanced Materials, 16[22] 2036-+.2004.
7. X. Wang, N. P. Padture, and H. Tanaka, "Contact-damage-resistant ceramic/single-wall carbon nanotubes and ceramic/graphite composites," Nat Mater, 3[8] 539-544.2004.
8. J. Tatami, T. Katashima, K. Komeya, T. Meguro, and T. Wakihara, "Electrically conductive CNT-dispersed silicon nitride ceramics," J. Am. Ceram. Soc., 88[10] 2889-2895.2005.
9. C. Balazsi, Z. Konya, F. Weber, L. P. Biro, and P. Arato, "Preparation and characterization of carbon nanotube reinforced silicon nitride composites," Materials Science & Engineering C-Biomimetic and Supramolecular Systems, 23[6-8] 1133-1137.2003.
10. R. Z. Ma, J. Wu, B. Q. Wei, J. Liang, and D. H. Wu, "Processing and properties of carbon nanotubes-nano-SiC ceramic," Journal of Materials Science, 33[21] 5243-5246.1998.
11. R. W. Siegel, S. K. Chang, B. J. Ash, J. Stone, P. M. Ajayan, R. W. Doremus, and L. S. Schadler, "Mechanical behavior of polymer and ceramic matrix nanocomposites," Scripta Materialia, 44[8-9] 2061-2064.2001.
12. A. Peigney, C. Laurent, E. Flahaut, and A. Rousset, "Carbon nanotubes in novel ceramic matrix nanocomposites," Ceramics International, 26[6] 677-683.2000.
13. A. Peigney, C. Laurent, O. Dumortier, and A. Rousset, "Carbon nanotubes Fe alumina nanocomposites. Part I: Influence of the Fe content on the synthesis of powders," Journal of the European Ceramic Society, 18[14] 1995-2004.1998.
14. C. Laurent, A. Peigney, and A. Rousset, "Synthesis of carbon nanotube Fe-Al2O3 nanocomposite powders by selective reduction of different Al1.8Fe0.2O3 solid solutions," Journal of Materials Chemistry, 8[5] 1263-1272.1998.
15. C. Balazsi, Z. Shen, Z. Konya, Z. Kasztovszky, F. Weber, Z. Vertesy, L. P. Biro, I. Kiricsi, and P. Arato, "Processing of carbon nanotube reinforced silicon nitride composites by spark plasma sintering," Composites Science and Technology, 65[5] 727-733.2005.
16. C. Balazsi, K. Sedlackova, and Z. Czigany, "Structural characterization of Si3N4-carbon nanotube interfaces by transmission electron microscopy," Composites Science and Technology, 68[6] 1596-1599.2008.
17. F. F. Lange, "Fracture-Toughness of Si3n4 as a Function of the Initial Alpha-Phase Content," Journal of the American Ceramic Society, 62[7-8] 428-430.1979.
18. F. F. Lange, S. C. Singhal, and R. C. Kuznicki, "Phase Relations and Stability Studies in Si3n4-Sio2-Y2o3 Pseudoternary System," Journal of the American Ceramic Society, 60[5-6] 249-252.1977.
19. G. Ziegler, J. Heinrich, and G. Wotting, "Relationships between Processing, Microstructure and Properties of Dense and Reaction-Bonded Silicon-Nitride," Journal of Materials Science, 22[9] 3041-3086.1987.
20. H. H. Lu and J. L. Huang, "Microstructure in silicon nitride containing beta-phase seeding: Part I," Journal of Materials Research, 14[7] 2966-2973.1999.
21. Z. L. Hong, H. Yoshida, Y. Ikuhara, T. Sakuma, T. Nishimura, and M. Mitomo, "The effect of additives on sintering behavior and strength retention in silicon nitride with RE-disilicate," Journal of the European Ceramic Society, 22[4] 527-534.2002.
22. J. Y. Park and C. H. Kim, "The Alpha-Si3n4 to Beta-Si3n4 Transformation in the Presence of Liquid Silicon," Journal of Materials Science, 23[9] 3049-3054.1988.
23. D. R. Messier, F. L. Riley, and R. J. Brook, "Alpha-Beta-Silicon Nitride Phase-Transformation," Journal of Materials Science, 13[6] 1199-1205.1978.
24. M. H. Lewis, B. D. Powell, P. Drew, R. J. Lumby, B. North, and A. J. Taylor, "Formation of Single-Phase Si-Al-O-N Ceramics," Journal of Materials Science, 12[1] 61-74.1977.

25. S. Rul, F.Lefèvre-schlick, E.Capria, C. Laurent, and A. Peigney, Percolation of single-walled carbon nanotubes in ceramic matrix nanocomposites, 2004 52: 1061.Acta Mater.
26. A. Peigney, Ch. Laurent, E. Flahaut, and A. Rousset Carbon nanotubes in novel ceramic matrix nanocomposites. Ceram Int 2000; 26:677–83.
27. G. Van Lier, C. Van Alsenoy, V. Van Doren, and P. Geerlings, Ab initio study of the elastic properties of single-walled carbon nanotubes and graphene. Chem Phys Lett 2000; 326:181–5.
28. M.M.J. Treacy, T.W. Ebbesen, and J.M. Gibson, Exceptionally high Young's modulus observed for individual carbon nanotubes. Nature 1996; 381:678–80.
29. A. Thess, R. Lee, P. Nikolaev, H. Dai, P. Petit, J. Robert, C. Xu, Y.H. Lee, S.G. Kim, A.G. Rinzler, D.T. Colbert, G.E. Scuseria, D. Tomanek, J.E. Fischer,and R.E. Smalley, Crystalline ropes of metallic carbon nanotubes. Science 1996; 273:483–7.
30. Ando Y, Zhaoa X, Shimoyama H, Sakai G, and Kaneto K, Physical properties of multiwalled carbon nanotubes. Int J Inorg Mater 1999; 1:77–82.
31. M.J. Biercuk, M.C. Llaguno, M. Radosavlijevic, J.K. Hyun, and A.T. Johnson, Carbon nanotube composites for thermal management. Appl Phys Lett 2002; 80:2767–9.
32. R.H. Baughman, A.A. Zakhidov, and W.A. de Heer, Carbon nanotubes—the route toward applications. Science 2000; 297:787–92.
33. K.M. Prewo, Fiber-reinforced ceramics: new opportunities for composite materials. Am. Ceram Soc Bull 1989; 68:395–400.
34. J. Tatami, T. Katashima, K. Komeya, T. Meguro, and T. Wakihara, Electrically Conductive CNT-Dispersed Silicon Nitride Ceramics. Journal of American Ceramic Society 2005; 88 [10]: 2889–93.
35. E.T. Thostenson, Z. Ren, and T.W. Chou, Advances in the science and technology of carbon nanotubes and their composites: a review. Compos Sci Technol 2001; 61:1899–912.
36. H. Dai, Carbon nanotubes: opportunities and challenges. Surf Sci 2002; 500:218–41.
37. J. Sun, and L. Gao, Attachment of Inorganic Nanoparticles onto Carbon Nanotubes. J Electroceram 2006; 17:91–94.
38. G.D. Zhan, J.D. Kuntz, J.E. Garay, and A.K. Mukherjee, Electrical properties of nanoceramics reinforced with ropes of single-walled carbon nanotubes. Appl. Phys. Lett. 2003; 83: 1228.
39. H.J. Dai, A.G. Rinzler, P. Nikolaev, A. Thess, D.T. Colbert, and R.E. Smalley, Single-wall nanotubes produced by metal-catalyzed disproportionation of carbon monoxide. Chem. Phys. Lett 1996; 260: 471.
40. W.Z. Li, J.G. Wen, M. Sennett, and Z.F. Ren, Clean double-walled carbon nanotubes synthesized by CVD. Chem Phys Lett 2003; 368:299–306.
41. X. Wang, N.P. Padture, and H. Tanaka, Contact-damage-resistant ceramic/single-wall carbon nanotubes and ceramic/graphite composites. Nat Mater 2004; 3:539–44.
42. W. Huang, Y. Wang, G. Luo, and F. Wei, 99.9% purity multi-walled carbon nanotubes by vacuum high-temperature annealing, Carbon 41 (2003) 2585–2590.
43. A.P. Naumenko, N.I. Berezovska, M.M. Biley, and O.V. Shevchenko, Vibrational analysis and Raman spectra of tetragonal Zirconia, Physics and Chemistry of Solid State, 2008, 121 – 125.
44. P.C. Eklund, J.M. Holden, and R.A. Jishi, Vibrational modes of carbon nanotubes; spectroscopy and theory. Carbon 1995, 959-972.

SYNTHESIS OF YTTRIA STABILIZED ZIRCONIA (3YTZP) - MULTI-WALLED NANOTUBE (MWNTs) NANOCOMPOSITE BY DIRECT IN-SITU GROWTH OF MWNTs ON ZIRCONIA PARTICLES

Amit Datye, Kuang-Hsi Wu, V. Monroy, S. Kulkarni, S. Amruthaluri
Mechanical and Materials Engineering
Florida International University
Miami, Florida 33174

H. T. Lin
Oak Ridge National Laboratory
Oak Ridge, TN 37831-6068.

J.Vleugels, K. Vanmeensel
Department of Metallurgy and Materials Engineering
Katholieke Universiteit Leuven
Leuven, Belgium

Wenzhi Li, Latha Kumari
Department of Physics
Florida International University
Miami, Florida 33174

ABSTRACT

Zirconia is one of the most widely investigated structural ceramics and is widely used as a refractory material in metallurgy and structural components for engine applications. Zirconia exhibits high strength, fracture toughness and wear resistance properties compared to other ceramics, but like all other ceramics, it is brittle. Various attempts have been made to reinforce Zirconia with reinforcing fibers[2-26]. Carbon Nanotubes (CNTs) are 100 times stronger than steel with only 1/6 of its mass and have thermal conductivity comparable with diamond. These remarkable properties make CNTs the ideal reinforcements for nanocomposites. The application of CNTs with ceramic based nanocomposites has been greatly hindered by its lack of interface bonding with matrix material and its poor dispersion in the matrix.

In this research, Yttria Stabilized Zirconia (3YTZP) - CNT composites are manufactured by direct in-situ growth of CNTs on the Zirconia particles followed by densification via Spark Plasma Sintering (SPS) technique. Detailed electron microscopy analysis of the 3YTZP-CNT powders shows uniform distribution of CNTs in the matrix without the formation of agglomerates frequently seen with traditional ex-situ mixing of CNTs in ceramic compositions. The samples were sintered to nearly 100% theoretical density and showed a finer grain size microstructure. Raman Spectroscopy confirms CNT retention in the sintered nanocomposite.

INTRODUCTION

Carbon nanotubes (CNTs) have been an attractive candidate for fundamental research since their discovery by Iijima[1]. Theoretical and experimental results have shown extremely high elastic modulus, greater than 1 TPa (the elastic modulus of diamond is 1.2 TPa)[2-6], and reported strengths 10–100 times higher than the strongest steel at a fraction of its weight. In addition to their extraordinary mechanical properties, carbon nanotubes also possess superior thermal properties. They are thermally stable up to

2800°C in vacuum, and the thermal conductivity is about twice as high as that of diamond[7-8]. CNTs with their outstanding mechanical properties and extraordinarily high aspect ratios are, therefore, the ideal reinforcing fibers for ceramic matrix composites.[9-10]

Zirconia (ZrO_2) is a technologically important material due to its stability at high temperatures, large energy bandgap, high breakdown electric field and low leakage current level. It has been extensively used in applications as a refractory material, in orthopedic implants, in common high temperature applications such as seals, valves and pump impellers and as synthetic gemstones. Recently, Zirconia has been used in solid oxide fuel cells (SOFC), oxygen sensors, ceramic membrane oxygen separation technology and high-temperature steam electrolysis[11,12]. Several applications were proposed for CNTs, many of which are concerned with conductive or high strength composites[13-15]. It is anticipated that the inclusion of CNTs in a ceramic matrix would allow one to produce composites with high stiffness and improved mechanical properties compared to that of the single phase ceramic material[14]. Several researchers have tried previously to incorporate CNTs into Zirconia matrix, but the results are far short of the predicted theoretical enhancement of properties in the CNT-ceramic composites. This is due to nonuniform dispersion of CNTs in the matrix and poor interface bonding between the ceramic and CNTs[13-15]. Lupo et al.[15] successfully fabricated ZrO_2-CNT composites by hydrothermal crystallization at 200°C for 8 h of zirconium hydroxide in the presence of carbon nanotubes. It was reported that CNTs acted as nucleation sites for the growth of ZrO_2 crystallites. SEM and TEM studies showed that the Zirconia particles tend to precipitate on and around the CNTs. Jing et.al[16] tried to overcome the difficulties regarding incorporation of CNTs in structural and functional ceramics by coating the CNTs with materials that will eliminate the undesirable attractive interactions between the nanotubes and facilitate their incorporation into composites. Ferlauto et al.[17,18] produced multi-walled carbon nanotubes (MWNTs) by chemical vapor deposition (CVD) using Yttria-stabilized Zirconia/nickel (YSZ/Ni) catalysts. The catalysts were obtained by a liquid mixture technique that resulted in fine dispersed nanoparticles of NiO supported in the YSZ matrix. The idea was to establish a direct route for the production of Ni/YSZ/CNTs composites. Shiw et al.[19] prepared MWNT/3Y-TZP composites by the spark plasma sintering (SPS) method and determined the DC electrical conductivity and dielectric behavior at room temperature. The relative density of all of the samples was higher than 95%, and the dielectric constant of the composites was greatly enhanced when the MWCNT concentration was close to the percolation threshold, which was attributed to the critical behavior of the dielectric constant near the percolation threshold as well as the polarization effects inside the composites. Even though Zirconia exhibits superior properties than most of the other ceramics, its mechanical properties decrease drastically at high temperature. This is due to thermally activated grain boundary (GB) sliding, which leads to plastic or even super-plastic deformation[20]. The 3Y-TZP reinforced CNT composite prepared by conventional sintering shows that this composite material exhibits better resistance to GB sliding at high temperature, due to CNT amorphization observed at the GBs, which might be due to the destruction of the crystalline structure of the carbon sheets by conventional sintering. The same was observed by Ionascu et al., when they prepared 3Y-TZP-MWNT composites using powder metallurgy[21]. Other researchers[22-28] who tried various conventional processing and sintering techniques have found experimentally that the CNTs enhance the electrical and the mechanical properties of the 3Y-TZP ceramics. All of the previous researches were carried out by mixing CNTs ex-situ with the ceramic powders. The major drawbacks of the ex-situ mixing technique are the nonuniform dispersion of nanotubes in the ceramic matrix and the formation of CNT bundles in the ceramic matrix, which would have a negative impact on the mechanical properties of the ceramic.

In this research, a unique approach to manufacture MWNT reinforced Yttria-Tetragonal Zirconia Polycrystals (3YTZP) nanocomposite by direct in-situ growth of MWNTs on the 3Y-TZP particles followed by spark plasma sintering is presented. Direct in-situ growth of MWNTs on ceramic particles has been tried before by A.Peigney et al.[2] on alumina and magnesium oxide using Iron as a catalyst with encouraging results. Enhancement of both the mechanical and thermal properties of the ceramic was observed. In this research, modifications to the CVD growth technique and optimization of catalysts and coating techniques for better CNT yield and uniform growth were made.

EXPERIMENTAL DETAILS

In-Situ Growth of MWNT on Spray Dried 3YTZP Powder

Previous research has shown that cobalt gives the highest yield of nanotubes per wt. percent addition. CNTs can be grown by using $Co(NO_3)\cdot 6H_2O$ as a catalyst precursor[29,30]. $Co(NO_3)\cdot 6H_2O$ (Cobalt(II) nitrate hexadydrate, 98+%, A.C.S. reagent, Sigma-Aldrich) and 3YTZP powder (HC Starck – spray dried with 10-100 micron agglomerates with individual average particle size of 40 nm) were mixed in ethanol, followed by sonication for 15 min. Then the mixture was dried overnight at 125°C in a furnace followed by ball milling into fine nanopowder. For CNT synthesis, 3YTZP powder containing three different cobalt contents (Co/3YTZP = 1, 2, and 5 wt%) was placed in a tube furnace fitted with a stirrer for all around growth of CNTs on the 3YTZP particles by CVD technique.

CNT Yield Characterization

The CVD grown powders were analyzed using a field emission scanning electron microscope (FESEM - JEOL JSM-633OF) and X-ray diffraction (XRD). XRD was performed using a Siemens D5000 Diffraktometer employing Cu Ka radiation operated at 40 kV and 40 mA and fitted with a graphite monochromator. The step size and scan rate were fixed at 0.01° and 1sec/step, respectively. The yield of the in-situ growth of CNTs on 3YTZP was characterized using TGA (TA Instruments: High–Res TGA 2950 Thermogravimetric Analyzer). The samples were heated in an oxidizing atmosphere with a heating rate of 20°C/min in the temperature range 50–800°C under 20ml/min ambient air and the weight change was recorded.

Spark Plasma Sintering (SPS)

Spark plasma sintering of the powders was carried out using equipment from FCT Systeme GmbH (Rauenstein, Germany) allowing several modes of operation (constant and pulsed DC current, pulse and pause times between 1 and 255 ms, different atmospheres: vacuum, N_2, Ar, high pressure forces: up to 250 kN). The base and in-situ CNT grown 3YTZP nanopowders were placed in a 30 mm diameter graphite die and sintered to full density under vacuum in SPS unit. The SPS conditions are as follows: Applied pressure 60 MPa, peak temperature 1200°C, 1400°C, and 1600°C with a dwell time of 2 and 4 min for the 3Y-TZP base powders. It was found that the samples sintered at 1400°C and 1600°C gave >99% theoretically dense samples, and, therefore, all the samples were sintered at 60MPa for 4 minutes at 1400°C and 1600°C.

Microstructural Characterization

Microstructure of the sintered samples was analyzed using the FESEM. The density of the sintered samples was measured by Archimedes method (BP210S balance, Sartorius AG, Germany) with isopropanol as the immersion medium. XRD analysis was carried out for the powders before and after sintering to determine the chemical composition of each nanocomposite. The fractured surface of the samples was also analyzed using the FESEM.

CNT Retention

CNT composites were characterized using Raman Spectroscopy to ensure the CNT retention in the composites after sintering. The SPS sintered samples were fractured, and Raman spectroscopy was carried out on the fractured surface. The Raman spectroscopy measurements were conducted at room temperature with a Raman spectrometer in the back scattering configuration. An argon ion laser tuned to 514 nm was used to collect the patterns. To avoid a heating effect, the laser power was operated at 3 mw after filter to excite the sample. Raman spectra were collected by using a high throughput holographic imaging spectrograph with a volume transmission grating, holographic notch filter, and thermoelectrically cooled charge coupled device (CCD) detector (Spectra Physics) with a resolution of 4 cm^{-1}. A 15-min exposure was used for each spectral collection.

RESULTS AND DISCUSSION

The sintered samples will follow a nomenclature 3YTZP-X-Y where X is the wt% of cobalt catalyst used and Y is the sintering temperature in celcius.

FESEM, HRTEM, XRD and TGA analysis of in-situ grown CNTs on Zirconia

Figure 1 shows the FESEM images of in-situ grown CNTs on 3YTZP particles after the CVD process. Figure 1 (a) shows the FESEM image of the raw powder of 3%Y-TZP acquired from HC Stark. This powder is in the form of spherical agglomerates of a size between 20-60 μm and the individual particle size is in the range of 40-50 nm. The growth conditions were the same for all except for the catalyst content. It is evident that the CNT content increases with an increase of the catalyst percentage, as illustrated in Figures 1 (b), (c) and (d). The CNTs grown are MWNTs with an average diameter of 10-40nm and lengths from 1- 5 microns.

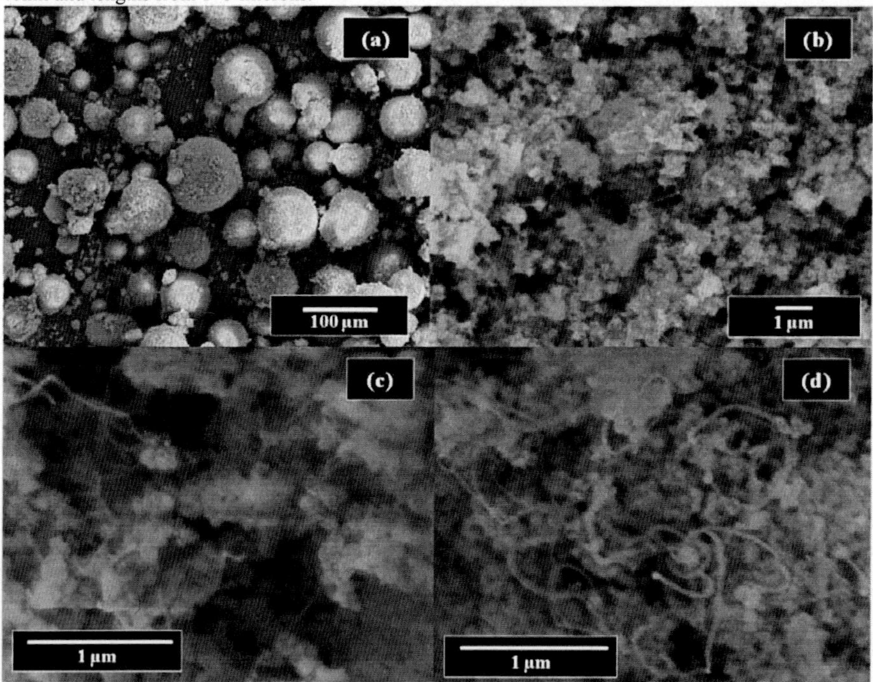

Figure 1: FESEM images of (a) Base spray dried 3YTZP (b) CNT on 3YTZP for 1% Co (c) CNT on 3YTZP for 2% Co and (d) CNT on 3YTZP for 5% Co

Figure 2 shows the TGA analysis on the zirconia base powder and the in-situ CNT grown 3YTZP with various catalyst contents. Huang et al. showed that MWNTs start decomposing above 400°C in an oxidizing atmosphere with a rapid loss at 600°C after which the curve flattens[34]. The weight percent loss of the sample from 400°C to 600°C is taken as the yield of the CNTs in the CVD process. It is important to note that a significant amount of amorphous carbon is also deposited after the CVD growth process, shown by the drop in the curve from 125°C to 400°C in Fig. 2 and must be removed prior to SPS sintering. It can be seen from the data that the CNT yield increases with an increasing catalyst content, as expected. The yield is ~4.2 wt% for 5% Co catalyst and ~3wt% for 2% Co catalyst, respectively. Since 3YTZP particles are heavy (Density ~ 6.2 gm/cm^3), a 4.2% by weight translates to a great number of nanotubes in the composite by volume. CNTs are hollow structures it has been proven experimentally by various researchers that an increase in the CNT/matrix ratio of more than 10% by volume has a negative effect on the mechanical properties of the ceramic composite; therefore, a higher catalyst ratio than 5% Co which gives around 5% by weight CNTs is not desirable.

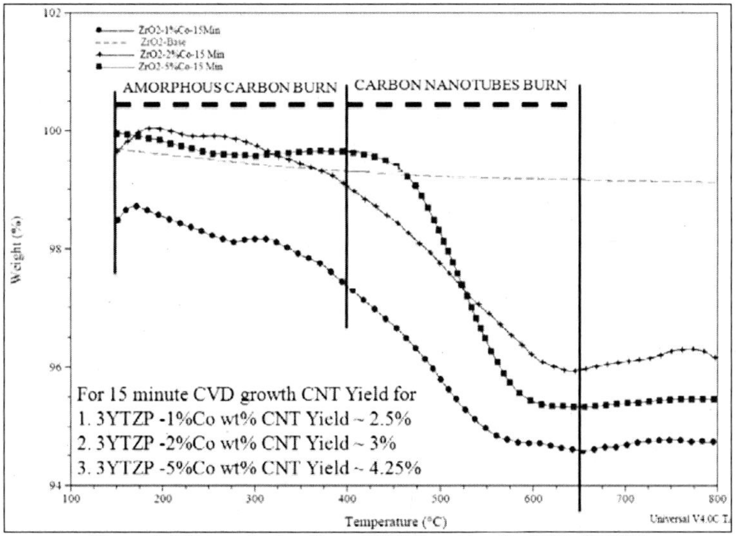

Figure 2: TGA analysis of base 3YTZP and in-situ CNT- 3YTZP

Figure 3 shows the XRD data for 3YTZP and the 3YTZP – CNT powders after heating them in an oxidizing atmosphere in an oven for 30 minutes at 300°C, to remove the amorphous carbon present in the after CVD powders. As evident from the XRD data, there is no change in the phase of 3YTZP of the base powders after the CVD process. In the processed powders after CVD, it was observed that the majority of the peaks corresponded to Zirconia in the tetragonal phase, which was expected since the nanopowders used were stabilized in the tetragonal phase with the 3% Yttria. Traces of the monoclinic phase might have been due to the processing methods used to synthesize the powder.

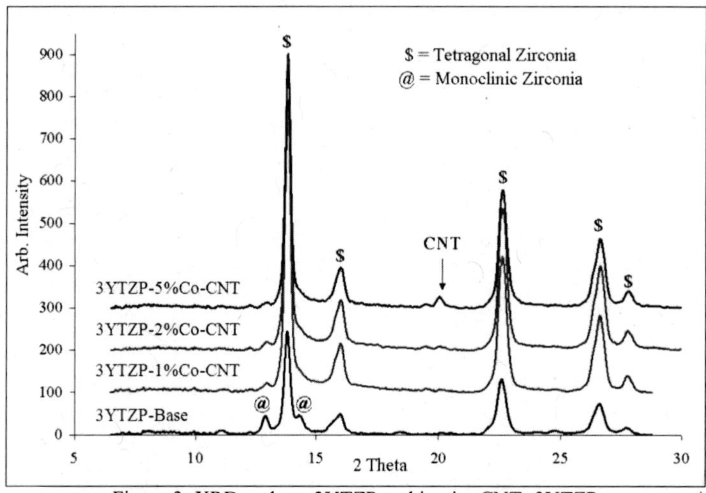

Figure 3: XRD on base 3YTZP and in-situ CNT- 3YTZP

Density Measurement

The density measurements for the samples with and without CNTs after SPS are shown in Table 1. It can be observed that there is a correlation between the amount of CNTs and density. At 0% CNT the density of the sample increases with the increasing temperature. The density of the pure 3YTZP sintered in these studies is in good agreement with the reported density of pure Zirconia of 6.04 g/cm^3. Based on the trend of the data in Table 1, it is clear that there is a strong correlation between the sintering temperature and the density of the final products. However, for the CNT-3YTZP samples, adding CNTs apparently slightly reduces the density when compared with the pure 3YTZP samples. There is a trade off in the density with an increased concentration of CNTs due to the hollow nature of CNTs.

Table 1: Density measurement of SPS samples

	Sample ID	Density gm/cm^3
w/o CNT	3YTZP-1200	6.02 ± 0.02
	3YTZP-1400	6.05 ± 0.02
	3YTZP-1600	6.05 ± 0.02
w/ CNT	3YTZP-1-1400	5.94 ± 0.02
	3YTZP-1-1600	5.99 ± 0.02
	3YTZP-2-1400	6.02 ± 0.02
	3YTZP-2-1600	6.03 ± 0.02
	3YTZP-5-1400	6.02 ± 0.02
	3YTZP-5-1600	6.03 ± 0.02

Microstructure of fractured surfaces of SPS sintered samples using FESEM

FESEM imaging of the fractured surfaces show that the CNTs have been uniformly distributed in the ceramic matrix due to the in-situ growth of the CNTs on the ceramic particles. Figure 4 shows the fractured surface of the sample with 5% wt catalyst sintered at 1600°C for 4 minutes. As seen from the

FESEM image, the CNTs are seen uniformly in the matrix, some of them broken and some of them intact on the fractured surface.

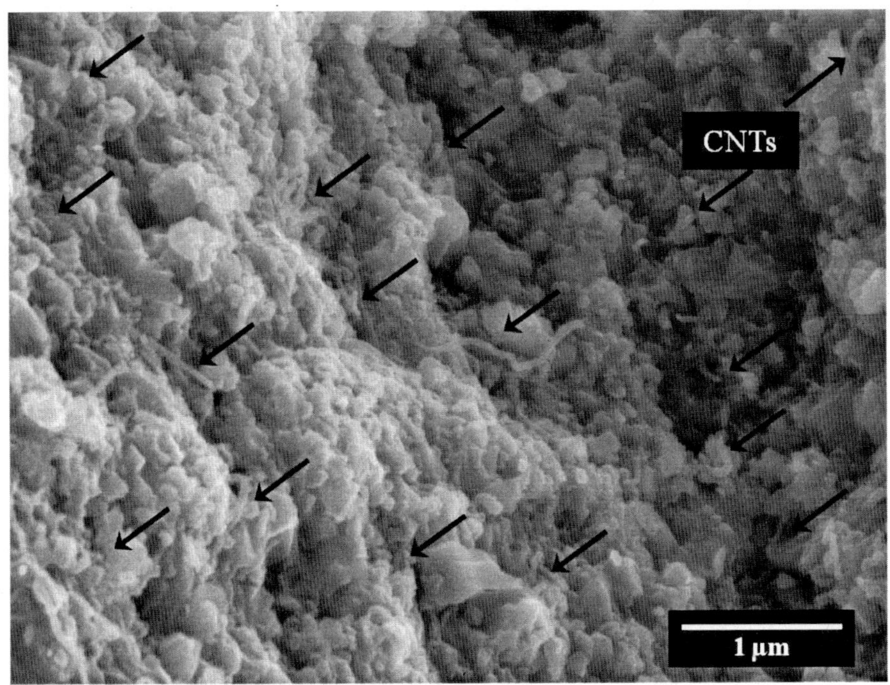

Figure 4: Fractured surface of sample 3YTZP-5-1600 showing presence of CNTs both broken and intact

Figure 5 (a), (b), (c) and (d) show the FESEM images of the fractured surfaces of the samples with 0%, 2.5%, 3 %, and 4 % CNTs in the matrix sintered at 1400°C for 4 minutes, respectively. The FESEM images confirm the survival of the CNTs during the SPS process in the Zirconia matrix. It can be seen from the FESEM images that these samples exhibit very low porosity, indicating that the CNTs have bonded effectively with the 3YTZP powders forming a dense ceramic matrix. The fractured surfaces show CNTs embedded in the matrix. The fracture of the pure Zirconia samples is intergranular, as evident from Figure 5(a), while the fracture of composite samples reinforced with CNTs in Figure 5(b), (c) and(d) exhibit a mixed mode of intergranular and transgranular feature. It can also be seen from Figure 4(a) that the average grain size of the sample sintered at 1400°C with no CNT content is approximately 0.3 microns, whereas the average grain size of the sintered compacts shows an inverse relation with the CNT content as seen in Figure 5(b), (c) and (d). The average grain size for the samples sintered at 1400°C with ~4% wt CNTs is approximately 0.15 microns. In Figure 4, which shows fractured surfaces of the samples, black arrows indicate the locations of CNTs, and a black rectangular box shows CNTs embedded in the matrix.

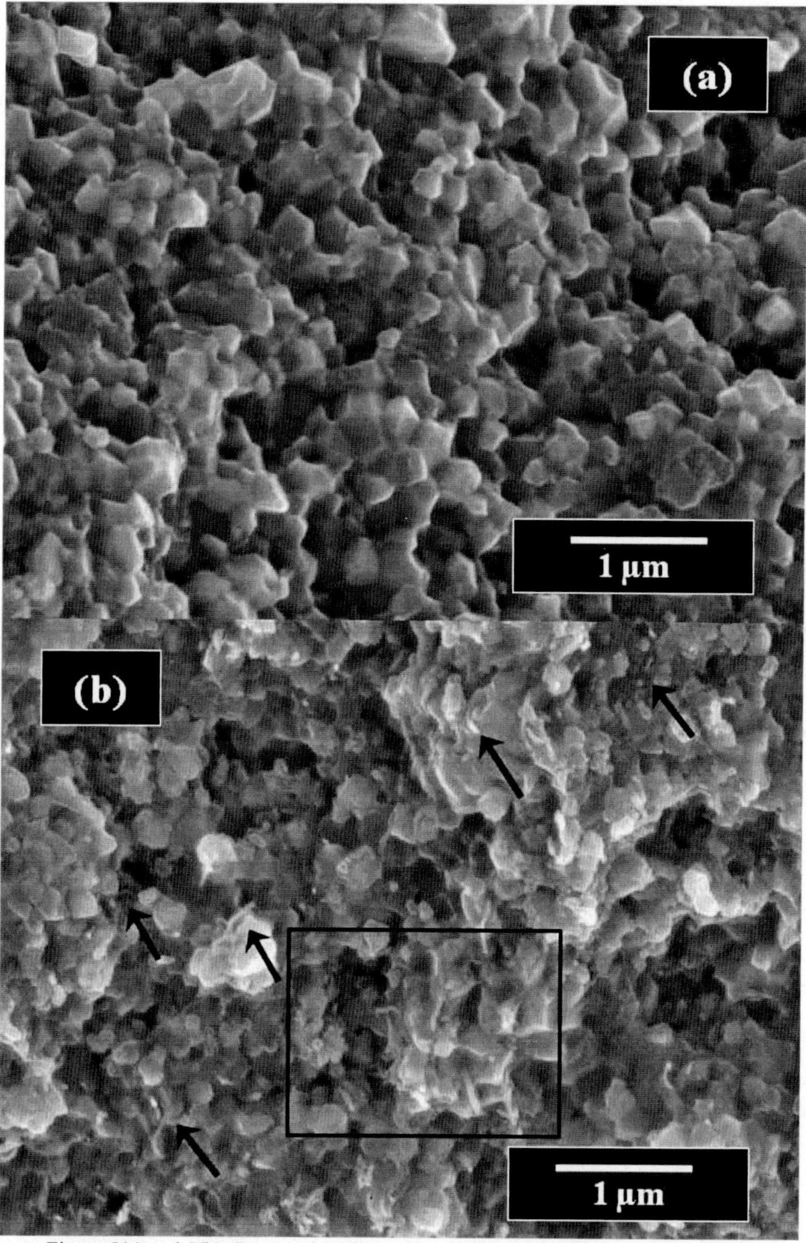

Figure 5(a) and 5(b): Fractured surface of (a) 3YTZP-0-1400 (b) 3YTZP-1-1400

Figure 5(c) and 5(d): Fractured surface of (c) 3YTZP-2-1400 (d) 3YTZP-5-1400

Raman Spectroscopic Analysis

Raman studies were performed on fracture surface of the sintered samples as shown in Figure 6, to get a confirmation of the presence of CNTs after sintering. The peaks at the lower wave numbers (254, 330, 467 and 638) are from the tetragonal Zirconia phase[35]. This suggests that Zirconia retains its tetragonal form after sintering and does not completely transform to monoclinic form. The Raman vibrational modes from the CNTs observed at 1373 cm-1 (D band) and 1565 cm-1 (G band) correspond to the MWNT modes reported in the literature[35]. Furthermore, the intensities and line profiles also suggest that these modes are from MWNTs and not from the other forms of carbon[36]. This observation thus suggests that the CNTs are retained after the SPS process.

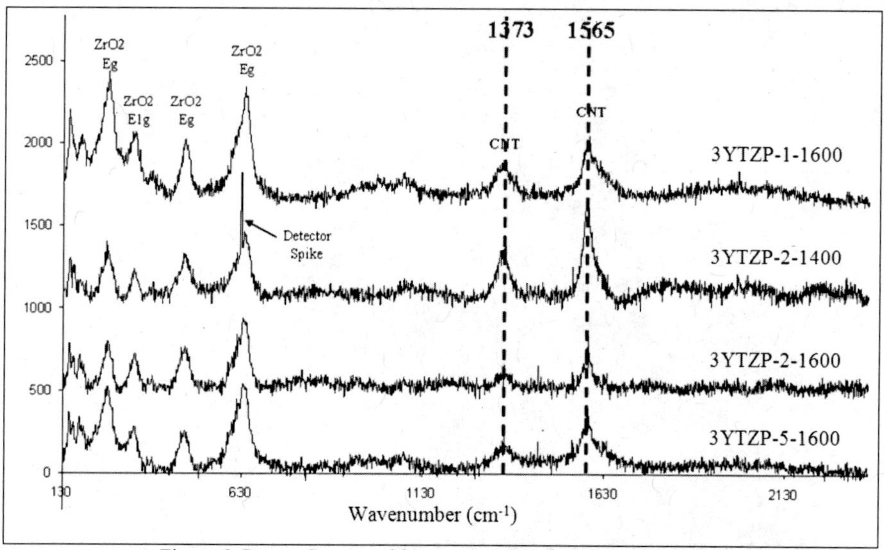

Figure 6: Raman Spectra of fractured surfaces of sintered samples

HRTEM studies

CNT retention is confirmed by HRTEM studies on the samples sintered at 1600°C. As seen from Figure 7, the CNTs are present in the sintered sample. The CNTs seem to be damaged due to the nature of the SPS process; however, the damage seems to be limited to the outer walls, and the inner layers of CNTs seem to be intact. The MWNTs do not seem to have undergone graphitization. The damage to the outer walls of the CNTs can also be explained by the sample preparation, where the sintered sample is crushed into powder and then put under an HRTEM

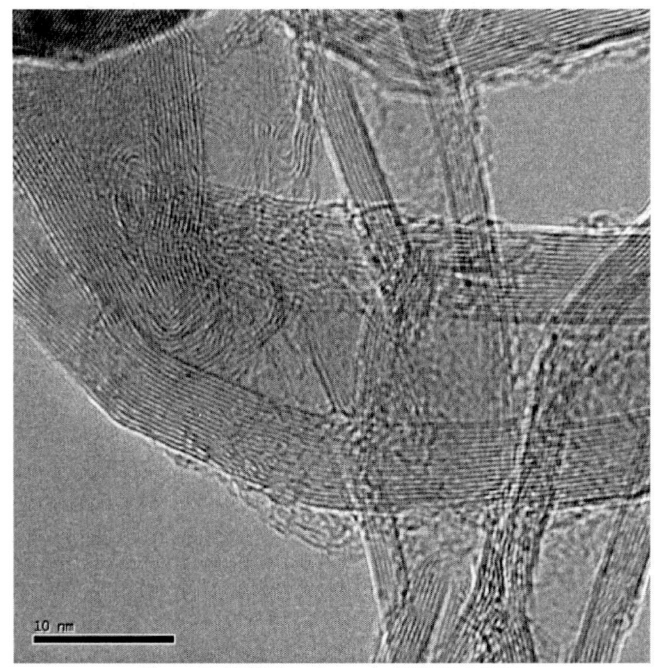

Figure 7: HRTEM of sample 3YTZP-5-1600

CONCLUSION

In this study, we have shown that using CVD technique a uniform distribution of CNTs by in-situ growth can be easily obtained. Further densification by SPS technique results in nearly 100% theoretical density and high CNT retention. The method outlined in this research is thus better suited for bulk manufacturing, with a more uniform distribution of MWNTs in the matrix, than conventional sintering methods. In addition, it was also found that average grain size of the sintered compacts showed an inverse relation with the CNT content; this suggests that the properties of these composites can be tailored by varying the CNT content, thus increasing the range of application of these composites.

In this research, multi-walled carbon nanotubes have been directly grown in-situ on Zirconia nanopowder. The grown MWNTs have been characterized using FESEM, and the yield with different catalyst ratios has been analyzed using TGA. The Zirconia powders with the in-situ grown nanotubes were then sintered to nearly 100% theoretical density using spark plasma sintering. Analysis of the fracture surfaces of the composite shows a maximum grain size of 0.15 microns when sintered at 1600°C. CNTs are uniformly distributed in the entire ceramic matrix due to the direct growth of CNTs on the powder particles.

REFERENCES

[1.] Iijima S. Helical microtubules of graphitic carbon. Nature 1991; 354:56–8.

[2.] A. Peigney, Ch. Laurent, E. Flahaut, and A. Rousset Carbon nanotubes in novel ceramic matrix nanocomposites. Ceram Int 2000; 26:677–83.

[3.] G. Van Lier, C. Van Alsenoy, V. Van Doren, and P. Geerlings, Ab initio study of the elastic properties of single-walled carbon nanotubes and graphene. Chem Phys Lett 2000; 326:181–5.

[4.] M.M.J. Treacy, T.W. Ebbesen, and J.M. Gibson, Exceptionally high Young's modulus observed for individual carbon nanotubes. Nature 1996; 381:678–80.

[5.] M.F. Yu, O. Lourie, M.J. Dyer, K. Moloni, T.F. Kelly, and R.S. Ruoff, Strength and breaking mechanism of multiwalled carbon nanotubes under tensile load. Science 2000; 287:637–40.

[6.] A. Thess, R. Lee, P. Nikolaev, H. Dai, P. Petit, J. Robert, C. Xu, Y.H. Lee, S.G. Kim, A.G. Rinzler, D.T. Colbert, G.E. Scuseria, D. Tomanek, J.E. Fischer,and R.E. Smalley, Crystalline ropes of metallic carbon nanotubes. Science 1996; 273:483–7.

[7.] Ando Y, Zhaoa X, Shimoyama H, Sakai G, and Kaneto K, Physical properties of multiwalled carbon nanotubes. Int J Inorg Mater 1999; 1:77–82.

[8.] M.J. Biercuk, M.C. Llaguno, M. Radosavlijevic, J.K. Hyun, and A.T. Johnson, Carbon nanotube composites for thermal management. Appl Phys Lett 2002; 80:2767–9.

[9.] R.H. Baughman, A.A. Zakhidov, and W.A. de Heer, Carbon nanotubes—the route toward applications. Science 2000; 297:787–92.

[10.] K.M. Prewo, Fiber-reinforced ceramics: new opportunities for composite materials. Am. Ceram Soc Bull 1989; 68:395–400.

[11.] J. Tatami, T. Katashima, K. Komeya, T. Meguro, and T. Wakihara, Electrically Conductive CNT-Dispersed Silicon Nitride Ceramics. Journal of American Ceramic Society 2005; 88 [10]: 2889–93.

[12.] S. Y. Lee, H. Kim, P. C. Mclntyre, K. C. Saraswat, and J. S. Byun, Atomic Layer Deposition of ZrO_2 on W for Metal–Insulator–Metal Capacitor Application. Appl. Phys Letter 2003; 82:2874–6.

[13.] E.T. Thostenson, Z. Ren, and T.W. Chou, Advances in the science and technology of carbon nanotubes and their composites: a review. Compos Sci Technol 2001; 61:1899–912.

[14.] H. Dai, Carbon nanotubes: opportunities and challenges. Surf Sci 2002; 500:218–41.

[15.] F. Lupo, R. Kamalakaran, C. Scheu, N. Grobert, and M. Rühle, Microstructural investigations on zirconium oxide—carbon nanotube composites synthesized by hydrothermal crystallization. Carbon 2004; 42:1995–9.

[16.] J. Sun, and L. Gao, Attachment of Inorganic Nanoparticles onto Carbon Nanotubes. J Electroceram 2006; 17:91–94.

[17.] A.S.Ferlauto, D.Z. Deflorio, F.C. Fonseca, V. Esposito, R. Muccillo, E. Traversa, and L.O. Ladeira, Chemical vapor deposition of multi-walled carbon nanotubes from nickel/yttria-stabilized zirconia catalysts. Appl. Phys 2006; A84:271–276.

[18.] A.S. Ferlauto, D.Z. Deflorio, F.C. Fonseca, V. Esposito, R. Muccillo, and E. Traversa, Composites of Nickel, Zirconia and Carbon Nanotubes. In Solid Oxide Fuel Cells VIII – Proceedings of the 8th International Symposium, ed. by S.C. Singhal, M. Dokiya. The Electrochemical Society Inc., New York, 2002; 643–654

[19.] S.L. Shiw, and J. Liang, Effect of Multiwall Carbon Nanotubes on Electrical and Dielectric Properties of Yttria-Stabilized Zirconia Ceramic. J. Am. Ceram. Soc 2006; 89 [11]:3533–3535.

[20.] M. Daraktchiev, B. Van De Moortele, R. Schaller, E. Couteau, and L. Farro. Effects of carbon nanotubes on grain boundary sliding in zirconia polycrystals. Advanced Materials 2005; 17:1.

[21.] C. Ionascu. High temperature mechanical spectroscopy of fine-grained zirconia and alumina containing nano-sized reinforcements. Ph.D. Thesis. 2008.

[22.] S. Rul, F. Lefèvre-schlick, E. Capria, Ch. Laurent, and A. Peigney, Percolation of single-walled carbon nanotubes in ceramic matrix nanocomposites. Acta Mater. 2004; 52: 1061.

23. K. Ahmad, W. Pan, and S.L. Shi. Electrical conductivity and dielectric properties of multiwalled carbon nanotube and alumina composites. Appl. Phys. Lett. 2006; 89: 133122.
24. G.D. Zhan, J.D. Kuntz, J.E. Garay, and A.K. Mukherjee, Electrical properties of nanoceramics reinforced with ropes of single-walled carbon nanotubes. Appl. Phys. Lett. 2003; 83: 1228.
25. T.Y. Luo, T.X. Liang, and C.S. Li, Stabilization of cubic zirconia by carbon nanotubes. Materials Science and Engineering 2004; A366: 206–209.
26. T. Ukai, T. Sekino, A. Hirvonen, N. Tanaka, T. Kusunose, T. Nakayama, and K. Niihara, Preparation and electrical properties of carbon nanotubes dispersed zirconia nanocomposites. Key Eng. Mat. 2006; 317–318:661–664.
27. J. Sun, L. Gao, M. Iwasa, T. Nakayama, and K. Niihara, Failure investigation of carbon nanotube/3Y-TZP nanocomposites. Ceram. Int. 2005; 31:1131–1134.
28. A. Duszova, J. Dusza, K. Tomasek, J. Morgiel, G. Blugand, and J. Kuebler, Zirconia/carbon nanofiber composite. Scripta Materialia, 2008; 58:520–523.
29. H.J. Dai, A.G. Rinzler, P. Nikolaev, A. Thess, D.T. Colbert, and R.E. Smalley, Single-wall nanotubes produced by metal-catalyzed disproportionation of carbon monoxide. Chem. Phys. Lett 1996; 260: 471.
30. W.Z. Li, J.G. Wen, M. Sennett, and Z.F. Ren, Clean double-walled carbon nanotubes synthesized by CVD. Chem Phys Lett 2003; 368:299–306.
31. Wichmann MHG, Sumflesh J, Fiedler B, Gojny FH, Schulte K. Multiwall carbon nanotube/epoxy composites produced by a masterbatch process. Mech Compos Mater 2006; 42:395–406.
32. Laurent Ch, Peigney A, Dumortier O, Rousset A. Carbon nanotubes–Fe–alumina nanocomposites. Part II: microstructure and mechanical properties of the hotpressed composites. J Eur Ceram Soc 1998; 18:2005–13.
33. X. Wang, N.P. Padture, and H. Tanaka, Contact-damage-resistant ceramic/single-wall carbon nanotubes and ceramic/graphite composites. Nat Mater 2004; 3:539–44.
34. W. Huang, Y. Wang, G. Luo, and F. Wei, 99.9% purity multi-walled carbon nanotubes by vacuum high-temperature annealing, Carbon 41 (2003) 2585–2590.
35. A.P. Naumenko, N.I. Berezovska, M.M. Biley, and O.V. Shevchenko, Vibrational analysis and Raman spectra of tetragonal Zirconia, Physics and Chemistry of Solid State, 2008, 121 – 125.
36. P.C. Eklund, J.M. Holden, and R.A. Jishi, Vibrational modes of carbon nanotubes; spectroscopy and theory. Carbon 1995, 959-972.

PROCESSING, MICROSTRUCTURE AND MECHANICAL PROPERTIES OF ULTRA HIGH TEMPERATURE CERAMICS FABRICATED BY SPARK PLASMA SINTERING

Amit Datye, Kuang-Hsi Wu, Srinivas Kulkarni
Department of Mechanical and Materials Engineering
Florida International University
Miami, Florida 33174

H. T. Lin
Oak Ridge National Laboratory
Oak Ridge, TN 37831-6068.

J.Vleugels
Department of Metallurgy and Materials Engineering
K.U.Leuven
Leuven, Belgium.

ABSTRACT

Recently, Zirconium based ultra high temperature ceramics (UHTC) — Zirconium Diboride (ZrB_2) and Zirconium Carbide (ZrC) — with Silicon Carbide (SiC) additions have been synthesized by various methods such as hot pressing, reactive hot pressing and pressureless sintering. Spark plasma sintering, also known as FAST (Field assisted sintering technique) or PECS (Pulse Electric Current Sintering), is a novel process characterized by high heating and cooling rates and is, therefore, ideally suited for bulk manufacturing. The high heating and the cooling rates coupled with pressure prevents grain coarsening and, therefore, enhances densification. In this research, UHTC composites with varying percentages of Zirconium Carbide (ZrC), Zirconium Diboride (ZrB_2) and Silicon Carbide (SiC) were sintered at 1950°C for varying hold times. Mechanical properties and microstructures of the composites thus synthesized were studied.

UHTCs have been synthesized using spark plasma sintering to more than 95% theoretical density and with mechanical properties (hardness, fracture toughness and bending strength) comparable to those reported previously by different researchers using other conventional sintering methods. This research gives a comparative analysis of the properties of various ZrC, ZrB_2 and SiC based compositions of UHTCs sintered under similar conditions.

INTRODUCTION

Currently, there is a great interest in developing advanced super-sonic and hypersonic flying vehicles. The continuous pushing for higher speeds has imposed an enormous demand to improve the thermal management for various heat-generating sources. The sharp leading edges of the hypersonic flying vehicles required for maneuvering at high speeds can experience temperatures greater than 2000°C. Current thermal protection materials are not sufficient for sustained operation at these extreme temperatures. Ultra high temperature ceramics are a unique class of materials with melting temperatures in excess of 3000°C[1, 2]. The borides, carbides and nitrides of the transition metals (Zr, Hf, Ta, etc.) are considered as ultra high temperature ceramics (UHTCs). Zirconium Diboride and Hafnium Diboride compounds are of ongoing interest in ultra high temperature applications because they have the highest melting points known, above 3200°C. Combined with their high melting temperatures, these materials have high hardness and chemical stability at elevated temperatures. UHTCs have a wide spectrum of applications, which includes thermal protection materials for advanced reentry vehicles, molten-metal crucibles, high-temperature electrodes and ultra-high-temperature aerospace applications[3-5]. In this research, ZrB_2-based ceramics have been particularly investigated because of the unique combination of thermal shock resistance and strength at high temperatures[6,7]. Pure ZrB_2 ceramic cannot be achieved with a high theoretical density due to the poor sinterability of ZrB_2 and, therefore, various sintering additives, such as Silicon Carbide [1,2,6], Boron Carbide[3], Aluminum Nitride[4], Silicon Nitride[5], Molybdenum Disilicide[7], Zirconium Nitride and Zirconium Silicide[8], have been used to enhance the density in the SPS process. These additives enhance densification by formation of an intergranular secondary phase. In particular, silicon containing additives are of interest, since silicon forms a stable oxide scale enhancing the oxidation resistance of the ZrB_2 based ceramic[7-11].

Most of the previous research has been aimed at fabrication of ZrB_2 based ceramics using conventional sintering techniques, such as hot pressing, pressure less sintering and reactive sintering. Spark plasma sintering, also known as FAST (Field assisted sintering technique) or PECS (Pulse Electric Current Sintering), is a novel process characterized by high heating and cooling rates and is, therefore, ideally suited for bulk manufacturing. The high heating and the cooling rates coupled with pressure prevent grain coarsening. The pulsed DC current used in this technique has been verified by various researchers experimentally to achieve a higher density in the sintered samples than conventional sintering. In this paper, the fabrication of bulk samples of ZrB_2 based ultra high temperature ceramics using spark plasma sintering is reported. The mechanical properties of the sintered samples are discussed.

MATERIALS AND METHODS

Zirconium Diboride (ZrB_2, average particle size = 0.7 micron), Zirconium Carbide (ZrC, average particle size = 50nm) and Silicon Carbide (SiC, average particle size = 0.5 microns, Grade UF-25– HC Starck) are mixed together in various proportions by volume percentage as shown in Table I below. The samples were first placed in a rolling mill using baffled bottles for 48 hours. The powder mixtures were then wet-mixed on a multidirectional mixer (Turbula) in ethanol using ZrO_2 milling balls (Ø 10 mm) for 24 hours. The ethanol was removed in a rotating evaporator at 65°C, and the powder was dried in a furnace at 90°C for 24 hours in air.

Table I: Powder compositions by volume percent and sintering conditions

Mixture	Sample ID	Composition (Volume %)			Hold Time (Min)	Pressure (MPa)
		ZrB_2	(ZrC)	SiC		
BZS	BZS4	60	30	10	4	60
	BZS8				8	60
BZ	BZ4	62	38	0	4	60
	BZ8				8	60
BS	BS4	80	0	20	4	60
	BS8				8	60
ZC	ZC4	0	80	20	4	60
	ZC8				8	60

. The final sintering temperature is 1950°C for all samples, and the hold time is varied from 4 to 8 minutes. The sintering profile is adjusted based on previous experiences to get rid of the B_2O_3 present on the surface of the ZrB_2 particles. The pressure on the die is adjusted so as to keep an open porosity up to a temperature of about 1500°C to allow evaporation of the B_2O_3 vapor. Application of high pressure might inhibit the evaporation of all B_2O_3 vapor, which will result in a dense material, but boron containing species might remain at the interphase destroying the ultra high temperature properties of the material. The dry powder mixture was poured into a cylindrical graphite die with an inner and outer diameter of 30 and 66 mm, and PECS with a heating rate of 200°C/min from 450 to 1450°C and 100°C/min from 1450°C to 1950°C. The maximum pressure of 60 MPa was applied at 1950°C and dwelled for 4 or 8 min.

The final densities of the sintered samples were measured using the Archimedes principle in distilled water. Vickers hardness testing was carried out using a Wilson Tukon Hardness tester. The sintered samples were then characterized using field-emission gun scanning electron microscopy (FEGSEM, JEOL, JSM – 6330F) in both the SEI (Secondary Electron) and the BSE (Back-Scattered Electron) mode. The X-Ray Diffraction (XRD) data for phase analysis was collected using Siemens D5000 XRD machine equipped with copper anode. The x-rays were generated by supplying 40 KV and 40 mA current to the anode. The scans were preformed from 10 to 90 degrees (2θ) with a step size of 0.01 degrees and a collection time of 1 deg/min. The samples for XRD were cut from the center of the spark plasma sintered samples using a diamond saw. The XRD data was collected on the surface parallel to the direction of the applied pressure during sintering. The Vickers hardness testing was conducted according to the ASTM C1327 standard at loads of 300gf and 1000gf with a dwell time of 15 seconds. Five indentations were taken, and the average is reported. The fracture toughness is measured by measuring the length of the cracks formed at 1000 gf of loading using an SEM. Three point bending tests were carried out on samples according to the ASTM C1161 standard at ambient temperature using a Bose Enduratec tester. The crosshead speeds of the tester are maintained using displacement control and the force measured with a load cell accurate up to 0.1N.

RESULTS AND DISCUSSION

The results from the density measurements, hardness testing and three point bending tests are summarized in Table II below and discussed in the subsequent sections.

Table II: Density, hardness, fracture toughness and bending strength of samples

Mixture	Sample ID	Measured Density (gm/cm^3)	Relative Density	Vickers Hardness (GPa)		Fracture Toughness ($MPa.m^{1/2}$)	Bending Strength (MPa)*
				300 gf	1000 gf		
BZS	BZS4	5.63	91.14	12.04	11.86	3.5 ± 0.4	570
	BZS8	5.88	95.15	18.95	17.76	3.8 ± 0.4	601
BZ	BZ4	5.25	80.97	7.88	7.13	2.8 ± 0.5	120
	BZ8	5.43	83.77	8.55	7.14	3.1 ± 0.3	133
BS	BS4	5.34	88.80	18.25	16.37	4.2 ± 0.3	628
	BS8	5.79	96.24	19.64	17.38	4.8 ± 0.4	640
ZC	ZC4	4.93	87.30	4.65	3.77	-	-
	ZC8	4.98	90.57	3.66	3.27	-	-

* The 3 point bending tests were carried out on only one sample, and, therefore, the data needs to be verified with more tests.

FEGSEM and HRTEM analysis of raw powders

The powders were first analyzed using SEM, TEM and XRD. The FEGSEM images show the average particle size of the ZrB_2 and SiC powders to be around 500 nm, whereas the ZrC powder average particle size is 50 nm. The HRTEM imaging confirms the presence of a thin layer of B_2O_3 coating on the ZrB_2 particles and a layer of ZrO_2 on the ZrC particles. Due to the nanometer size of the ZrC particles, this ZrO_2 coating can prove to be deleterious in the sintering of the ZrC based ceramics.

Figure 1: FEGSEM image of raw as received ZrB_2 powder

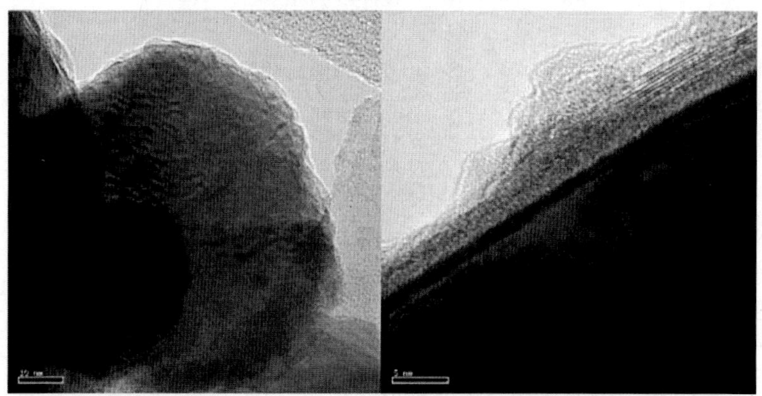

Figure 2: HRTEM image of raw ZrB_2 powder showing B_2O_3 oxide layer

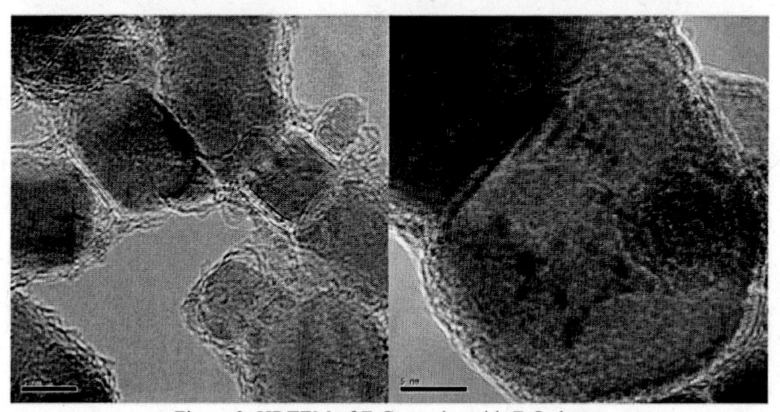

Figure 3: HRTEM of ZrC powder with ZrO_2 layer

Density Measurements

The density measurements show that the samples with longer hold time at 1950°C achieved a higher relative density compared to the one with shorter hold time as expected for all samples. The samples from Sample BZ are not expected to have a higher than 85% density due to the constituents being ZrB_2 and ZrC, both of which do not sinter well at 1950°C. Samples BZS4 and BZS8, which are mixtures of ZrB_2-ZrC-SiC, show an increase in density with longer hold time and achieves close to 95% density. The density of Sample BS (BS4 = 5.29 gm/cm^3 and BS8 = 5.77 gm/cm^3) and is higher than that reported by Tian et al.[12] (5.25 gm/cm^3) for similar samples hot pressed at 1900°C for one hour. This increase can be attributed to a higher sintering temperature and the kinetics due to the high pulsed DC current used during the SPS process. Sample ZC did not achieve more than ~90% theoretical density. Since there was no interaction between the ZrC and the SiC in the sintered samples, the samples were not sintered due to the presence of ZrO_2 on the surface of the ZrC nanoparticles.

FEGSEM analysis of microstructure of SPS sintered samples

Mixture BZS

FEGSEM (BSE and SEI mode) images of the ZrB_2-ZrC-SiC samples (BZS4 and BZS8) are shown in Figures 4 and 5, respectively. It can be seen from the FEGSEM images that the distribution of the SiC, ZrC and ZrB_2 is uniform throughout the sample. There is also a reduction in the porosity in Sample BZS8 compared to Sample BZS4 due to a longer hold time, as expected.

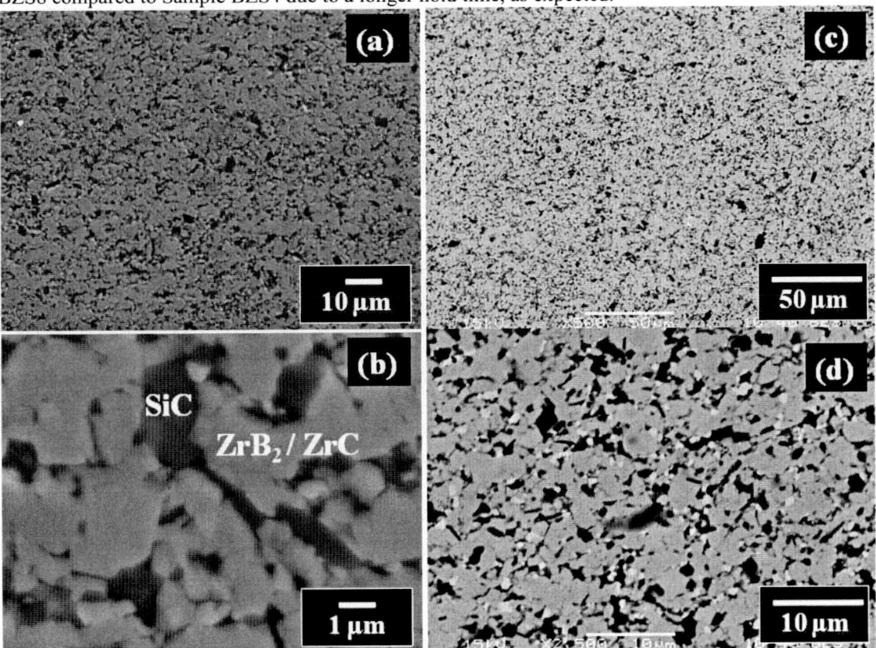

Figure 4: (a) SEI image of polished Sample BZS4. The grey areas are ZrC or ZrB_2 and the darker areas are SiC. The black areas are the pores (b) High magnification SEI image of BZS4 (c) BSE low magnification image (d) BSE image -- the grey areas are ZrB_2, the white particles are ZrC and the black areas are SiC or pores.

The microstructure seems to be free of any other residual foreign phases usually present after hot pressing or other conventional sintering methods. It can be seen from the microstructure that dense (~90% TD) samples can be obtained of the ZrC-ZrB_2-SiC system using the spark plasma sintering method without additives. The microstructures in Figures 4 and 5 for Samples BZS4 and BZS8 also show that the average grain size is less than 1 micron. This is due to the nanometer sized ZrC particles present at the grain boundary which prevented the diffusion related grain growth of the ZrB_2 along with SiC particles.

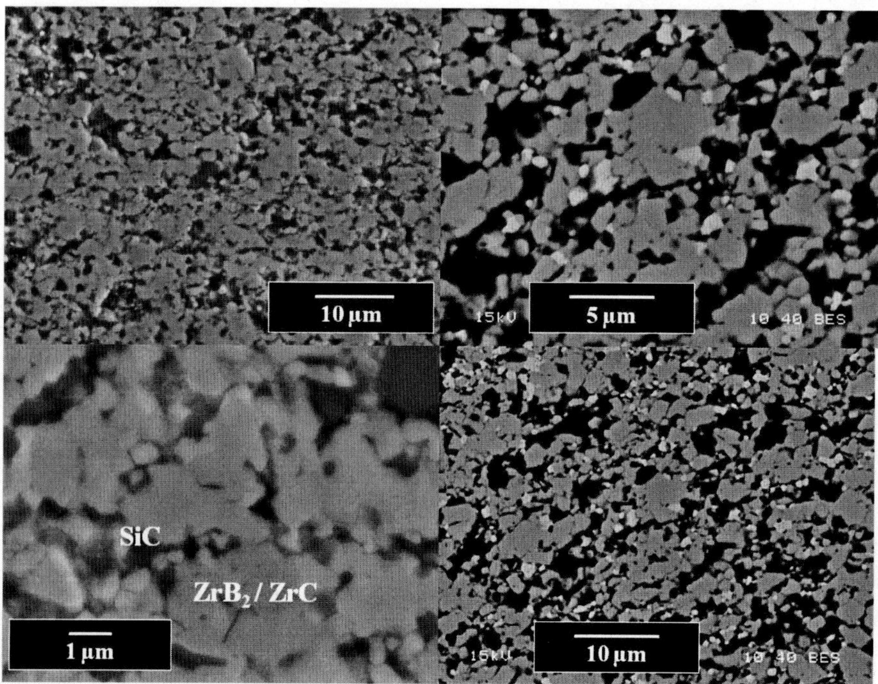

Figure 5: (a) SEI image of polished Sample BZS8 showing grey areas (ZrC or ZrB_2) and darker areas (SiC) The black area are the pores. (b) High magnification SEI image of BZS8 (c) BSE low magnification image (d) BSE image: the grey areas are ZrB2; the white particles are ZrC and the black areas are SiC or pores.

Mixture BZ

Figure 6 shows the representative microstructure of Sample BZ sintered at 1950°C. Microstructure examination of Sample BZ4 sintered for 4 minute shows that it has not yet completely densified (80.97 and 83.77% TD). The sample sintered for 8 minutes shows a better densification with an average grain size of around 2 microns for the ZrB_2 particles the ZrC particles form around the ZrB_2 particles with an average grain size of 1 micron. Unlike Samples BZS4 and BZS8, where fine particle of ZrC more or less uniformly coated on the surface of ZrB_2, in this set of samples, ZrC powders fused together to form a major second, coexisting with the major ZrB_2 phase, as clearly evident in Figure 6(b). However, these two ZrC and ZrB_2 were not completely fused into one phase. They, in fact, remain as two phases, lacking strong interface bonding. This implies that without the additive of SiC, ZrC and ZrB_2 cannot bond well to form a strong UHTC. It is also noticed that in the as received ZrC powder, there is a very thin layer of ZrO_2 film on the surface of the powder. This oxide film might have played important role on preventing the fusion of ZrC and ZrB_2 phases. The relatively weak bonding between the two major phases was also clearly reflected in the Vicker's hardness values and bending strength, as can be seen in Table II.

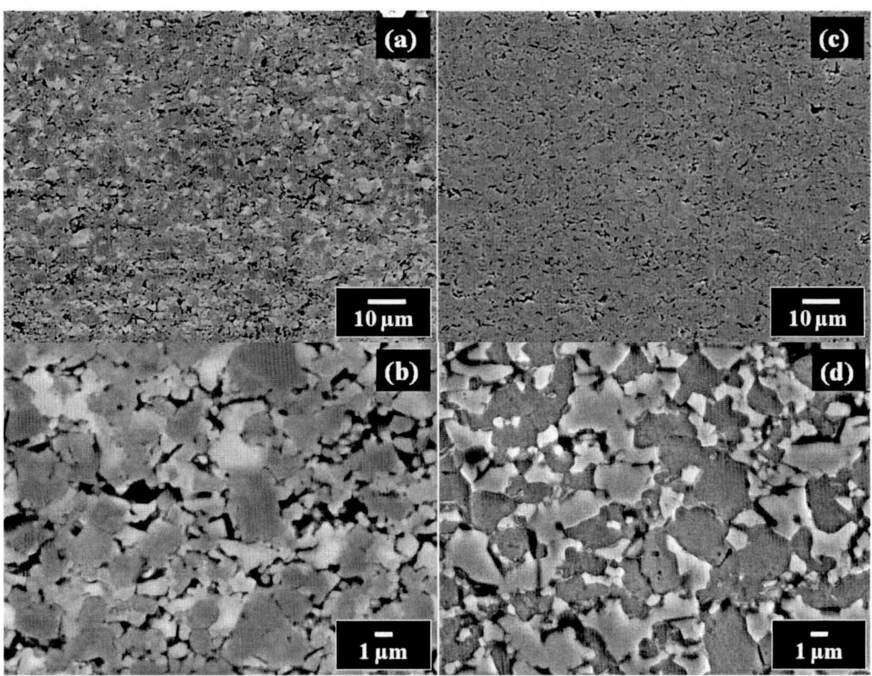

Figure 6: a) SEI image of polished Sample BZ4 showing grey area (ZrB_2), lighter areas (ZrC). The black areas are pores (b) Higher magnification image of Sample BZ4 showing ZrB_2 (grey), ZrC (lighter) and pores (darker) (c) SEI image of polished Sample BZ8 showing grey areas (ZrC or ZrB_2) and black areas (pores) (d) Etched sample showing ZrB_2 (darker) and ZrC (lighter)

Mixture BS

The microstructures of mixture BS are illustrated in Figures 7(a), (b), (c) and (d). It is obvious from these FEGSEM images that SiC appears to be able to assist the sintering of ZrB_2. As a consequence, with a four minute hold time a density of 88.8% TD is achieved for sample BS4 as shown in Figures 7(a) and (b). Increasing hold time to eight minutes significantly increases the density, to 96.2% in BS8, as shown in Figures7(c) and (d). It is also noticed that the ZrB_2 phase is well connected together, instead of separated regions as in Samples BSZ and BZ. The average grain size of BS4 and BS8 also appear to be larger compared to those of Samples BSZ and BZ. The well-connected microstructure of the ZrB_2-SiC samples might have been the reason that both Samples BS4 and BS8 achieve higher hardness and higher bending strength values, as indicated in Table II. It can also be seen that the porosity present in the sintered samples is closed porosity.

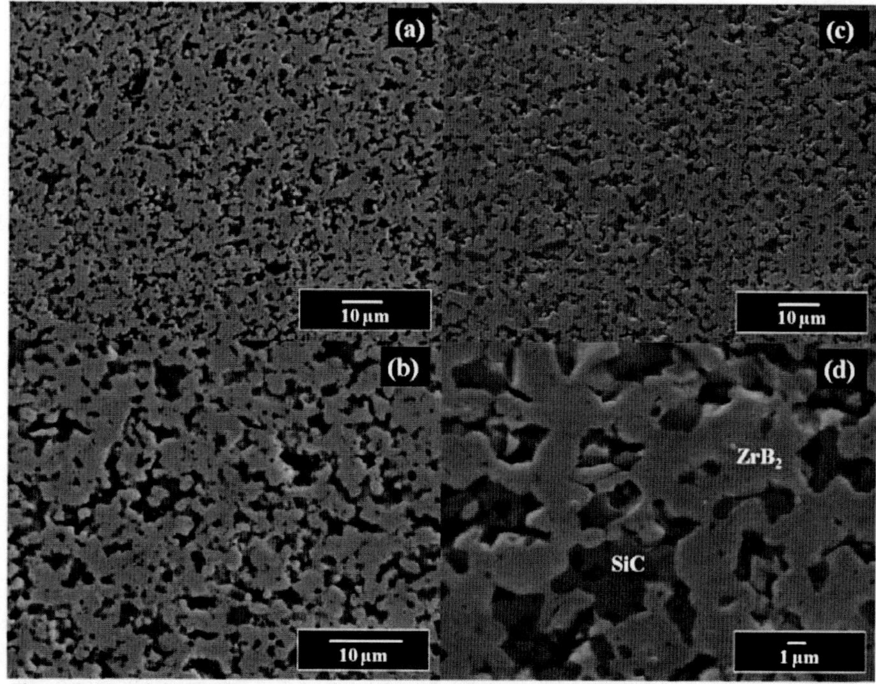

Figure 7: (a) and (b) SEI images of polished Sample BS4 showing lighter grey area (ZrB_2) and darker grey area (SiC) and black area is pores (c) and (d) SEI images of polished sample BS8 showing lighter grey area (ZrB_2) and darker grey area (SiC) and dark black area are pores.

Mixture ZS

The microstructures for samples sintered of mixture ZS are shown in Figures 8 (a), (b), (c) and (d). Samples ZS4 and ZS8 consist of ZrC (80 vol%) and SiC (20 vol%), without ZrB_2. These two samples were not sintered well as demonstrated by the very low hardness values out of four compositions attempted, as evident in Table II. The reason for the poor sintering of this set of samples is not clear and requires further studies. The poor sintering of the samples also manifested by the fact that both ZS4 and ZS8 samples crumpled immediately upon the application of load in the three-point bending test, therefore, registering no valid bending strength values. Figure 8 illustrates the SEM images of Samples ZS4 and ZS8. Instead of nicely sintered structure, as shown in previous samples, it is clear that the powders were not well fused together. Further studies are required for this composition.

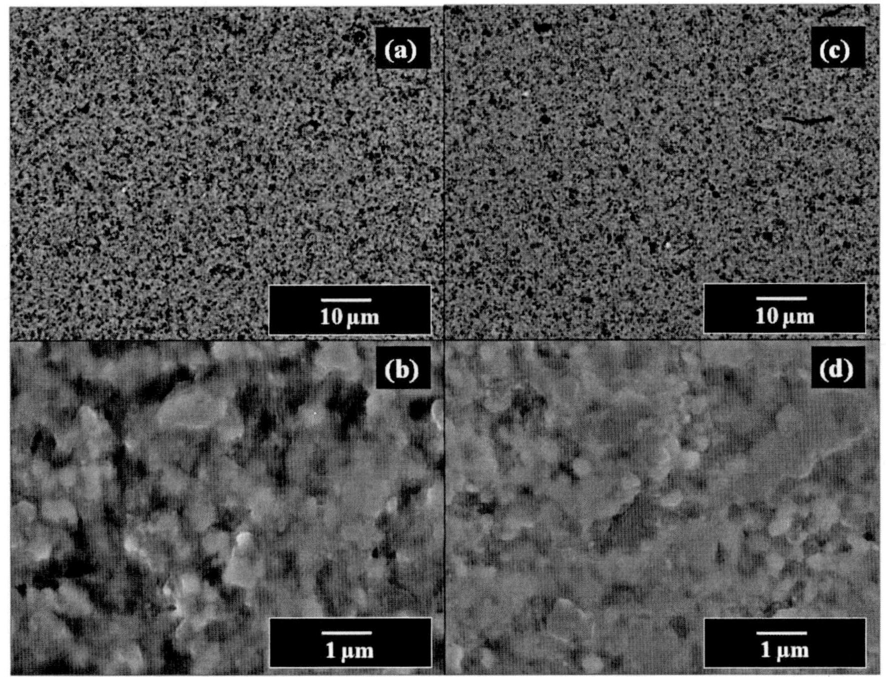

Figure 8: (a) BSE image of polished sample ZS4 – dark area is SiC lighter areas (ZrC), (b) SEI image at higher magnification of ZS4, (c) BSE image of polished sample ZS4 - dark areas are SiC; the lighter areas are ZrC, (d) SEI image at higher magnification of ZS4

XRD Analysis

Figure 9 shows the XRD patterns of the base powders (ZrC, ZrB_2, SiC) and the various powder samples sintered by SPS method. It can be seen from the patterns that all the base powders are single phase, remaining intact after sintering, and there is no reaction between the respective mixed powders. The only additional phase that is observed in these samples after sintering is ZrO_2 (Baddelyte). However, it should be noted that this phase is present in only ZrC containing samples and it is almost absent in samples containing ZrB_2 and SiC only. The amount of ZrO_2 is higher in the samples that have a higher amount of ZrC. This suggests that, during sintering, a certain amount of ZrC is oxidized, even though a very strong non oxidizing atmosphere was maintained. This may be because of the small particle size of ZrC or the presence of the amorphous ZrO_2 that was not detected in the XRD studies of the raw powder.

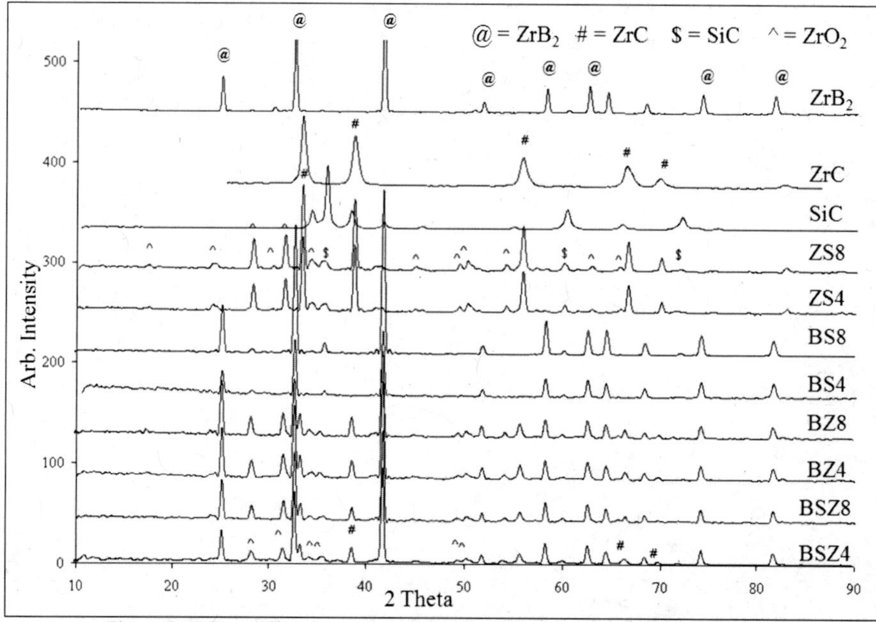

Figure 9: X-Ray diffraction patterns of SPS sintered samples and base powders

Hardness and Fracture Toughness by the Vickers Indentation Method

The testing results listed in Table II show an expected trend with samples from the same powder mixture sintered at the same temperature but with a higher hold time showing an increase in the fracture toughness and hardness. The Vickers indentation method was used for assessing the fracture toughness of the SPS samples. This method involves using a Vickers diamond microhardness indenter to induce radial cracks in the material. Those radial cracks are assumed to emanate from the indent as a result of residual tensile stresses, which develop during the tip unloading and arrest when the near-tip stress intensity equals the materials toughness, K_{IC}. This fracture toughness is calculated shown below,

$$K_{IC} = 0.026\left(\frac{E^{1/2}P^{1/2}a}{C^{3/2}}\right) \quad (1)$$

where P is the indentation load, a the half length of the indent, C the half length of the crack, and E the elastic modulus of the composite. E is calculated from the elastic moduli of the components (E_{ZrB2} = 450 GPa and E_{SiC} = 414 GPa), according the rule of mixtures.

The Vickers hardness (BZS4 = 12 GPa and BZS8 = 19 GPa) and fracture toughness (BZS4 and BZS8 = 3.5 and 3.8 MPa-m$^{1/2}$, respectively) obtained for BZS8 is comparable to the hardness and fracture toughness obtained by Bellosi et al.[10]. Even though the sintering temperature (1950°C) is less than the sintering temperature used (2150°C) by Bellosi et al.[10], the high hardness and fracture toughness can be due to a higher pressure 60MPa applied during the SPS sintering process. The modified sintering profile also helps in better densification of the sample and prevents boron species from interfering at the grain boundaries. The Vickers hardness and fracture toughness (2.8 and 3.1 MPa-m$^{1/2}$ for BZ4 and BZ8) obtained for Sample BZ8 is less than a hardness of ~17.8 GPa, and fracture toughness of 3.8 MPa-$m^{1/2}$ was reported by Takeshi et al.[16], who sintered a ZrB_2-ZrC composite using SPS by mechanical alloying Zr-C-B together and then SPS sintering at 1800°C. This difference is attributed to the presence of ZrO_2 coatings on the ZrC particles, which prevent proper sintering of the sample; therefore, the hardness of the sample is much less than reported by Takeshi et al.[16]. Samples BS4 and BS8 with a hardness (~19 GPa and ~18 GPa, respectively) and a fracture toughness of (4.2 and 4.8 MPa-$m^{1/2}$) are comparable to fracture toughness reported by other researchers for ZrB_2-20%vol.SiC[1] produced by hot pressing and pressure-less sintering.

Three-Point Bending Tests
Samples BZS4 and BZS8 with a flexural strength of 570 and 601 MPa, respectively, are comparable to that reported by Bellosi et al.[10] and more recently by Guo et al.[21] but at lower sintering temperatures. This is due to the optimized sintering profile, which helps to reduce the formation of intergranular species that cause a decrease in the flexural strength. The flexural strength in bending for BS4 and BS8 (>600MPa) is considerably less than that reported by Chamberlain et al.(>1000MPa)[1]. It is, however, comparable or even higher than the values reported by other researchers who have obtained flexural strength of ZrB_2-SiC based composites up to 580 MPa[12,20]. Another possible influential factor is the fact that the grain size of the sintered samples obtained in the current research (less than 5 microns) is smaller compared to the grain sizes obtained by other researchers.

CONCLUSION

ZrB_2 based ultra high temperature ceramics have been consolidated to a greater than 90% theoretical density using the spark plasma sintering process. An optimal pressure and heating cycle during spark plasma sintering have been developed to remove the boron species at the interphase to form single-phase ZrB_2 based ceramics with a homogeneous and a fine grained microstructure. The SPS process requires much shorter time for sintering these ceramics to a high theoretical density compared to the conventional sintering methods.

Mechanical properties comparable to those reported by other researchers for the various mixtures have been obtained at comparatively lower sintering temperatures and pressures.

ACKNOWLEDGEMENTS

The authors also wish to acknowledge the effort of George Gomes, Kevin Hernandez, S. Amruthaluri and C.Ferrer in running the experiments and collecting information.

REFERENCES

1. Adam L. Chamberlain, William G. Fahrenholtz, Gregory E. Hilmas, Donald T. Ellerby. High-Strength Zirconium Diboride-Based Ceramics. Journal of the American Ceramic Society 2004;87:1170.
2. Adam L. Chamberlain, William G. Fahrenholtz, Gregory E. Hilmas. Pressureless Sintering of Zirconium Diboride. Journal of the American Ceramic Society 2006;89:450.
3. Adrian Goldstein, Ygal Geffen, Ayala Goldenberg. Boron Carbide-Zirconium Boride In Situ Composites by the Reactive Pressureless Sintering of Boron Carbide-Zirconia Mixtures. Journal of the American Ceramic Society 2001;84:642.
4. F. Monteverde and A. Bellosi, ''Beneficial Effects of AlN as Sintering Aid on Microstructure and Mechanical Properties of Hot-Pressed ZrB2,'' Adv. Eng. Mater., 5, 508–12 (2003).
5. F. Monteverde and A. Bellosi, ''Effect of the Addition of Silicon Nitride on Sintering Behavior and Microstructure of Zirconium Diboride,'' Scr. Mater., 46, 223–8 (2002).
6. F. Monteverde and A. Bellosi, ''Development and Characterization of Metal Diboride Based Composites Toughened with Ultra-Fine SiC Particulates,'' Solid State Sci., 7, 622–30 (2005).

7. D. Sciti, M. Brach, and A. Bellosi, ''Oxidation Behavior of a Pressureless Sintered ZrB2–MoSi2 Ceramic Composites,'' J. Mater. Res., 20 [4] 922–30 (2005).
8. S. Q. Guo, Y. Kagawa, T. Nishimura, and H. Tanaka, ''Pressureless-Sintering and Physical Properties of ZrB2-Based Composites with ZrSi2 Additive,'' Scr. Mater., 58, 579–82 (2008).
9. W. C. Tripp, H. H. Davis, and H. C. Graham, ''Effect of an SiC Addition on the Oxidation of ZrB2,'' Ceram. Bull., 52 [8] 612–6 (1973.
10. Alida Bellosi, Frederic Monteverde, Diletta Sciti. Fast Densification of Ultra-High-Temperature Ceramics by Spark Plasma Sintering. International Journal of Applied Ceramic Technology 2006; 3:32.
11. Alireza Rezaie, William G. Fahrenholtz, Gregory E. Hilmas. Oxidation of Zirconium Diboride-Silicon Carbide at 1500 oC at a low Partial Pressure of Oxygen. Journal of the American Ceramic Society 2006; 89:3240.
12. Wu-Bian Tian, Yan-Mei Kan, Guo-Jun Zhang and Pei-Ling Wang, "Effect of carbon nanotubes on the properties of ZrB2–SiC ceramics", Materials Science and Engineering: A, Volume 487, Issues 1-2, 25 July 2008, Pages 568-573.
13. Bruno R. Miccioli, Peter T. B. Shaffer. High-Temperature Thermal Expansion Behavior of Refractory Materials: I, Selected Monocarbides and Binary Carbides. Journal of the American Ceramic Society 1964;47:351.
14. C. C. Sorrell, H. R. Beratan, R. C. Bradt, V. S. Stubican. Directional Solidification of (Ti, Zr) Carbide-(Ti, Zr) Diboride Eutectics. Journal of the American Ceramic Society 1984;67:190.
15. Charles C. Sorrell, Vladimir S. Stubican, Richard C. Bradt. Mechanical Properties of ZrC-ZrB2 and ZrC-TiB2 Directionally Solidified Eutectics. Journal of the American Ceramic Society 1986;69:317.
16. T.Takeshi and Y.Satoshi, "Spark plasma sintering of ZrB2-ZrC powder mixtures synthesized by MA-SHS in air," Journal of Materials Science, 42(3): 772-778
17. D. W. Lee, I.S. Haggerty. Plasticity and Creep in Single Crystals of Zirconium Carbide. Journal of the American Ceramic Society 1969; 52:641.
18. F. G. Keihn, E. J. Keplin. High-Temperature Thermal Expansion of Certain Group IV and Group V Diborides. Journal of the American Ceramic Society 1967; 50:81.
19. George H. Beall, Linda R. Pinckney. Nanophase Glass-Ceramics. Journal of the American Ceramic Society 1999;82:5.
20. Guo-Jun Zhang, Zhen-Yan Deng, Naoki Kondo, Jian-Feng Yang, Tatsuki Ohji. Reactive Hot Pressing of ZrB2-SiC Composites. Journal of the American Ceramic Society 2000; 83:2330.
21. Guo, S. Q., Kagawa, Y., Nishimura, T. and Tanaka, H., Elastic properties of spark plasma sintered (SPSed) ZrB2–ZrC–SiC composites. Ceram. Int., 2008, 34(8), 1811–1817.
22. Hasselman D. P. H. Experimental and Calculated Young's Moduli of Zirconium Carbide Containing a Dispersed Phase of Graphite. Journal of the American Ceramic Society 1963;46:103.
23. Jihong She, Jian-Feng Yang, Naoki Kondo, Tatsuki Ohji, Shuzo Kanzaki, Zhen-Yan Deng. High-Strength Porous Silicon Carbide Ceramics by an Oxidation-Bonding Technique. Journal of the American Ceramic Society 2002;85:2852.
24. Jinkwan Jung, Shinhoo Kang. Advances in Manufacturing Boron Carbide-Alumium Composites. Journal of the American Ceramic Society 2004;87:47.
25. Kantesh Balani, Gabriela Gonzalez, Arvind Agarwal, Robert Hickman, J. Scott O'Dell. Synthesis, Microstructural Characterization, and Mechanical Property Evaluation of Vacuum Plasma Sprayed Tantalum Carbide. Journal of the American Ceramic Society 2006; 89:1419.
26. Marschall D. John. Oxidation and Catalytic Efficiency of ZrB2 and HfB2 based Ultra-High Temperature Ceramic (UHTC) Composites Exposed to Supersonic Air Plasma. 2004:1.
27. S. C. Zhang, G. E. Hilmas, W. G. Fahrenholtz. Zirconium Carbide-Tungsten Cermets Prepared by In Situ Reaction Sintering. Journal of the American Ceramic Society 2007; 90:1930.

28. S. H. Shim, K. Niihara, K. H. Aug, K. B. Shim. Crystallographic orientation of ZrB2-ZrC composites manufactured by the spark plasma sintering method. Journal of Microscopy 2002; 205:238.
29. Shiro Shimada, Tadao lshii. Oxidation Kinetics of Zirconium Carbide at Relatively Low Temperatures. Journal of the American Ceramic Society 1990; 73:2804.
30. Wen-Wen Wu, Guo-Jun Zhang, Yan-Mei Kan, Pei-Ling Wang. Reactive Hot Pressing of ZrB2-SiC-ZrC Ultra High-Temperature Ceramics at 1800C. Journal of the American Ceramic Society 2006; 89:2967.
31. Wen-Wen Wu, Guo-Jun Zhang, Yan-Mei Kan, Pei-Ling Wang. Reactive Hot Pressing of ZrB2-SiC-ZrC Composites at 1600 oC. Journal of the American Ceramic Society 2008:1.
32. 1R. A. Cutler. Engineering Properties of Borides. Ceramics and Glasses, Engineered Materials Handbook, Vol. 4. Edited by S. J. Schneider Jr. ASM International, Materials Park, OH, 1991: 787–803.
33. P. T. B. Shaffer. Engineering Properties of Carbide. Am. Ceram. Soc. Bull 1994; 73 [6]; 141–42.
34. K. Upadhya, J.-M. Yang, and W. P. Hoffman. Materials for Ultrahigh Temperature Structural Applications. Am. Ceram. Soc. Bull 1997; 58 [12]:51–56.
35. F. Monteverde, S. Guicciardi, and A. Bellosi. Advances in Microstructure and Mechanical Properties of Zirconium Diboride-Based Ceramics. Mater. Sci. Eng 2003; 346 [1–2]:310–19.
36. S. Norasetthekul, P. T. Eubank, W. L. Bradley, B. Bozkurt, and B. Stucker. Use of Zirconium Diboride – Copper as an Electrode in Plasma Applications. J. Mater. Sci 1999; 34 [6]:1261–70.
37. E. V. Clougherty, R. J. Hill, W. H. Rhodes, and E. T. Peters. "Research and Development of Refractory Oxidation-Resistant Diborides, Part II, Vol. II: Processing and Characterization," Tech. Rept. No. AFML-TR-68–190, Air Force Materials Laboratory, Wright–Patterson Air Force Base, OH, 1970.
38. L. Kaufmann and H. Nesor, "Stability Characterization of Refractory Materials under High-Velocity Atmospheric Flight Conditions, Part I, Vol. I, Summary," Tech. Rept. No. AMFL-TR-69–84, Air Force Materials Laboratory, Wright–Patterson Air Force Base, OH, 1970. 17.

FABRICATION OF CARBON FIBER REINFORCED CERAMIC MATRIX COMPOSITES POTENTIAL FOR ULTRA-HIGH-TEMPERATURE APPLICATIONS

S. M. Dong[1], Z. Wang[1,2], Y. S. Ding[1], X. Y. Zhang[1], P. He[1], L. Gao[1]
[1] Shanghai Institute of Ceramics, Chinese Academy of Sciences,
Shanghai 200050, PR China
[2] Graduate University of Chinese Academy of Sciences,
Beijing 100049, PR China

ABSTRACT

In order to improve the properties of carbon fiber reinforced ceramic matrix composites at ultra-high temperatures, both ultra high temperature ceramic (UHTC) matrix and UHTC coatings were applied. The influences of carbon fiber types on mechanical properties and microstructures of C_f/SiC-ZrC composites as well as the microstructures of $(ZrB_2\text{-}SiC)_4$ coating were investigated. Composite with M40JB carbon fibers as fiber reinforcements possesses better mechanical properties though the dispersion of carbon fibers is less homogenous, obvious fiber pull-outs can also be found on the fracture surface of this composite. FPCVD-derived SiC can fill the open pores formed in the slurry pyrolysis process and connect slurry-derived coatings to CVD-derivrd SiC, which will result a high bonding strength between the coatings and substrates.

INTRODUCTION

Carbon fiber reinforced ceramic matrix composites, mainly comprised of carbon or SiC fibers in a carbon or SiC matrix(C/C, C/SiC, SiC/SiC) are potential candidates for a variety of applications in aerospace fields, including rocket nozzles, heat shields and aeronautic jet engines.[1,2] A common aspect of the future hypersonic flight vehicles include sub-orbital and earth-to-orbit vehicles for rapid global and space access missions is the need for new high-temperature materials. Hypersonic vehicles with sharp aerosurfaces, such as engine cowl inlets, wing leading edges (LEs) and nosecaps, have projected needs for 2000°C to 2400°C materials which must operate in air and be re-usable.[3] However, the oxidation of C/C composites above 500°C limits their application at high temperatures[4] and the maximum use temperatures of Silicon-based ceramics is limited to ~1600°C duo to the onset of active oxidation and lower temperature in water vapor environment which will result in gaseous SiO. Therefore, the development of structural materials for use in oxidizing and rapid heating environments at temperatures above 1600°C is therefore of great engineering importance.[5] In order to develop fiber-reinforced ceramic composites that perform at ultra-high temperatures (greater or more than 1500 °C) under oxidative conditions, especially for hypersonic vehicles, two potential approaches are: (a) Utilizing existing carbon-fiber-reinforced carbon-matrix composites (C/C) or carbon-fiber-reinforced silicon carbide-matrix composites (C/SiC) coated by thick (>100μm) UHTC coatings, or (b) Replacing the C and SiC matrices of such composites with an ultra-high temperature matrix.[6] UHTCs based on the several carbides, nitrides, and borides of group IV_B and V_B transition metals have been identified as a class of materials with the potential to withstand extreme aerothermal heating environments.[7]

In this work, Zirconium based UHTCs were applied to modify the matrix of C_f/SiC composites and fabricate ultra-high-temperature coatings for C_f/SiC composites. The properties and microstructures were studied as well.

EXPERIMENTAL PROCEDURE

C_f/SiC-ZrC composites

ZrC, SiC and polycarbosilane (PCS) with the ratio of 25wt%:25wt%:50wt% were ball milled for 4h to form homogenous slurries, using xylene as solvent. Carbon fibers (M40JB, 6K, and T700SC, 12K,Toray, Tokyo, Japan) were used as the reinforcements. The fiber tows were wound on the graphite frame to form aligned UD sheets and then impregnated with the aforementioned slurry. After drying, the prepregs were cut and stacked in a graphite die. The graphite die was then put into the hot-pressing furnace and after being heated to ~200°C, pressures were applied to adjust the thickness of the composites according to the desired fiber volume fraction (~40%). The samples were then further densified by several cycles of polymer impregnation and pyrolysis (PIP) using PCS as the polymer precursor and the pyrolysis temperature was 1100°C. The detailed procedures were described by Wang[8]

UHTC coating

ZrB_2, SiC and PCS were mixed to form homogenous dispersed slurries with a weight ratio of 70:10:20. The 2D C_f/SiC substrates used in this work were prepared by PIP in our laboratory. $(ZrB_2\text{-}SiC)_4$ coatings were fabricated by a slurry dipping plus forced pressure-pulsed chemical vapor deposition (FPCVD) which is similar to FPCVI[9] but with a higher pressure(15KPa) and a longer hold-time (20s). After being machined into 5×2.5×20mm bars, the specimens which were first deposited with ~20μm SiC as adhering coating were dipped into the afore-mentioned slurries. FPCVD were performed after the specimens were pyrolyzed at 1300°C to fill the open pores and enhance the bonding strength between adhering coating and pyrolyzed ZrB_2-SiC coating. The above mentioned experimental procedures can be repeated for several times to get multilayered coatings with desired thickness.

Sample characterization

Densities and open porosities of composites were measured by the Archimedes method. The mechanical properties of C_f/SiC-ZrC composites were tested through three-point-bending tests in an Instron-5566 machine, operated at a crosshead speed of 0.5mm/min and a span of 24mm, the dimensions of the testing bars were 4mm×2mm×40mm. The fracture surfaces and polished cross-sections of the composites as well as the surfaces and polished cross-sections of the UHTC coatings were observed by electron probe microanalyzer (EPMA, JXA-8100, Joel, Tokyo, Japan) to characterize the microstructures.

RESULTS AND DISCUSSION

Mechanical properties and microstructures of C_f/SiC-ZrC composites

Some physical and mechanical properties of composites with different fiber reinforcements were listed in table 1. Bulk density of composite with T700SC fibers (S_T, density:2.66±0.05g/cm^{-3})is some higher than that of composite with M40 fibers (S_M, density: 2.47±0.05 g/cm^{-3}) though there is just little differences in the open porosities. This may be mainly caused by the different ZrC volume fractions in the composites. It can be observed from the back-scattered electron micrographs shown in Fig.1 that the dispersion of ZrC particles (white phase) as well as carbon fibers is more homogenous in composite S_T than that in S_M. There are only very few ZrC particles be found in the intra-bundle areas in composite S_M while there are much more ZrC dispersed in the intra-bundle area of S_T. This phenomenon may be accounted by the difference in the structure between fiber tows used in this study. The structure of T700SC carbon fiber tows is flat while that of M40JB carbon fiber tows is round, so it is easy to handle T700SC carbon fiber tows without twisting, which will result in a higher efficiency for the ZrC particle to infiltrated into the inner part of the fiber tows. More pores can also be found in composite S_M.

Table 1. Effects of fiber type on the properties of Cf/SiC-ZrC composites

Sample	Fiber Volume fracture/%	ZrC volume fraction /%	Density g/cm^3	Open porosity/%	Bending stress /MPa	Elastic modulus /GPa
S_T	36	20	2.66±0.05	4.3±0.7	81±24	91±12
S_M	39	14	2.47±0.05	6.6±0.03	178±77	106±13

The bending stress/displacement curves obtained from bending test are shown in Fig.2. It

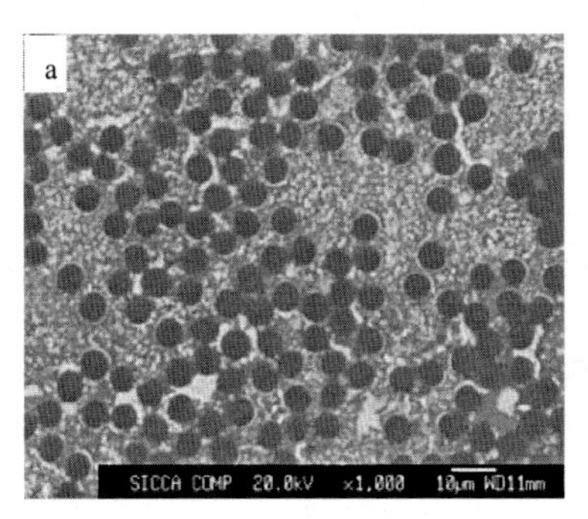

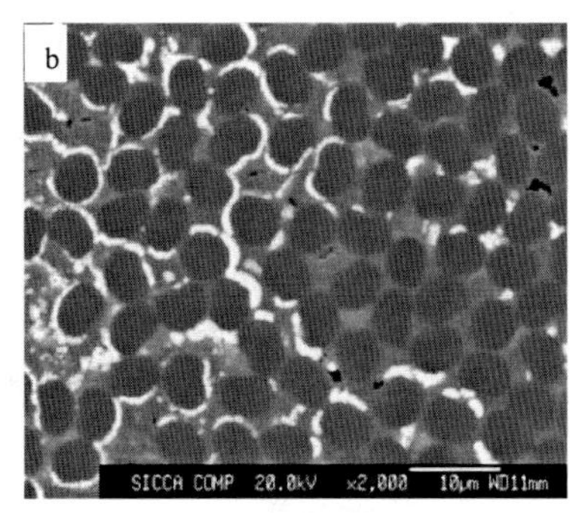

Fig.1 Back-scattered electron graphs for the polished cross-sections of Cf/SiC-ZrC composites With different fiber reinforcements (a. T700SC, b.M40JB)

indicates that composite S_M shows typical non-brittle fracture behaviors meanwhile it also displays higher bending stress and elastic modulus as summarized in table 1. However, composite S_T shows brittle fracture behavior, lower bending stress as well as smaller displacement. This may be induced by the erosion of carbon fibers. T700SC fibers are a grade of carbon fiber of high strength and standard modulus while M40JB is a grade of carbon fiber with high modulus. Meanwhile, M40JB fibers have a higher stability at high temperature due to high graphitization degree. In consistence with the brittle fracture behaviors of S_T, there is almost no fiber-pull-out observed on the fracture surface shown in Fig.3a but some pulled-out fibers can be found on fracture surface of composite S_M (Fig.3b). Some matrix can also be found on the surfaces of pulled-out fibers of composite S_M.

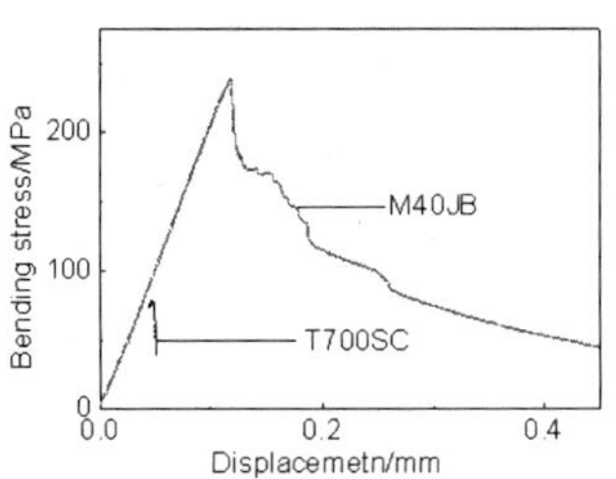

Fig.2 Bending stress versus displacement curves for C_f/SiC-ZrC composites

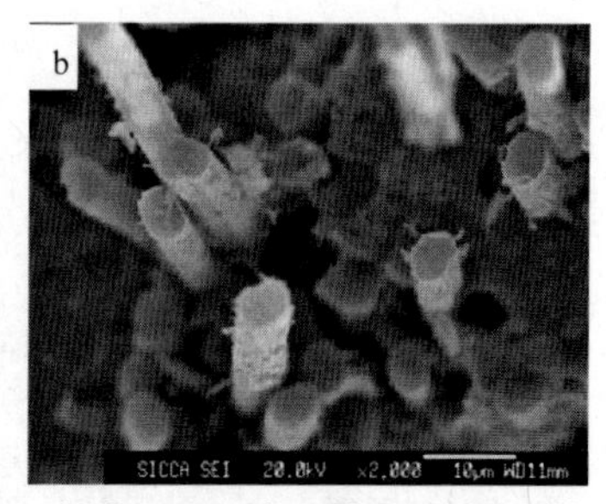

Fig.3 Scanning electron microscope graphs on the fracture surface of composites with different fiber reinforcements (a. T700SC, b.M40JB)

Microstructure evolution of UHTC coatings

Fig.4 shows the microstructure of SiC adhering coating. The thickness of the coating is about 20μm. No cracks can be found between the composite substrate and the SiC coating. Some large particles as well as some small particles can be observed from the surface of the pyrolyzed

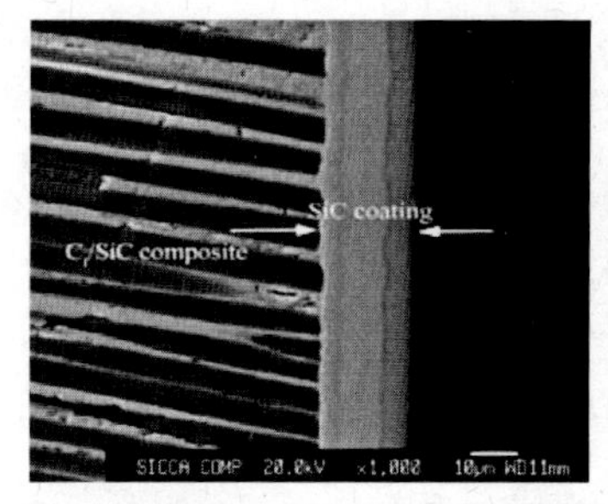

Fig.4 Microstructures of the SiC adhering coating

Fig.5 Surface morphology of pyrolyzed ZrB_2-SiC coating

surface (Fig.5), which can correspond to ZrB_2 and SiC respectively according to the particle size of the raw materials. Furthermore, there are still some pores existed, which can be induced by the

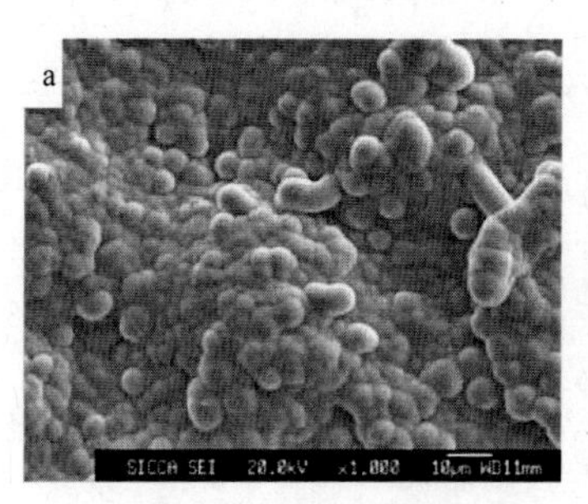

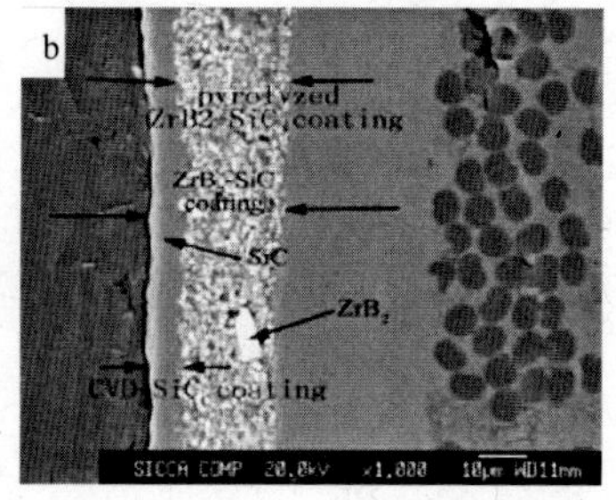

Fig.6 SEM graphs of ZrB_2-SiC coating after FPCVD (a. surface; b. polished cross-section)

volume shrinkage accompanied in the PCS pyrolysis. Compared to the surface microstructrures of pyrolyzed ZrB_2-SiC coating, coating after FPCVD is much denser, shown in Fig 6(a). As it is shown in Fig. 6(b), the thickness of FPCVD SiC is about 7μm and only a small amount of closed

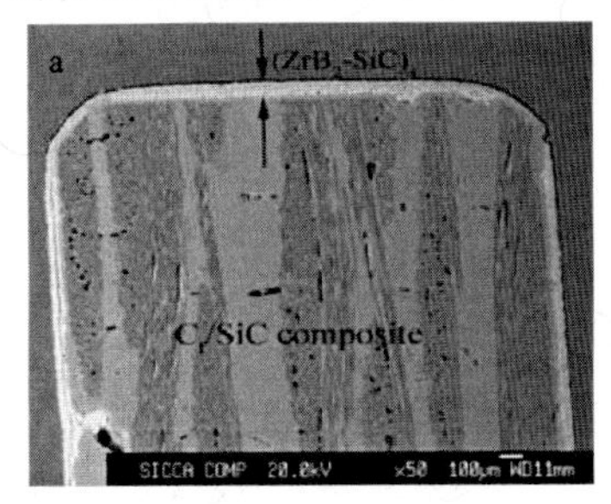

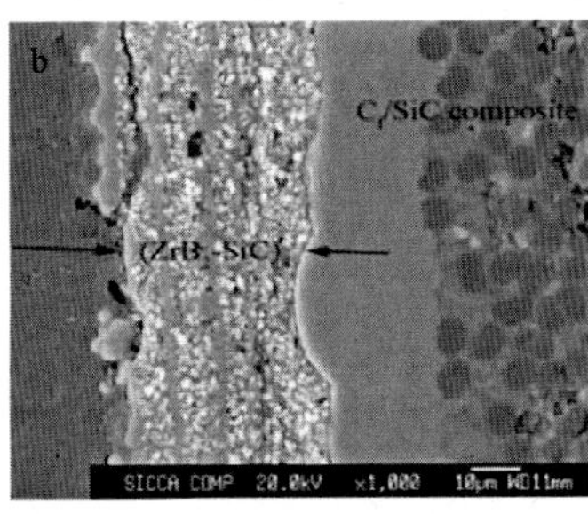

Fig.7 Microstructures of $(ZrB_2\text{-}SiC)_4$ coating

pores can be found in the ZrB_2-SiC coatings. ZrB_2 particles dispersed homogenously in the pyrolyzed coating and were wrapped by either PCS-derived SiC or FRCVD SiC. The advantage of FPCVD is the exhausts are pumped out of the furnace after each pulse, so the source gases can easily infiltrated into the open pores existed in the samples at the beginning of the pulse, therefore the formed SiC can fill the open pores. FPCVD formed SiC can fill the throughout holes in the pyrolyzed ZrB_2-SiC coatings and connect them to the FPCVD-derived SiC layers, which will enhance the bonding strength between of the resulted UHTC coatings. It can be concluded from the microstructures of $(ZrB_2\text{-}SiC)_4$ (Fig.7) that the thickness of the as-prepared coatings was homogenous and combination between the coating and the substrate is excellent.

CONCLUSIONS

The density of composite S_T is higher than that of composite S_M while the mechanical properties of composite S_M are better. Composite S_M shows non-brittle fracture behavior. The dispersion of ZrC particles and carbon fibers are much more homogenous in composite S_T but pulled-out fibers can be observed on the fracture surfaces of composite S_M.

ZrB_2-SiC and (ZrB2-SiC)$_4$ UHTC coatings were fabricated through slurry dipping and FPCVD. $(ZrB_2\text{-}SiC)_4$ coating is homogenous around the samples and the surface of the coating is dense due to the filling of open pores by FPCVD-derived SiC, meanwhile, the combination of coatings to the substrate is excellent.

ACKNOWLEDGEMENT

This work was supported by the National High Technology Research and Development Program of China (863) under Project No. 2006AA03Z565.

REFERENCES

[1] R. Naslain, Design, preparation, and properties of non-oxide CMCs for application in engines and nuclear reactors: An overview. Compos. Sci. Technol., 64, 155-170 (2004).

[2] S. Dong, Y. Katoh, S. Schwab, and L. Snead, Microstructural evolution and mechanical performances of SiC/SiC composites by polymer impregnation/microwave pyrolysis (PIMP) process, Ceram. Int., 28, 899-905 (2002).

[3]M. Opeka, I. Talmy, J. Zaykoski, Oxidation-based materials selection for 2000°C+ hypersonic aerosurfaces: theoretical considerations and historical experience, J. MATER. SCI. 39,

5887-5904 (2004).
[4]E. Corral, R. Loehman, Ultra-high-temperature ceramic coatings for oxidation protection of carbon-carbon composites, J. Am. Ceram. Soc., 91, 1495-1503 (2008).
[5] E. Wuchina, M. Opeka, S. Causey, et al. Designing for Ultra-high-tempperature applications: The mechanical and thermal properties of HfB_2, HfCx, HfNx and αHf(N), J. Mater. Sci., 39, 5939-5949 (2004)
[6] Y. Blum, J. Marschall, H. Kleebe, et al, Low temperature, low pressure fabrication of ultra high temperature ceramics (UHTCs), Technical reports. ADA456577 (2006)
[7] R. Lochman, E. Corral, H. Dumm, and P. Kotula, Ultra high temperature ceramics for hypersonic vehicle applications, Technical Report No. SAND 2006-2925, Sandia National Laboratories, (2006).
[8]Z. Wang, S. Dong, X. Zhang, Fabrication and properties of Cf/SiC-ZrC composites, J. Am. Ceram. Soc., 91, 3434-3436, (2008).
[9]Q. Zhou, S. Dong, X. Zhang, Y. Ding, Z. Huang, and D. Jiang, Synthesis of the fiber coating by FP-CVI, Key Engineering Materials, 336-338, 1307-1309(2007).

ESTIMATION OF SINTERING WARPAGE OF A CONSTRAINED CERAMIC FILM

Kais Hbaieb
Institute of Materials Research and Engineering (IMRE)
3 Research link
Singapore, 117602

ABSTRACT:

A sintering model describing the stress evolution in a sintering film on a pre-sintered thick ceramic substrate is developed. The model is bench-marked by measuring the development of curvature during sintering of multiple thin films on a thin pre-sintered ceramic strip for three different constant heating rates 1°C/min, 3°C/min and 5°C/min. A theoretical analysis was conducted assuming that the strip deforms under plane strain conditions. The predicted curvature results are compared with the experimental results. It was found that the viscosity of the substrate has a large effect on the curvature. The curvature increases with decreasing viscosity of the substrate. Since the substrate viscosity was not measured it was assumed such that the model prediction is forced to agree well with the experimental results for the specific heating rate of 3°C/min. When the same expression was used for the other two heating rates, the model prediction agreed fairly well with the measurement.

INTRODUCTION

Constrained sintering model has been developed to model sintering of thin film on a pre-sintered substrate[1]. This model adopted the linear viscous constitutive relationship to simulate the mechanical response of the sintered thin film[2-6]. Some input parameters, such as free sintering rate and viscosity, have to be measured experimentally. While standard shrinkage experiment is conducted to measure free sintering rate, a recently developed cyclic dilatometry technique[7] is used to quantify the viscosity. Using these experiments, fully empirical relationships for these quantities are developed assuming temperature and relative densities of sintered material as the only intrinsic properties affecting these two input parameters.

Although the model is semi-empirical, it has not been yet validated. A possible approach for the model validation would be to predict the stresses developed in the sintering film and compare the predicted results to the experiment. Thus, a theoretical framework is developed based on the recently developed semi-empirical model to predict curvature developed during sintering of thin film deposited on pre-sintered thin film. To quantify the curvature experimentally, an optical dilatometer is used to measure the curvature developed during sintering of a thin film screen-printed onto a thin dense substrate. The material used is zirconia partially stabilized by 3 mol% yttria (3YSZ). Three heating rates were used in both experiment and modeling, namely 1°C/min, 3°C/min and 5°C/min. It is shown that the results strongly depend on viscosity of the substrate. By using the same relationship for viscosity for the three different heating rates, good agreement could be found between predicted and measured curvature.

EXPERIMENTAL PROCEDURE

The purpose of the experiments is to check the capability of the recently developed sintering model[1] in predicting the stresses/strains developed in the 'constrained' sintering films. The substrate used in these experiments is 200 μm 3YSZ Kerafol electrolyte which has a coefficient of themal expansion of $\alpha_s = 11\text{x}10^{-6}$ /deg C. Multiple 3YSZ films were screen printed on the substrate on one side and sintered at three different heating rates- 1°C/min, 3°C/min and 5°C/min- with one hour dwell at 1350°C. An optical dilatometer was used that allowed in-situ monitoring of sample warpage. If the film coated substrate is roughly square shaped, the stresses developed during sintering will be biaxial. Such a coated substrate will start to deform into a spherical surface during the initial stage of sintering, but as the displacements become large – in this context that means displacements greater than the thickness of the substrate – significant stretching of the substrate as a membrane will occur and gradually the deformation will transform from a spherical to a cylindrical surface which is energetically more favorable. The complications that will arise by the change in the character of the deformation from small to large scale deformation can be avoided if a thin strip of the coated substrate is fired. In this case, because the strip is thin, the full biaxial stress distribution will not develop and the strip will deform into a near cylindrical form. Hence the following analysis assumes that the strip deforms under plane strain conditions. The validity of this assumption could be tested by sintering strips of different widths whose behavior should asymptotically approach plane strain as the width is diminished. Fig. 1 shows a typical deformation undergone by the film coated substrate during firing.

Fig. 1: In-situ monitoring of thin dense substrate bonded to a sintering film and warping at high temperature.

THEORETICAL FRAMEWORK

In this section we describe the approach for predicting curvature development during sintering.

Sintering Model for the film and deformation behaviour of the substrate

(a) Sintering model for the film

The free sintering strain rate can be given by

$$\dot{\varepsilon}_f = A\frac{(1-\rho)^2}{\rho}\exp\left(-\frac{Q}{T}\right) = \exp(18.3)\frac{(1-\rho)^2}{\rho}\exp\left(\frac{-40000}{T}\right) \tag{1}$$

where ρ is the relative density of the film, A is a constant, Q is the "activation energy" for sintering, T is the absolute temperature. The free sintering rate is measured using a standard shrinkage experiment. A curve fitting to the experimental data suggests the values *exp(18.3)* and *40000* for A and Q, respectively. Note the free sintering strain rate is negative, but $\dot{\varepsilon}_f$ is defined here as a positive quantity.

The coefficient of viscosity of the film is given by

$$\eta = \frac{E_{fv0}\exp\left(\frac{Q_{fv}}{T}\right)}{\left[1+CF(\rho)\right]} = 19.2\frac{\exp\left(\frac{14335}{T}\right)}{(1+13.888F)} \tag{2}$$

where Q_{fv} is the activation energy for viscous flow and $E'_{fv0}\exp\left(\frac{Q_{fv}}{T}\right)$ is the viscosity of the fully sintered dense film, and C is a constant. F is a function of density, namely:

$$F(\rho)=\frac{1-\rho}{1-\rho_0} \quad (3)$$

Where ρ_0 is the initial density. Cyclic dilatometry technique[7] is used to measure viscosity. After curve fitting the parameters C, E_{fv0}, Q_v are determined to be 13.888, 19.2 and 14335, respectively. The viscous Poisson's ratio is assumed to be given by

$$\nu_{fv}=\nu_{v0}+(0.5-\nu_{v0})\left[1-F(\rho)\right] \quad (4)$$

The Young's modulus, E_f, is assumed to be independent of temperature and given as a function of relative density by

$$\frac{E_f-E_{f0}}{E_{fd}}=\frac{\rho-\rho_0}{1-\rho_0} \quad (5)$$

where E_{fd} is the Young's modulus of the fully sintered dense film, and E_{f0} is the Young's modulus of the film before sintering. The elastic Poisson's ratio is assumed to be a constant, ν_{fe}.

(b) Viscoelastic model for substrate

Young's modulus of the substrate, E_s, is assumed to be a constant independent of temperature. The elastic Poisson's ratio is assumed to be a constant, ν_{se}.

The coefficient of viscosity of the substrate is given by

$$E_{sv}=E_{sv0}\exp\left(\frac{Q_{sv}}{T}\right) \quad (6)$$

The viscous Poisson's ratio for the substrate is assumed to be 0.5.

Deformation Equations

The variation of the density in the film is given by

$$\dot{\rho}=-\rho\dot{\varepsilon}_v \quad (7)$$

where $\dot{\varepsilon}_v$ is the volumetric strain.

Here we only consider viscous stresses and omit the elastic terms. Thus, we calculate the volumetric strain rate under plane strain conditions to be:

$$\dot{\varepsilon}_v=(1-2\nu_{fv})(1+\nu_{fv})\sigma_f/\eta_f+2\alpha_f(1+\nu_{fv})\dot{T}-2(1+\nu_{fv})\dot{\varepsilon}_f \quad (8)$$

The in-plane strain rate in the film is given by:

$$\dot{\varepsilon}^f=\frac{\sigma_f(1-\nu_{fv}^2)}{\eta_f}-(1+\nu_{fe})\dot{\varepsilon}_f+(1+\nu_{fv})\alpha_f\dot{T} \quad (9)$$

The strain rate in the substrate is given by:

$$\dot{\varepsilon}^s=\frac{0.75\sigma_s}{\eta_s}+1.5\alpha_s\dot{T} \quad (10)$$

(A7.5)

Equations of bending

The sign convention for the curvature and the definition of the reference strain and curvatures, ε_0, κ_0 and the z coordinate are shown in Fig. 2.

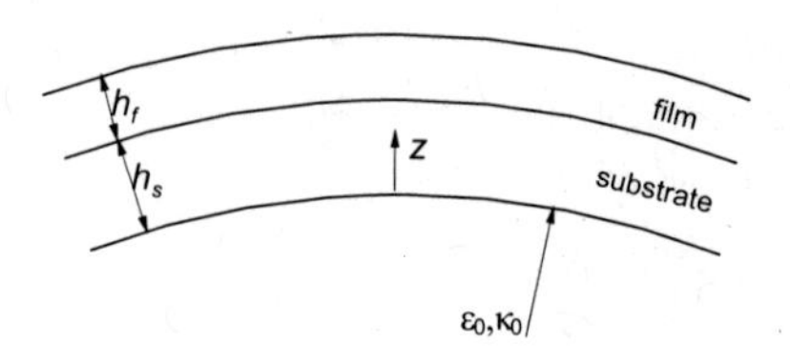

Fig. 2: Schematic illustration of the curved substrate coated by a sintering film.

The conditions for equilibrium are:

$$\int \sigma dz = 0$$
$$\int \sigma z dz = 0 \tag{11}$$

Method of solution using Fortran

The film is portioned into N elements each of thickness $h_i = h_f / N$, with centre at $z = z_i$. It is found that 5 elements give three figures accuracy. The average stress in each element across the film is:

$$\sigma_f^i = \frac{\eta_f}{\left(1 - v_{fv}^2\right)} \left[\dot{\varepsilon}_0 + z_i \dot{\kappa}_0 + \left(1 + v_{fv}\right) \dot{\varepsilon}_f - \left(1 + v_{fv}\right) \dot{T} \right] \tag{12}$$

The stress in the substrate is:

$$\sigma_s = \frac{4}{3} \eta_s \left[f_1 + z f_2 \right] = \frac{4}{3} \eta_s \left[\dot{\varepsilon}_0 + z \dot{\kappa}_0 - \frac{3}{2} \alpha_s \dot{T} \right] \tag{13}$$

The equilibrium equations then become:

$$\sum_1^n \sigma_i h_i + \frac{4 \eta_s h_s}{3} \left[f_1 + \frac{h_s}{2} f_2 \right] = 0$$
$$\sum_1^n \sigma_i h_i z_i + \frac{4 \eta_s h_s^2}{3} \left[\frac{f_1}{2} + \frac{h_s}{3} f_2 \right] = 0 \tag{14}$$

The quantities f_1 and f_2 are constant coefficients. A Fortran program has been written that solves the $2N + 4$ simultaneous first order differential equations for $\dot{\sigma}_i$, $\dot{\rho}_i$, $\dot{\varepsilon}_0$, $\dot{\kappa}_0$, f_1, f_2 and integrates. In the program, temperature profile with steady heating at dT deg. C per minute up to a the sintering temperature, hold at the sintering temperature for 1 hour is adopted. It is assumed that

- The dense elastic modulus of the film is the same as that of the substrate, 200000MPa.
- The elastic Poisson's ratio of the film and substrate is 0.23

RESULTS AND DISCUSSION

Fig. 3 shows the measured and calculated curvature for a sample comprising a thin film deposited on thin pre-sintered substrate and heated at 3°C/min to 1350°C. The results show that the curvature largely varies with the viscosity of the substrate. Therefore, it is essential to treat the substrate as viscous material. The viscosity of the substrate could be measured by conducting 4-pt bending at high temperature. Moreover, it is also noticeable that the curvature is much larger for plane strain that for the case of biaxial stress state. It is therefore useful to conduct curvature measurement on a substrate with much smaller width, or to calculate the curvature using three dimensional (3D) finite element simulating.

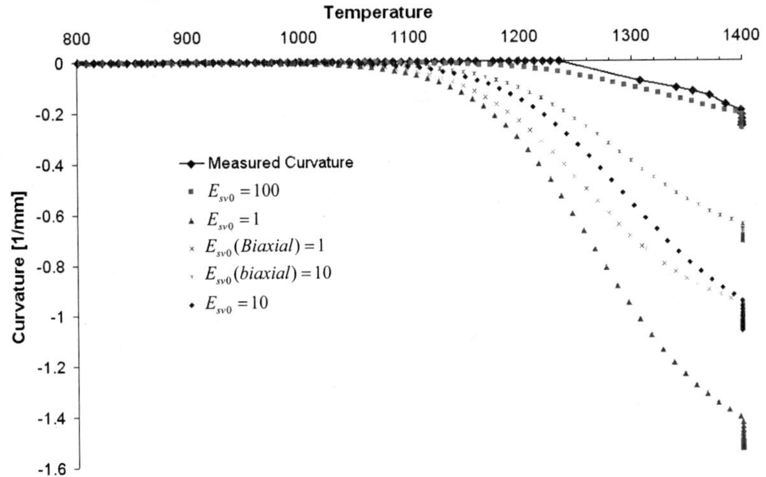

Fig. 3: Comparison between measured and calculated curvature. The different curves for the different calculated curvatures are given for different values of the pre-constant in the substrate viscosity equation (6). The calculated curvature varies significantly with the viscosity of the thin substrate.

As mentioned above the free sintering rate is measured experimentally and a mathematical expression is derived that closely fit the experimental data. The same procedure is followed for the viscosity. Three heating rate are chosen for these analyses and a universal expression for free sintering strain rate and viscosity can be approximated. Since we have not been able to measure the viscosity of the substrate, which we have shown that it has a considerable effect on the development of warpage, we estimate this quantity for a particular heating rate, e.g. 3°C/min, and then use the same expression for the other two heating rate (1°C/min and 5°C/min). We choose the same expression for viscosity as proposed for the sintering YSZ material but we change the pre-constant preceding the exponential term. The universal equation for both sintering rates and viscosities are given in equation (1) and (2). Fig.4 and 5 show comparison between curves corresponding to universal equations and experimental data at the 3 different rates for sintering rates and viscosities, respectively.

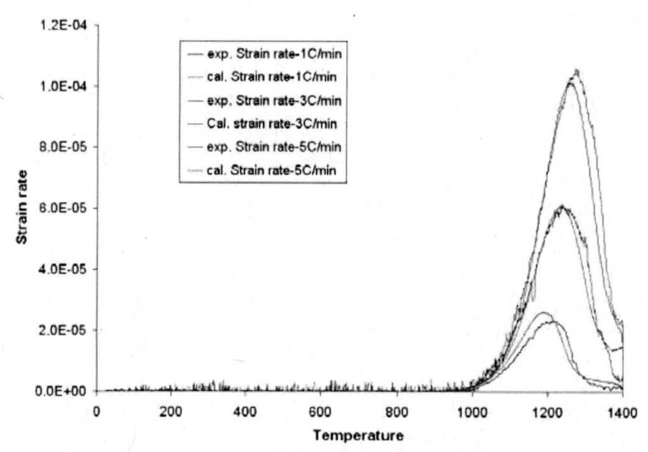

Fig. 4: Comparison between modeling and experimental sintering rates at different heating rates. A unique universal equation for modeling the sintering rates is used.

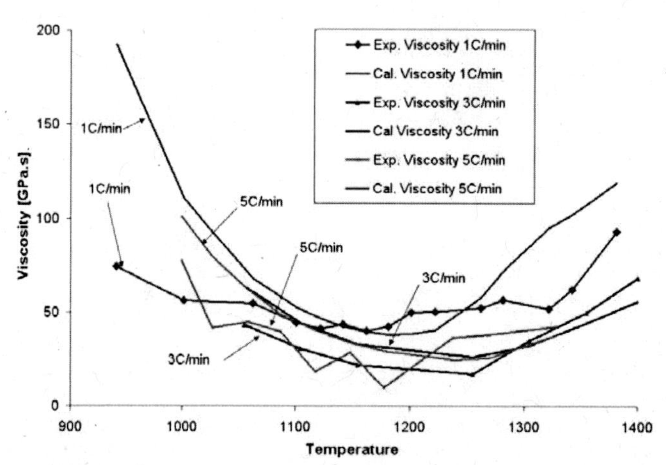

Fig. 5: Comparison between modeling and experimental data for film viscosity at different heating rates. A unique universal equation for modeling the viscosity is used.

The curvature developments at different heating rates are shown in Fig. 6-8 for both modeling prediction and experimental measurements. The combination of all experiments and model predictions are shown in Fig. 9. The results suggest that the viscosity of the substrate shall have a pre-constant of 220. If this constant is adopted in all three warpage estimations, the model gives an acceptable estimation of the curvature and compares well with the experimental results.

Note that the sintering model applies plane strain conditions in predicting curvature. However, transversal strain in the sample can affect the results and the model here may overestimate the curvature. It is therefore important to vary the width of the sample and study how does curvature change (increase) with reducing the width of the sample.

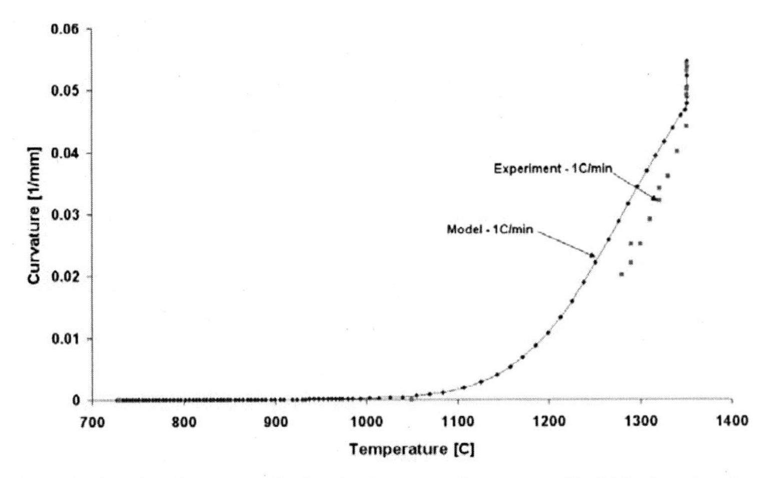

Fig. 6: Comparison between calculated and measured curvature of a thick sintering electrolyte film on a Kerafol 3YSZ substrate and at constant heating rate of 1°C/min.

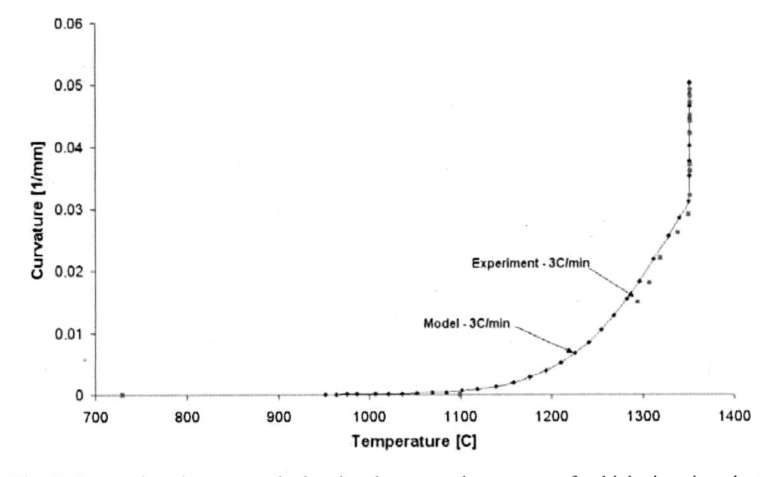

Fig. 7: Comparison between calculated and measured curvature of a thick sintering electrolyte film on a Kerafol 3YSZ substrate and at constant heating rate of 3°C/min.

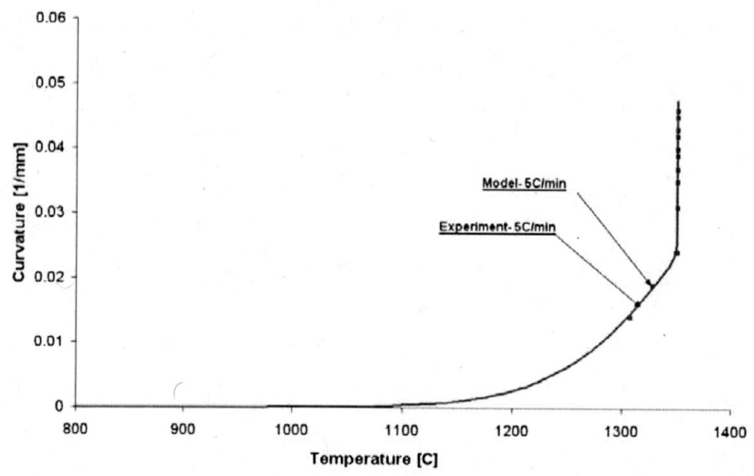

Fig. 8: Comparison between calculated and measured curvature of a thick sintering electrolyte film on a Kerafol 3YSZ substrate and at constant heating rate of 5°C/min.

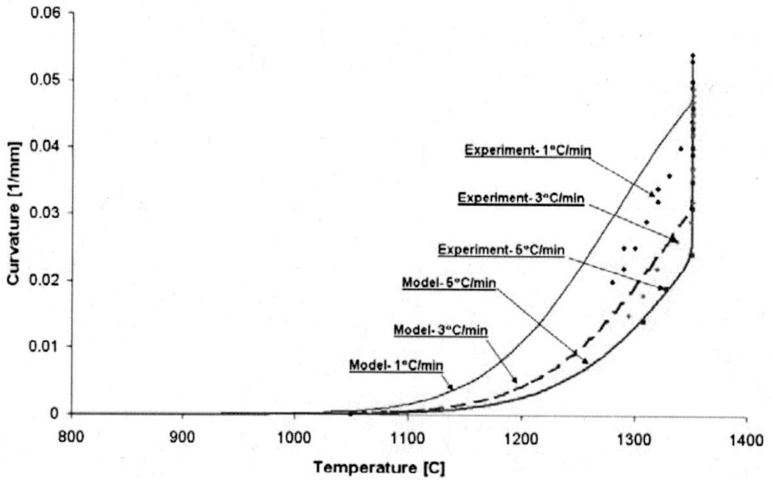

Fig. 9: Comparison of the curvatures developed in the electrolyte layer at different heating rates.

CONCLUSION

A theoretical framework is developed to predict curvature of sintering thin film deposited on pre-sintered substrate for three different heating rates as a way for validating a constrained sintering model developed recently. Using universal expression for free sintering rate and viscosity and estimating viscosity of substrate at one specific heating rate, the model can fairly predict the curvature for the other two heating rates.

REFERENCES

[1]K. Hbaieb, B. Cotterell, Model constrained sintering, Mechanical Properties and Performance of Engineering Ceramics and Composites III, Ceramic Engineering and Science Proceedings, **28** (2), 379-88 (2007), Jonathan Salem and Dongming Zhu, General Editors; Edgar Lara-Curzio, Editor.
[2]R.K. Bordia and G. W. Scherer, On Constrained Sintering-I. Constitutive model for a sintering body; and II. Comparison of constitutive models, Acta Metall., 36, 2393-09 (1988).
[3]T.J. Garino, H.K. Bowen. Kinetics of Constrained Film Sintering, *J. Am. Ceram. Soc.*, **73**, 251-57 (1990).
[4]D. Green, P.Z. Cai and G.L. Messing. Residual Stresses in Alumina-Zirconia Laminates, *J. Eur. Ceram. Soc.*, **19**, 2511-17 (1999).
[5]A. Petersson and J. Agren. Constitutive behaviour of WC-Co materials with different grain size sintered under load, *Acta Mater.*, **52**, 1847-58 (2004).
[6]A. Mohanram, S.H. Lee, G.L. Messing and D.J. Green. A Novel Use of Constrained Sintering to determine the Viscous Poisson's Ratio of Densifying Materials, *Acta Mater.*, **53**, 2413-18 (2005).
[7]P.Z. Cai, G.L. Messing and D.J. Green, Determination of the mechanical response of sintering compacts by cyclic loading dilatometry, *J. Am. Cer. Soc.*, **80**, 445-52 (1997).

LONG-TERM TEMPERATURE GRADIENT STRESS RELAXATION TESTING AND MODELING OF CERAMIC INSULATION MATERIALS

James G. Hemrick
Oak Ridge National Laboratory
Oak Ridge, TN, USA

Edgar Lara-Curzio
Oak Ridge National Laboratory
Oak Ridge, TN, USA

James F. King
Oak Ridge National Laboratory
Oak Ridge, TN, USA

ABSTRACT

Testing was carried out to characterize and predict the long-term thermomechanical properties of Thermal Ceramics Min-K 1400TE thermal insulation material, hereafter referred to as Min-K. In particular, the high temperature gradient stress relaxation behavior of Min-K up to 700°C was evaluated under various thermal gradients and a helium atmosphere for test times of up to over two years. Additionally, finite element and mathematical models were formulated to predict the mechanical behavior exhibited by Min-K out to 50,000 hours based on this testing. This paper discusses the design and construction of unique test equipment to carry out this testing, along with the results of the testing and the subsequent modeling.

INTRODUCTION

The thermomechanical properties of Thermal Ceramics Min-K 1400TE, hereafter referred to as Min-K, were characterized in support of its use for structural applications under a gradient temperature regime in an inert atmosphere.* Min-K is a compression resistant fibrous microporous, material widely used in applications requiring lightweight insulation. The TE line is specifically designed to be high purity and non-contaminating when in contact with other materials and to have high strength due to its fibrous content[1]. Because of its good mechanical properties this material is often used as an insulator in structural applications.

When this material is used as a thermal insulator it is subjected to temperature gradients. In this paper we discuss the design and construction of unique test equipment to evaluate the compressive stress relaxation behavior of structural components when subjected to temperature gradients. Selected results are presented for temperature gradient stress relaxation tests of Min-K and an empirical model is presented to describe the gradient temperature stress relaxation behavior.

Other tasks performed under this project consisted of isothermal evaluation of the Min-K material in compression as a function of temperature (including - assessment of size and geometric effect on the distribution of compressive strength of Min-K, determination of distribution of compressive strength of Min-K as a function of temperature, and statistical analysis of monotonic compressive strength results), isothermal stress relaxation testing, and associated modeling efforts. Results from these activities can be found in a previously published ORNL Technical Report[2].

GRADIENT STRESS RELAXATION EXPERIMENTAL PROCEDURES

A test set-up similar to that shown in Figure 1 was used for the stress relaxation tests. This set-up consists of an electromechanical testing machine equipped with load and displacement digital controllers, a heated metallic platen (Inconel, 304 stainless steel or S-7 tool steel encasing a nichrome/alumina ceramic heater) above and below the test specimen, and a single zone furnace or refractory box surrounding the sample/heated platen assembly.

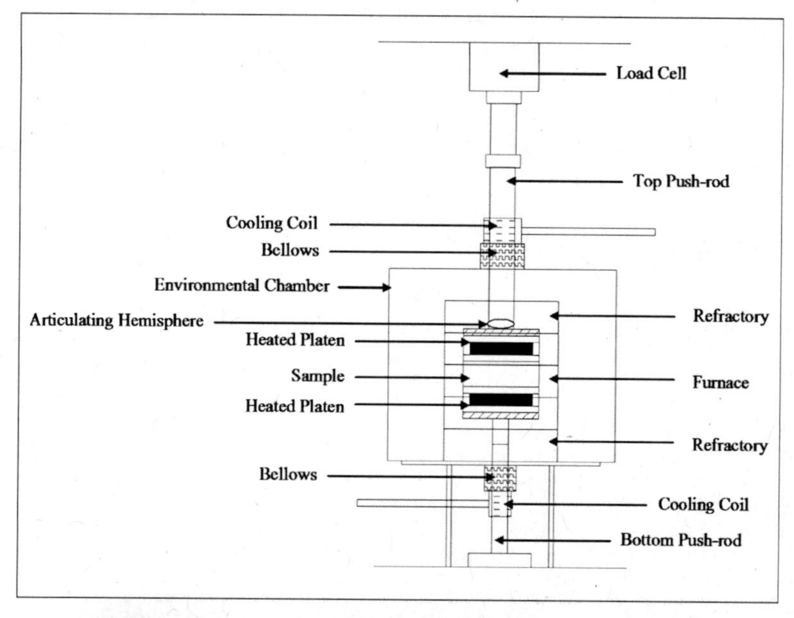

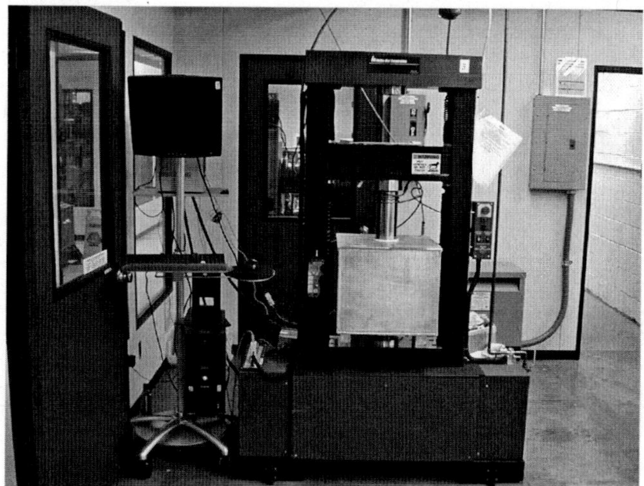

Figure 1. Schematic and image of gradient stress relaxation test fame set-up.

Heated platens capable of 1000°C and sustained loads (≈8000 lbs.), which were needed for this testing were not commercially available. Therefore, platens were custom built at ORNL by encasing a commercially obtained nichrome/alumina ceramic heater** in a metallic ring with metallic disks metal above and below to act as loading surfaces. Inconel or steel (304 stainless steel or S-7 tool steel) was used for the metallic rings and disks depending on the test temperatures. Such arrangement ensured that linear temperature gradients along the height of the test specimen could be achieved while the temperature was independent of the radial position.

Cooling coils were installed on the loading and support push-rods to protect frame electronics and load cells from high-temperatures. An aluminum environmental chamber, connected to the push-rods by a bellows assembly, was used to maintain a controlled environment. Test frames were connected to a back-up-power supply system to provide uninterrupted power during the event of a power failure.

Long-term gradient stress relaxation testing was performed on cylindrical Min-K test specimens with two different diameters and at various gradient temperature conditions as indicated in Table I. The tests were performed under a continuous flow of helium (99.999% purity).

Table I. Gradient Stress Relaxation Testing Matrix

Sample Geometry	Gradient (°C)	Initial Stress (kPa)	Test Duration (hours)
6" dia., 3" thick (152 mm dia., 76 mm thick)	Top – 700 Bottom – 100	1379 (200 psi) (twelve-step loading)	18,135
6" dia., 3" thick (152 mm dia., 76 mm thick)	Top – 700 Bottom – 100	1379 (200 psi) (twelve-step loading)	17,935
8" dia., 1.856" thick (203 mm dia., 47 mm thick)	Top – 593 Bottom - 149	1069 (155 psi) (single-step loading)	19,650
8" dia., 1.856" thick (203 mm dia., 47 mm thick)	Top – 538 Bottom – 71	1089 (158 psi) (single-step loading)	17,375
8" dia., 1.856" thick (203 mm dia., 47 mm thick)	Top – 482 Bottom – 25	1082 (157 psi) (single-step loading)	18,455

Loading was performed under strain control utilizing a twelve-step loading scheme with loading every half hour at a rate of 5.56% strain/hour or in a single-step at a rate of 0.4 mm/minute. An articulating hemisphere was installed between the top push rod and top heated platen to improve loading contact and transfer. Loading was followed by stress relaxation under constant strain with testing planned for in excess of two years (17,520 hours).

TEMPERATURE GRADIENT STRESS RELAXATION TESTING RESULTS

Characteristic results of the temperature gradient stress relaxation testing are shown in Figure 2, representing one of the testing conditions for each of the sample geometries. Magnitude of the stress reduction over the two year test period varied from 54% (18,135 hours with a 700/100°C gradient) to 26% (18,455 hours with a 482/25°C gradient), directly related to the temperature gradient to which the sample was subjected. Relaxation behavior took on a form with rapid stress loss initially (during the first 10 hours) followed by a decaying rate of stress relaxation. Although the rate of stress relaxation was seen to decay, a steady-state stress was never realized over the current test durations in excess of two years.

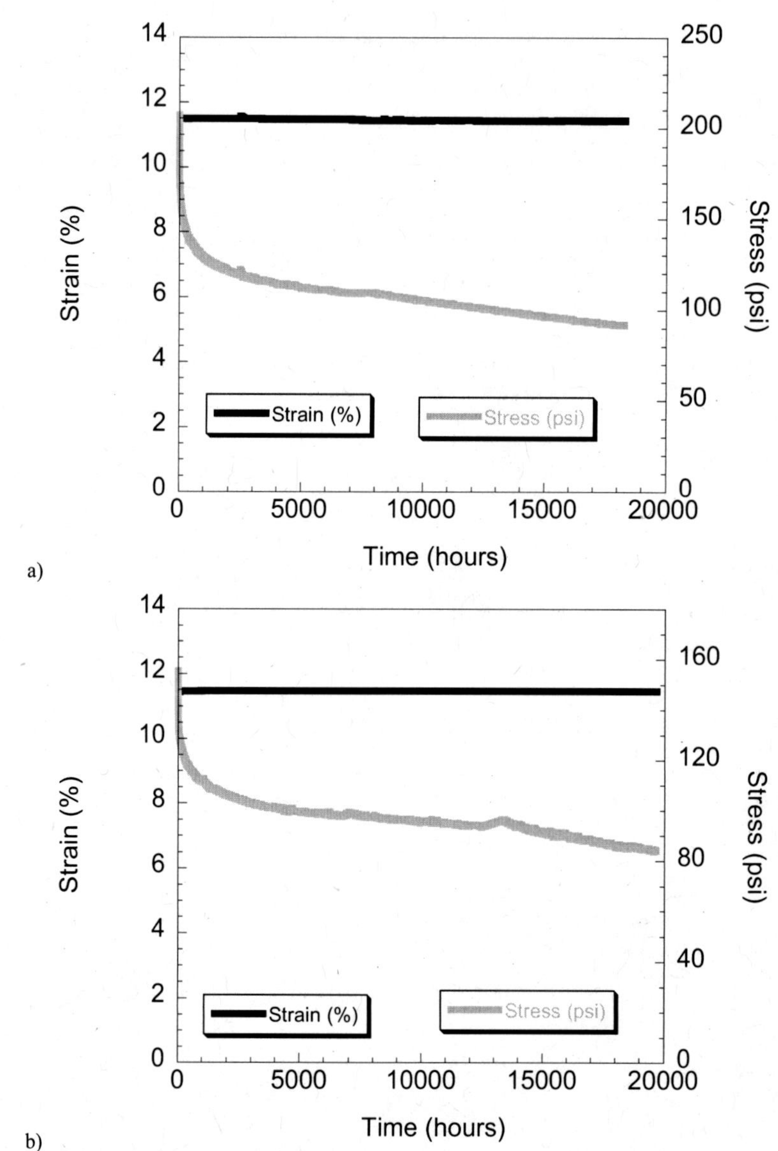

Figure 2. Characteristic gradient stress relaxation results showing an 18,135 hour test with a 700/100°C gradient (a) and a 19,650 hour test with a 593/149°C gradient (b).

MODELING OF GRADIENT STRESS RELAXATION BEHAVIOR

To analyze the stress-relaxation behavior of a Min-K component subjected to a constant axial strain under a temperature gradient, several approaches were taken. The initial approach consisted of discretizing the component into isothermal sections and analyzing the stress-relaxation of each section using Equation 1, where σ is the instantaneous stress at time t, σ_o is the initial stress at the onset of stress relaxation (at the end of mechanical loading), a_i are the temperature-dependent constants and t_i constitute a spectrum of relaxation times. It was assumed that the spectrum of relaxation times is logarithmically-distributed and that the temperature dependence of a_i is described according to an Arrhenius relationship. A schematic representation of such an approach is shown in Figure 3.

$$\frac{\sigma}{\sigma_0} = \sum_{i=1}^{n} a_i e^{-\frac{t}{t_i}} \qquad (1)$$

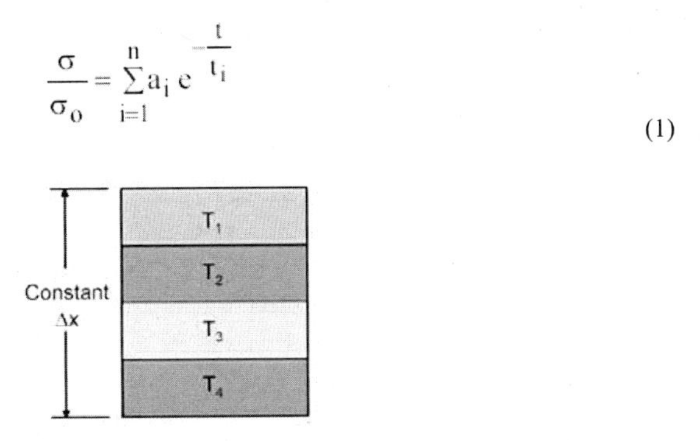

Figure 3. Schematic of discretization of temperature gradient stress relaxation sample into isothermal sections (T_i).

Relaxation times were selected arbitrarily so that they would represent the time span over which the model needs to be applicable. Fitting of this model to experimental isothermal data led to values for the constants a_i and t_i that could be expressed as shown in Table II, where T is temperature in °C.

Table II. Closed-form solution constants

a_1	$0.053 + 7.6x10^{-10}\,T^3$
a_2	$0.88 - \left(\frac{T}{880}\right)^4$
t_1	10 hours
t_2	10,000 hours

In another approach, the stress relaxation behavior was analyzed using the finite element program ANSYS by modeling the isothermal, time-dependent behavior of Min-K using Equation 2, where $\dot{\varepsilon}_{cr}$ is the change in creep strain with respect to time, σ is equivalent stress, t is time, T is temperature, and a, b, c, and d are temperature dependent parameters.

$$\dot{\varepsilon}_{cr} = a\sigma^b t^c e^{\frac{-d}{T}} \qquad (2)$$

However, it was found that the temperature distribution was not radialy isothermal and the results of this analysis predicted more stress relaxation than what had been determined experimentally. Even after specimen temperatures were adjusted to more closely match the actual test temperature distribution observed during an actual gradient stress relaxation test, results were still found to over predict the stress relaxation behavior. Therefore, this modeling approach was abandoned for the long-term prediction of the stress relaxation behavior.

Finally, a simple log function $((\sigma/\sigma_o) = a - b * \log(t))$, where a and b are numerical parameters, was fit to the temperature gradient stress relaxation data. As data in excess of one-year duration were obtained, it was found that the simple log function was sufficient to fit the long-term gradient stress relaxation data. This observation was further confirmed after obtaining two year duration data. Characteristic log fits of temperature gradient stress relaxation data are shown in Figure 4. For all cases an R value of greater than 0.90 was obtained for these fits.

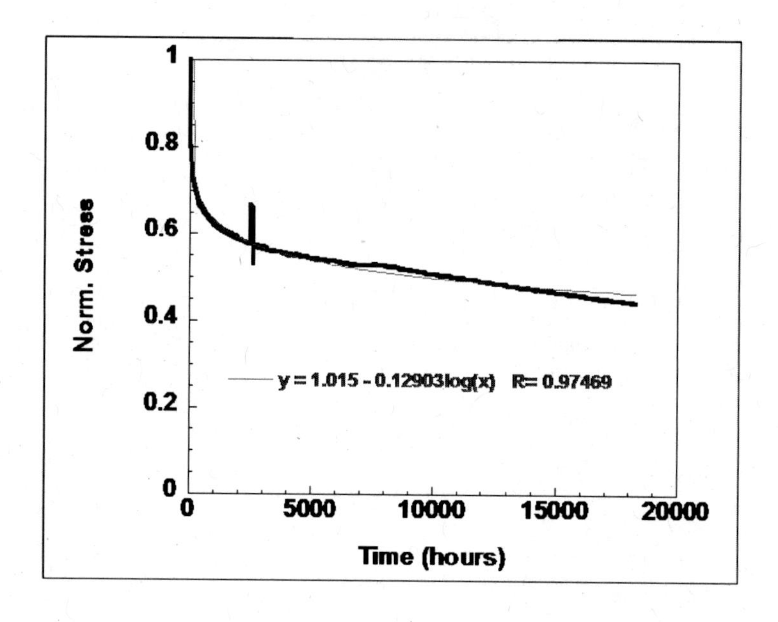

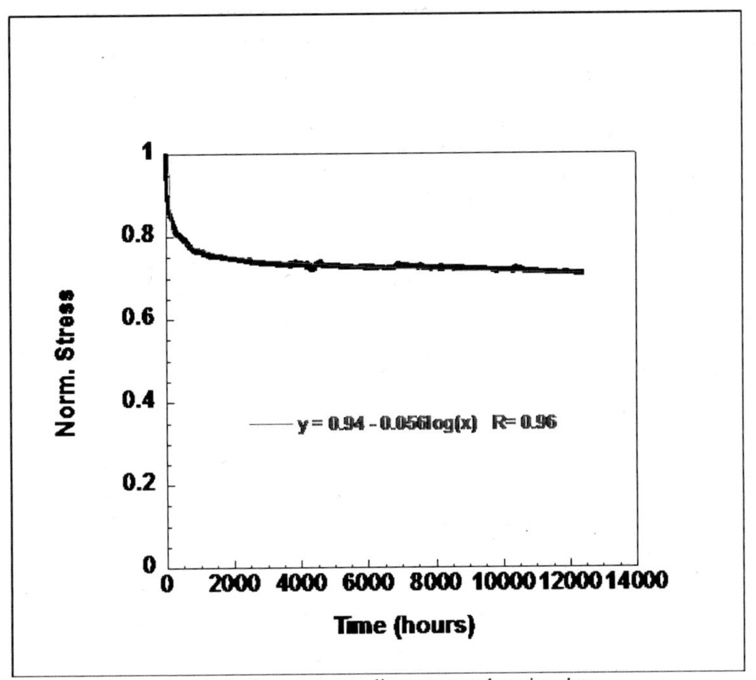

Figure 4. Log-function curve fits for long-term gradient stress relaxation data.

Also, the effect of fitting various time scales of data for predicting long term behavior was investigated. Fits were made using Equation 4 using data from 1,000, 2,500, 5,000, 7,500, 10,000, 12,500, 15,000 and 17,500 hours. In each case, predictions were made out to 17,500 and 50,000 hours. Through this analysis, it was found that for the 700/100°C gradient tests, the predictions of the 7,500 and 12,500 hour fits were the best.

For the lower temperature gradient tests, results varied. For the 593/149°C gradient test, all predictions over estimated the remaining stress in the sample with the 5,000 and 7,500 hour predictions being the closest (errors of 4.89 and 5.17%, respectively when comparing the last predicted and actual data point for each time scale), followed by the 17,500, 10,000 and 12,500 hour predictions (errors of 5.28, 5.51, and 5.54%, respectively). For the 538/71°C gradient test, all predictions were very close, but the prediction made with 12,500 hours of data was the closest (error of 0.02%), followed by predictions made with 10,000 hour and 1,000 hour data (errors of -0.16 and 0.32%, respectively). For the 482/25°C, predictions were again close with the 7,500 hour and 15,000 hour predictions being the closest (errors of 0.05 and 0.12%, respectively), followed by the 10,000, 5,000 and 17,500 hour predictions (errors of -0.27, -0.30, and 0.39%, respectively). No clear trends were evident in this analysis.

CONCLUSIONS

- Methods of platen construction were developed to produce heated platens capable of 900°C and continuous operation in excess of two years (17,520 hours) by utilizing Inconel and various steels for the platen body, dependent on temperatures, and nichrome resistive heaters encased in ceramic insulation[**].
- Data from testing were fitted with various empirical models but through modeling with data in excess of one-year and two-year duration, it was found that a simple log function $(((\sigma/\sigma_o) = a - b * \log(t))$ best describes the long-term temperature gradient stress relaxation data.
- Through analysis of the effect of fitting various time scales of data for predicting long term behavior, it was found that for the 700/100°C gradient tests, the predictions of the 7,500 and 12,500 hour fits were the best. For the lower temperature gradient tests, results varied.

FOOTNOTES

[*]This work was sponsored by the Office of Radioisotope Power Systems (NE-34) of the U.S. Department of Energy and performed at the Oak Ridge National Laboratory, managed by UT-Battelle, LLC, under contract DE-AC05-00OR22725.

[**]International Ceramics & Heating Systems, Inc., Circleville, New York

REFERENCES

[1]Thermal Ceramics Technical Data Sheet "Boards and Shapes Molded Min-K Insulation", 09.02/6 14-125, Thermal Ceramics Inc., Augusta, Georgia 30906.

[2]J.G. Hemrick, E. Lara-Curzio, and J.F. King, "Characterization of Min-K TE-1400 Thermal Insulation", ORNL Technical Report, ORNL/TM-2008/089, (2008).

FROM CONVENTIONAL TO FAST SINTERING OF ZIRCONIA TOUGHENED ALUMINA NANOCOMPOSITES

Enikö VOLCEANOV [1*], Gheorghe Virgil ALDICA [2]; Adrian VOLCEANOV [3], Dan Mihai CONSTANTINESCU [4] and Ştefania MOTOC [1]

[1]Laboratory of Refractories and Advanced Ceramics, Metallurgical Research Institute- ICEM SA, Bucharest, Romania,[*]Corresponding author: evolceanov@gmail.com
[2] Laboratory of Low Temperature Physics and Superconductivity, National Institute of Materials Physics, Magurele-Bucharest, Romania.
[3] Faculty of Applied Chemistry and Material Science, University Politehnica of Bucharest, Romania.
[4] Department of Strength of Materials, University Politehnica of Bucharest, Romania.

ABSTRACT

The aims of this study are to clarify the effect of different fast sintering techniques on densification behavior, microstructure and mechanical properties of a ZTA type composite obtained with 15 % vol. (Ca-Mg)-ZrO_2 nanosized zirconia dispersed in 85% vol. Al_2O_3 matrix. Three kind of heat consolidation techniques were applied respectively on coprecipitated nanosized alumina-zirconia powders by: 1) conventional sintering; 2) microwaves field sintering at 2.45 GHz and 3) external pulsed electrical field (Spark Plasma Sintering), making possible very high heating rates. Microwave sintering of ceramics can offer certain advantages over conventional firing methods, including fast firing times and improved properties. Heat generated by the interaction of microwaves with the ceramic material results in volumetric heating of the ceramic, thus high heating rates is possible. On the other hand, a newly developed rapid sintering method, the Spark Plasma Sintering (SPS) which involves the application of electrical field and an external pressure is currently used to consolidate ceramic, metal and composite powders. SPSed high density zirconia toughened alumina (ZTA) ceramics show excellent mechanical strengths. Microstructural coarsening within ZTA is found to produce a fracture toughness increment, mainly associated with the effect of zirconia particles phase transformability. The obtained mechanical properties and grain growth kinetics are discussed in terms of microstructural features. The effects of heating technique on the far infrared and Raman spectra were investigated. Additional bands development was observed on spectra as the crystal symmetry is lowered, and certain of these bands are highly sensitive to strain-induced distortions of the crystal lattice.

INTRODUCTION

Grain growth during sintering is an inherent problem in the sintering and densification of nanoceramics[1]. For structural applications, fine grain sizes on the order of microns, and ideally nanometers, are desirable to achieve higher hardness and strength[2]. Alumina ceramic has very high hardness and good strength[3] but relatively low toughness in comparison to zirconia, which exhibits very high strength and good toughness but shows relatively poor hardness. The strength and toughness of alumina may be enhanced by dispersion of zirconia solid solutions, mainly due to the tetragonal–monoclinic transformation[4-6]. Strength and toughness of zirconia dispersed ceramics is strongly affected by: 1) the amount of tetragonal zirconia phase (t-ZrO_2) retained in the body, 2) compositional variation and 3) critical size of dispersed zirconia, above which these particles transform to the monoclinic structure [7]. On the other hand, zirconia powders stabilization in binary systems is a very widely approached process, but in many cases the temperature needed for its achievement is too high. In order to decrease the stabilization temperature of zirconia, the simultaneous use of two stabilizer cations is effective in ternary systems[8]. The thermal gradients which are produced during conventional firing process, through radiative and conductive heat transfer can be minimized by the interaction of microwaves with the ceramic material in volumetric heating and thus, high heating rates are possible [9-13]. By temperature gradients attenuation, it is possible to reduce internal stresses to create a more uniform microstructure by avoiding the cracking of parts during sintering and finally may lead to improved mechanical properties. The SPS techniques is a consolidation technique, that uses a combination of pulsed electrical field application with high heating rates to densify, while preserving ultra-fine grain size, a large variety of powder materials that proved to be difficult to sinter by other methods [14-19]. SPS sintering method applies at the same time with the pulsed electrical current, an external pressure which is maintained until 600^{0}C cooling after sintering. The pulsed current is claimed to promote electrical discharges at powder particle surfaces, thus activating them for subsequent bonding [20-21]. All these requirements for improved mechanical properties of such structural ceramics have made the need for fast firing methods crucial.

In this context, the aims of this study are to clarify the effect of different fast sintering techniques such as: microwaves field at 2.45 GHz and external pulsed electrical field (Spark Plasma Sintering-SPS) on densification behavior of zirconia dispersed alumina nanopowders, on the microstructure and mechanic properties and to enhance the mechanical properties of the sintered composites.

EXPERIMENTALS

ZTA powder synthesis

One composition of alumina-zirconia composite with 85%Al_2O_3 and 15%ZrO_2 (by volume) was considered. Ternary zirconia particles with 92% ZrO_2, 5%CaO and 3% MgO were obtained starting from a mixture of aqueous salt solutions (2M) of: $ZrOCl_2.8H_2O$ (Merck) with M = 322.25 g/mol.g; $MgCl_2.6H_2O$-bishofite (Chimopar, Romania) with M = 203.31 g/mol.g, solubility in water of 54.2 g/100 cm³ at 20 °C; $CaCl_2$.$6H_2O$, M = 219.08 g/mol.g, solubility in water of 74.5 g/100 cm³ at 20 °C precipitated with a solution of ammonia hydroxide NH_4OH (c=25%, Chimopar) at pH = 13. The obtained precipitate was filtered, rinsed with distilled water precipitate and added to monoclinic $Al(OH)_3$ (ALOR-Oradea, Romania) having a density of 2.42 g/cm^3 and a calcination loss of 34.5 % and an average grain size of 120 nm and homogenized for 8 hours in corundum ball planetary mill. The obtained hydroxide mixture was filtered, rinsed with distilled water and calcinated at 600°C in electric furnace for 4 h, followed by sieving to avoid the particles agglomeration.

ZTA powders sintering

Two kind of fast consolidation techniques were applied on alumina-zirconia powder mixture: 1) microwave field (2.45 GHz) using SiC susceptors to sinters at 1350 °C and 1450 °C with a heating rate of 60°C/min and a 30 minutes soaking time and 2) external pulsed electrical field (FAST) in 1200-1250 °C apparently temperature range making possible very high heating rates (~ 125°C/min.).
The same composition alumina-zirconia mixture was conventional pressureless sintered in a kiln with methane gas burners at 1550 °C and 1710 °C with a heating rate of 20 °C/min. for 1 hour at maximum temperature. After approximately 16 hours of cooling, the specimens were removed from the oven. The batches submitted to microwave and to conventional sintering were cold uniaxialy pressed as cylinders (ϕ =h =12 mm) in hardened steel die to 15 MPa .

Microwaves field sintering

The microwave sintering was performed in a kitchen oven, commercially available, that has been modified by the addition of a temperature controller and an insulation box and silicon carbide susceptors. The modifications include a thermocouple (Pt-PtRh) and a power controller (with both manual and automatic options) for temperature control. The insulation box consists of a small chamber fabricated from low-density insulation board. Low density and very low dielectric loss is required for

the box to make it microwave transparent [9]. Microwaves pass through the material with little interaction, allowing the contents to heat. The box, in essence, acts as an oven within the microwave chamber, as it allows microwaves to pass through but contains the heat generated by the contents. Many ceramic materials do not absorb microwaves (2.45 GHz) well at room temperature. Susceptors are useful for initial heating of these ceramics[12]. Susceptors are made of a material that absorbs microwaves at room temperature and act as heating elements, which "boost" the temperature until the dielectric loss in the ceramic is high enough that the ceramic couples directly with the field[13].
To sinter the specimens, two specimens were stacked one on top of the other and centered on the floor of an insulation box fabricated from low-density insulation board. Once the insulating box was positioned in the microwave, a power level was selected from the controller and the timer on the microwave set. During the run up to 1350 °C and 1450 °C (and hold for 30 minutes), the power was manually stepped and the temperature measured every minute. When the run was complete, the microwave door was opened and it was possible to remove the box. After approximately 40 minutes of cooling, the specimens were removed from the oven. The sintering time as function of microwave heating temperature is given in Figure 1.

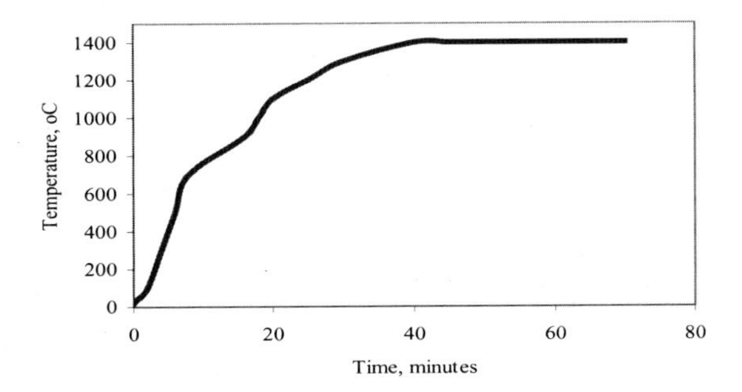

Figure 1 Sintering time as function of microwave heating temperature for a 1350°C running

External pulsed electrical field (Spark Plasma Sintering-SPS)

For SPS each experiment about 5 g of alumina -zirconia powder was loaded into a graphite die with 1.9 cm diameter punches. After loading the powder into the die, samples were processed using a "Dr Sinter" (Sumitomo Coal Mining Co, Japan) sintering machine. Sintering was performed in vacuum (6-15 Pa). The temperature was measured by a pyrometer on the outside wall of the die. A uniaxial pressure of 63 MPa was applied during sintering for all samples. In the SPS apparatus we utilized the default 12:2 (on:off) current pulsed pattern. The waveform is not square and, in fact, is composed of several spikes (pulses) separated by a current-free interval [20]. Regardless of the pattern, each pulse has the same period of about $3 \cdot 10^{-3}$ s. Thus, the pattern of 12:2 has a sequence of 12 pulses "on" and 2 pulses with no current (off). The total time of one sequence (cycle) is about 0.04 s. The operating parameters - voltage and the peak current - were below 5 V and 2000 A, respectively. The SPS machine provides shrinkage and shrinkage rate measurements. The FAST processing parameters were following: controlled computer heating from 20 °C to 600 °C - 3 min, controlled pyrometer heating from 600 °C to 650 °C - 1 min, heating from 650 °C to 1000 °C - 1 min, heating from 1000 °C to 1225 °C – 1 min, the average heating rate 200 °C / min; $t_{dwelling}$ - 4:40 min. The real temperatures in central point of the sample have been evaluated about 1350 °C. The rescale procedure from the theoretical simulation and the direct measurements were taken from reference [21].

Characterization methods

The density was calculated from measured mass and specimen size using a 0.1 mg Mettler Toledo balance (AB204-S type) and a 0.001 mm digital Mitutoyo micrometer. The surface area of powders was determined by BET nitrogen adsorptive method. The sintering bulk density was measured by the hydrostatic method according to SR EN 993 – 1:1997. X-ray diffraction (XRD) patterns were obtained with a DRON diffractometer using CuKα radiation. Computerized Vickers Hardness machine (Buehler 5114) for a test load of 20 kgf equipped with high-resolution digital cameras was used. The fracture toughness, hardness and mode of crack of ZTA composites have been evaluated by using Vickers indentation method at room temperature. The coprecipitated ZTA powder was submitted to thermogravimetric and differential thermal analysis (TGA/DTA) by using a Shimadzu DTG-TA-51H device. A Hitachi S 2600N scanning electron microscope (SEM) incorporating a microanalysis detector for energy disperse X-rays (EDS) was used for the microstructural analysis of the fracture surfaces of the investigated ZTA composites. Additionally, the fracture surface was observed with a FEI, QUANTA INSPECT "F" high resolution scanning electron microscope (HREM). An infrared grating spectrophotometer (BOMEM DA-8) with DTGS (FIR) and HgCdTe (MIR) detectors and appropriate beam splitters was used for pellet spectra over the 700 to 30 cm^{-1} region and this instrument was used for specular reflection measurements of solid samples. Raman spectra were collected at room temperature by a monochromator (model U 1000, Jovin Yvon/Horiba Group, Kyoto, Japan) equipped with Ar, Kr, He-Ne and He-Cd ion lasers, and classical RCA photomultiplier as a detection system and analyzed.

RESULTS AND DISCUSSIONS

Thermal analysis of co-precipitated ZTA powders. DTA, TGA and DTG curves for the coprecipitated alumina-zirconia precipitated mixture are shown in Figure 2 revealing the loss of

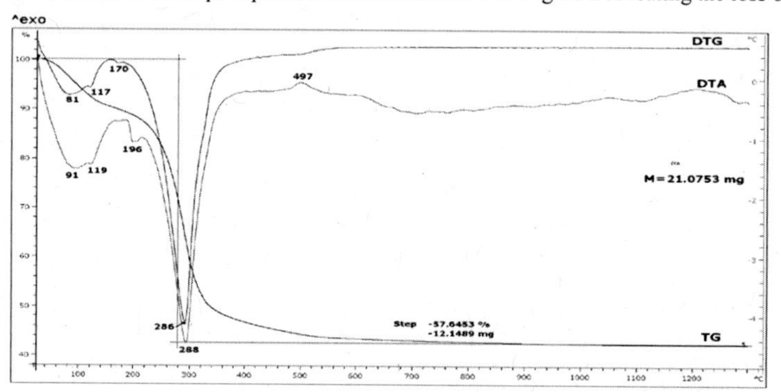

Figure 2. DTA, DTG and TG curves of alumina-zirconia coprecipitated powder

physically bonded water up to 196 °C and of chemically bonded water at 288 °C, attributable to the gibbsite-to-boehmite transition occurred over the temperature range of 220°-330°C followed by a crystallization process at 497 °C.

ZTA powders sintering

In Figure 3 the shrinkage, d, and the shrinkage rate, r, (and others parameters) versus temperature plots of a SPS processed sample are displayed. All curves are fitted in the same graph. Therefore the absolute values of some parameters were reduced by a number showing on the graph.

The maximum measured temperature was 1235 °C. The apparent onset temperature (T_d) of the densification process was observed at 932 °C. The process is composed by two steps (curve d) corresponding to two maxima on the densified rate (curve r) after 300 s and, respectively, 350 s from the heating beginning, because to apply a multiple consecutive rate of heating between 600 and 1225 °C. In the heating cycle were occurred two intervals of gas evolution from powder (curve v) between 150 and 650 °C and, respectively, 730 and 1200 °C.

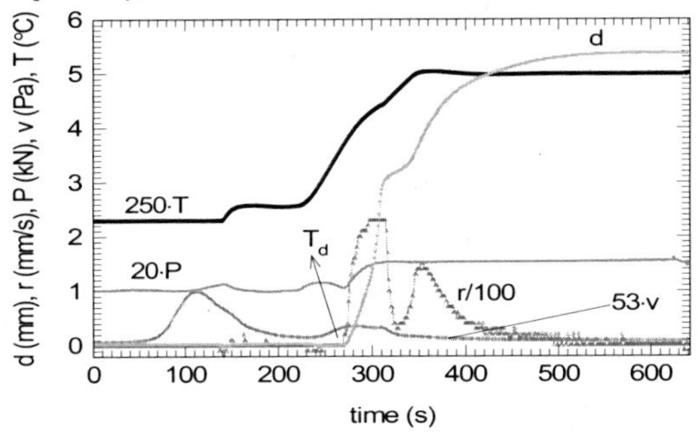

Figure 3 SPS-processed parameters versus temperature for ZTA sample.
The arrow shows the onset temperature (T_d) of the densification process.

Densification

Comparative density and open porosity data for the samples sintered by conventional presureless technique at 1550 °C and 1710°C and by unconventional SPS at 1235 °C and 63 MPa and by microwave field (at 1350 °C and 1450 °C) are given in Figure 4 (a, b). One can be notice that the SPSed specimens were more compact than the same specimens sintered in microwave field and even than the conventionally sintered sampled, cured at ~ 215 °C higher temperature.

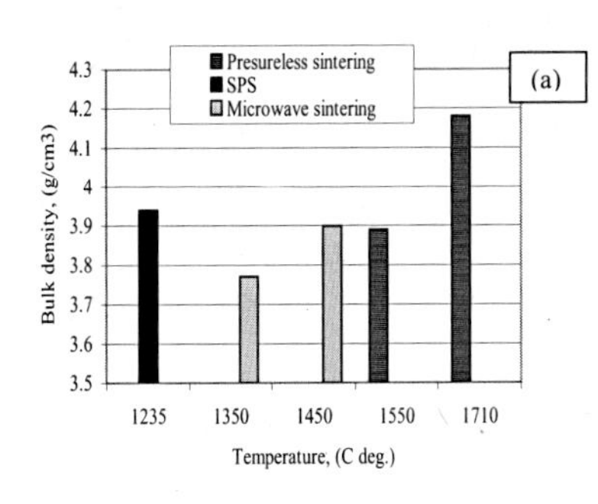

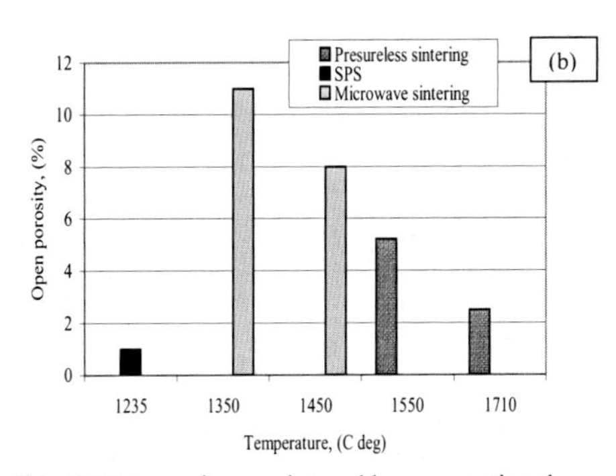

Figure 4. Bulk density-(a) and of open porosity-(b) of ZTA specimens sintered by conventional pressureless technique (at 1550 and 1710°C), SPS (at 1235 °C and 63 MPa) and 2.45 GHz microwave field sintering (at 1350 °C and 1450 °C).

Vickers hardness and toughness

Vickers hardness and toughness (for a indentation load of 20 kgf) for the samples sintered by conventional presureless technique at 1550 °C and 1710°C and by unconventional SPS at 1235 °C and 63 MPa and respectively, by microwave field at 1450 °C are given in Figure 5 (a, b).
Fracture toughness was calculated from the following equation (1) [22]:

$$K_{IC} = \frac{0.203(HV.a^{0.5})(\frac{c}{a})^{-1.5}}{\phi} \quad (1)$$

where, HV is Vickers hardness, a is half-diagonal length in mm, c is crack length from the centre of indentation in mm, and φ is a constraint factor (≈3).

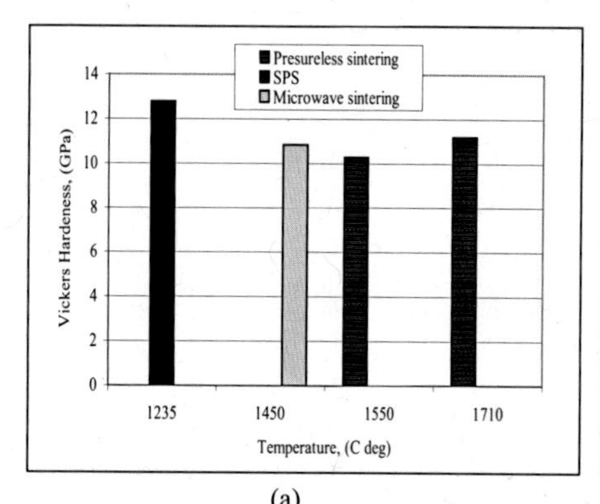

(a)

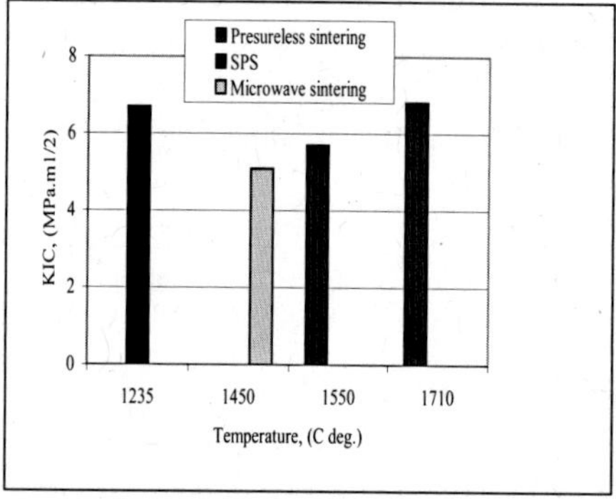

(b)

Figure 5. Vickers hardness and toughness of ZTA specimens sintered by conventional pressureless technique (at 1550 °C and 1710°C), SPS (at 1235 °C and 63 MPa) and microwave field sintering (at 1450 °C).

Various ratio and distribution among the toughening zirconia polymorphs as result of the different sintering techniques could partially explain the obtained mechanical properties (Vickers hardness and fracture toughness by indentation). This assessment would be sustained by microstructural features analysis, XRD, far infrared and Raman spectra as follows.

Microstructure

X-ray diffraction patterns of ZTA specimens sintered by conventional pressureless technique (at 1710°C), by SPS (at 1235 °C and 63 MPa) and microwave field sintering (at 1450 °C) are shown in Figure 6. The microstructures of ZTA are characterized by the presence of major α-Al_2O_3 and ZrO_2 which do not react with each other to form a solid solution [7]. The monoclinic polymorphs of ZrO_2 might be observed on all sample. However, for the SPSed samples, some weak peaks corresponding to tetragonal and cubic polymorphs are revealed.

In figure 7 are shown the microstructure of ZTA composites sintered by conventional pressureless technique at 1550 °C (a) and at 1710°C (b) . The grain size of Al_2O_3 matrix increases from 5-8 μm after sintering at 1550°C (as seen on Figure 7 a) to 10-13 μm at 1710°C (Figure 7 b). Microstructural coarsening within conventionally sintered ZTA composite produces a fracture toughness increment (Figure 5 b), mainly associated with the effect of zirconia particle phase transformability. The high sintering temperature of 1550 °C or 1710 °C required by conventional sintering leads to rapid diffusion and variation in the stabilizer content of the zirconia phase from grain to grain.

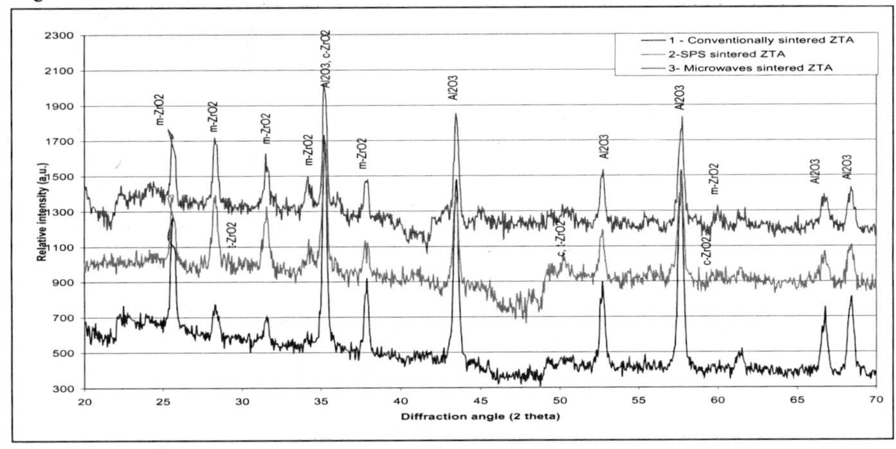

Figure 6 X-ray diffraction patterns of ZTA specimens sintered by conventional pressureless technique (at 1710°C), by SPS (at 1235 °C and 63 MPa) and microwave field sintering (at 1450 °C).

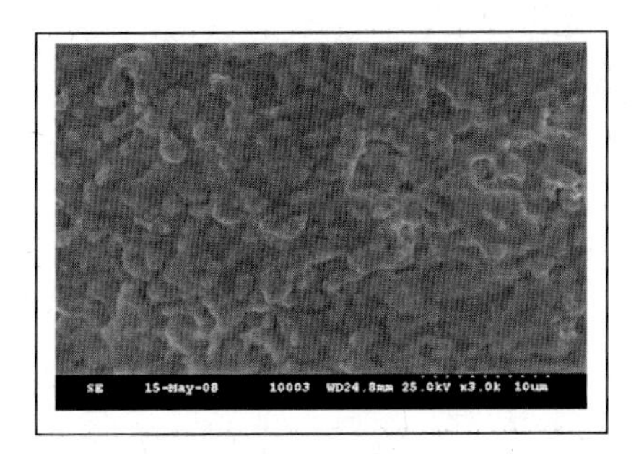

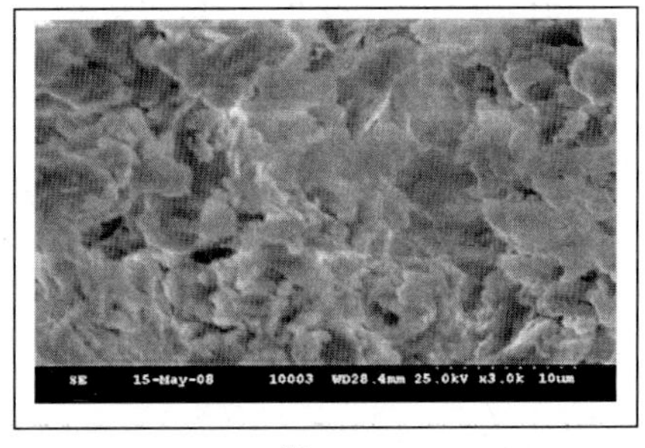

(a) (b)

Figure 7. SEM micrograph of ZTA specimens sintered by conventional pressureless technique at: 1550 °C (a) and at 1710°C (b)

In figure 8 are presented the EDS signals (a), SEM micrograph (b) and HREM micrograph on ZTA specimen surface after sintering in microwaves field at 1450 °C. Microwave sintering at 1450 °C of composites leads to a finer grain size in the matrix (3-7 μm) and to a lower toughness than that obtianed by conventionally sintering (Fig. 5b). Nevertheless, the hardness of composites sintered in microwaves at the same temperature is higher comparatively with those of samples sintered conventionally at 1550 °C.

SEM and HREM micrographs of ZTA ceramics sintered by SPS at 1235 °C in conjunction with energy dispersive signals (EDS) are presented in Figure 9.

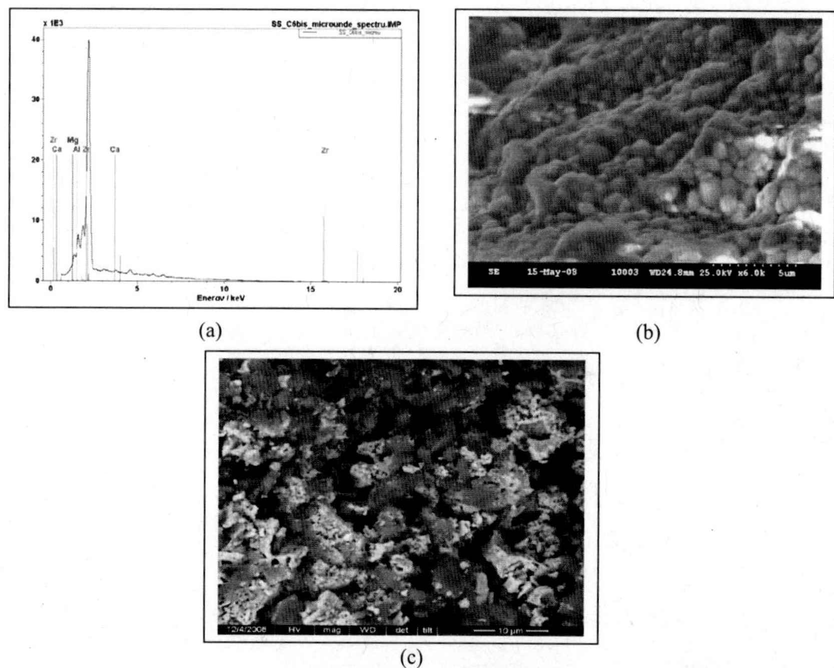

(a) (b)

(c)

Figure 8. EDS signals (a) SEM micrograph (b) and HREM micrograph on ZTA specimen surface after sintering in microwaves field at 1450 °C.

The microstructure consists of somewhat elongated very fine grains with a broad size distribution from 0.6 μm to 3 μm. For the SPSed samples, some weak peaks corresponding to tetragonal and cubic polymorphs are revealed on X-ray diffraction pattern (Figure 6). These microstructure features contribute probably to the development of the best Vickers hardness and fracture toughness in the investigated ZTA composites. However, the compositional variations usually lead to decreases and variations in the strength and toughness of the composite [7].

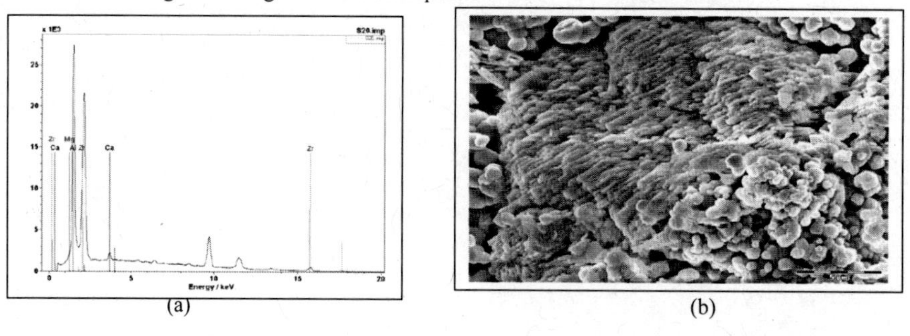

(a) (b)

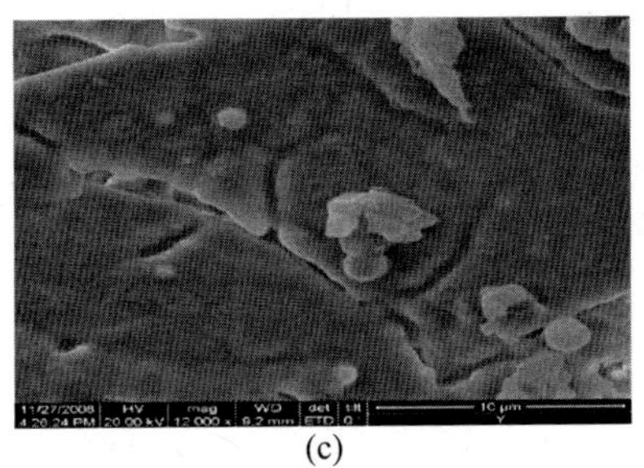

(c)

Figure 9. EDS signals (a), SEM micrograph (b) and HREM micrograph (c) on ZTA specimen surface after SPS sintering at 1235 °C and 63 MPa

Infrared powder transmission spectra analysis

IR spectra for ZTA composites sintered by SPS and in microwaves field at 2.45 GHz are presented in Figure 10 (a, b). The overall contour of the bands, exact frequencies of peaks, and band contrast differ slightly for the two examined samples. A band around 435 cm^{-1} was apparent corresponding to tetragonal ZrO_2[23], while for the sample sintered in microwaves the maximum of this frequency is shifted to 445$cm^{-1}$and for the sample sintered by SPS the bands became more pronounced shifted to 450 cm^{-1}, and which can be assigned to monoclinic ZrO_2. The stabilizing cations produce a structure with certain sites distorted sufficiently that the "tetragonal" modes appear around 602 and 604 cm^{-1}, instead of 575 cm^{-1} mentioned in literature for pure tetragonal zirconia policrystalls[20]. The IR spectrum of a material depends on its crystal structure; if it is inhomogeneous, a different spectrum is associated with each local structure. The appearance of a shifted band from 480 cm^{-1} (acc. to ref.[23]) to around 492 cm^{-1} and 494 cm^{-1} might be assigned to cubic zirconia, which became more pronounced for SPSed specimen. Thus, IR spectra might be more sensitive than X-ray diffraction patterns to small changes in the crystal lattice of the phases of ZrO_2 and may yield practical information on ZTA phase equilibrium after firing because can be used nondestructively.

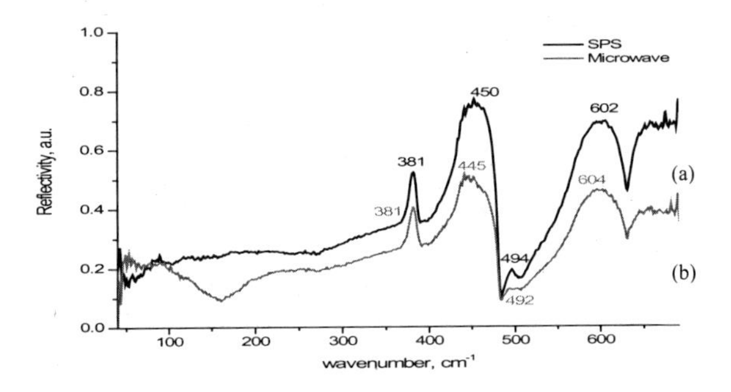

Figure 10. Reflection spectra of polycrystalline ZTA composites.
(a) Reflection from ZTA specimen sintered by SPS at 1250°C; (b) Reflection from ZTA sample sintered at 1450 °C microwaves field (2.45 GHz) sintered

In contrast with IR spectra, in which the overall band contours provide as much information as the exact band frequencies, Raman spectra ordinarily consist of discrete lines. The Raman spectrum of a ZTA sinterized in microwaves field is given in Figure 11, showed weak broad lines at 106 cm^{-1}, 225 cm^{-1}, 505 cm^{-1} and a medium line at 386 cm^{-1} assigned to the monoclinic ZrO_2. A distinct strong asymmetrical line at 179 cm^{-1}, 236 cm^{-1}, 475 cm^{-1}, 641 cm^{-1} and several other weak broad lines at 536, 561 cm^{-1} might be attribute for a tetragonal or a highly strained cubic sample. Several of the weaker lines were better developed for the sample sintered by SPS (Figure 12). The tetragonal and monoclinic phases have many lines in common, but two strong lines at 263 cm^{-1} and 148 cm^{-1} are peculiar to the tetragonal phase, whereas the strong line at 348 cm^{-1} and two weaker ones are peculiar to the monoclinic phase[23]. Additional bands development (150 cm^{-1}, 360 cm^{-1} 492 cm^{-1}, 494 cm^{-1} and 602 cm^{-1}, 604 cm^{-1}) was observed on spectra (being more pronounced in the case of specimens sintered by SPS) as the crystal symmetry is lowered, and certain of these bands are highly sensitive to strain-induced distortions of the crystal lattice.

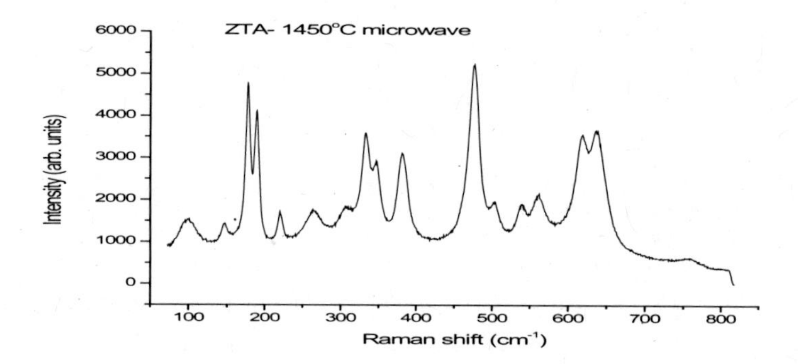

Figure 11. Raman spectra of ZTA ceramic sintered by SPS and in microwaves field at 2.45 GHz

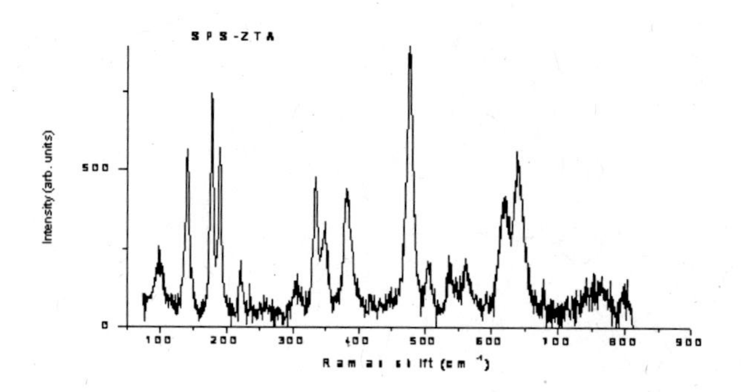

Figure 12. Raman spectra of ZTA ceramic sintered by SPS and in microwaves field at 2.45 GHz

CONCLUSIONS

The results highlighted the possibility of developing dense, high hardness ZTA ceramics by chemical route, using simultaneously Ca^{2+} and Mg^{2+}, starting from $CaCl_2x6H_2O$ or $MgCl_2x6H_2O$, $ZrOCl_2.8H_2O$ as precursors for zirconia dispersed in $Al(OH)_3$ matrix composite.

The effect of different fast sintering techniques on densification behavior of 15 % vol. ZrO_2 nanosized zirconia dispersed in 85% vol. Al_2O_3 matrix, on the microstructure and mechanical properties was investigated. Three kind of heat consolidation techniques were applied respectively on coprecipitated nanosized alumina-zirconia powders by: 1) conventional sintering 2) microwaves field sintering at 2.45 GHz and 3) external pulsed electrical field (SPS), making possible very high heating rates.

With uniform, volumetric temperatures by microwave field sintering at 2.45 GHz, the grain growth due to associated sintering gradients can be regulated.

Microstructural coarsening within ZTA produces a fracture toughness increment, mainly associated with the effect of zirconia particle phase transformability.

The effects of heating technique on the far infrared and Raman spectra were investigated. Additional bands development was observed on spectra as the crystal symmetry is lowered, and certain of these bands are highly sensitive to strain-induced distortions of the crystal lattice.

The high density and fine grain size of the SPS processed samples makes them at least as hard as conventionally sintered materials. It is therefore possible to sinter in a fraction of the time it takes for conventional firing.

The density and Vickers hardness of alumina-zirconia composite sintered by SPS at 1225 °C, with an average heating rate 125 °C/min; a dwelling time of 40 seconds were comparable to dense conventionally sintered specimens with conventional firing cycles at 1550 °C for 1 hours soaking time.

With short sintering runs, the time for grain growth is diminished and rapid sintering of nanosized zirconia powders dispersed in alumina matrix can offer certain advantages over conventional firing methods, including fast firing time and improved mechanical properties.

REFERENCES

[1] M.J. Mayo, D.-J. Chen, D.C. and Hague, Consolidation of Nanocrystalline Materials by Compaction and Sintering; in *Nanomaterials: Synthesis Properties and Applications*. Edited by A.S. Edelstein and R.C. Cammarata. Institute of Physics, Philadelphia, p.165-197, (1996)

[2] J.R Weertman and R.S. Averback, "Mechanical Properties"; in *Nanomaterials: Synthesis Properties and Applications*. Edited by A.S. Edelstein and R.C. Cammarata. Institute of Physics, Philadelphia, p.323-345, (1996).

[3] E.Volceanov, A.Volceanov, S. Stoleriu. Assessment on mechanical properties controlling of alumina ceramics for harsh service conditions, *Journal of European Ceramic Society*, vol.27, p.759-762, (2007), Elsevier Publisher.

[4] S. Hori, R. Kurita, M. Yoshimura, S. Somiya, Suppressed grain growth in final stage sintering of Al_2O_3 with dispersed ZrO_2 particles, J. Mater. Sci. Lett. **4** (9), p.1067–1070, (1985).

[5] G.L. Messing, M. Kumangi, Low-temperature sintering of seeded sol–gelderived, ZrO_2-toughened Al_2O_3 composites, J. Am. Ceram. Soc. **72** (1) p. 40–44, (1989).

[6] V.V. Srdic, L. Radonjic, Transformation toughening in sol–gel-derived alumina-zirconia composites, J. Am. Ceram. Soc. 80 (8), p.2056–2060, (1997)

[7] J. Wang, R. Stevens, *Review Zirconia-toughened alumina (ZTA) ceramics*, Journal of Materials Science **24,** p.3421-3440, (1989)

[8] E.Volceanov, A.Volceanov, S. Stoleriu, C. Plăpcianu, Influence of Ca^{2+} and Mg^{2+} as twinned ions on zirconia powders stability, *Journal of Optoelectronics and New Materials*, vol. 8, Nr. 2, p.589-592, (2006),

[9] A. Goldstein, N. Travitzky, A. Singurindy, and M. Kravchik, "Direct Microwave Sintering of Yttria-stabilized Zirconia at 2.45 GHz," *Journal of the European Ceramic Society*, 19: (1999), 2067-72
[10] Z. Xie, J. Yang, X. Huang, and Y. Huang, "Microwave Processing and Properties of Ceramics with Different Dielectric Loss," *Journal of the European Ceramic Society*, **19**, p.381-87, (1999)
[11] D.D. Upadhaya, A. Ghosh, K.R. Gurumurthy, R. Prasad, "Microwave Sintering of Cubic Zirconia", *Ceramics International* **27,** p . 415-418, (2001),
[12] D. Agrawal, J. Cheng and R. Roy, "Microwave Sintering of Ceramics, Composites and Metal Powders: Recent Development." Innovative Processing/Synthesis: Ceramics, Glasses, Composites IV, Am. Ceramic Soc.Publ., p. 273 – 284, (2000).
[13] J. Lasri, P.D. Ramesh, L. Schachter, "Energy Conversion during Microwave Sintering of a Multiphase Ceramic Surrounded by a Susceptor", *Journal of the American Ceramic Society*, **83** (6), p. 1465-68, (2000),
[14] J.R. Groza: Field Activated Sintering. ASM International Handbook, Vol. **7**: "Powder Metallurgy Technologies and Applications", eds. P.W. Lee, W.B. Eisen, R.M. German, Materials Park, OH, p. 583-589, (1998)
[15] M. Omori: Sintering, consolidation, reaction and crystal growth by the spark plasma system (SPS). Mater. Sci. Eng. A, **287, p.**183-188, (2000)
[16] M. Tokita: Development of Large-Size Ceramic/Metal Bulk FGM Fabricated by Spark Plasma Sintering, Mater. Sci Forum, **308-311,** p. 83-88, (1999)
[17] D.S. Perera, M. Tokita, S. Morricca: Comparative study of fabrication of Si_3N_4/SiC composites by spark plasma sintering and hot isostatic pressing. J. Eur. Ceram. Soc. **18**, p. 401-404, (1998)
[18] Z. Shen, M. Johnsson, Z. Zhao, M. Hygren: Spark Plasma Sintering of Alumina. J. Am. Ceram. Soc. **85,** p.1921-1927, (2002)
[19] C. Gras, F. Bernard, F. Chalot, E. Gaffet, Z.A. Munir: Simultaneous Synthesis and Consolidation of Nanostructured $MoSi_2$. J. Mater. Res. **17**, p. 542-549, (2002).
[20] W. Chen, U. Anselmi-Tamburini, J.E. Garay, J.G. Groza, Z.A. Munir: Fundamental Investigations on the Spark Plasma Sintering/Synthesis Process I. Effect of DC Pulsing on Reactivity. Mater. Sci. Eng. A **394**, p. 132-138, (2005)
[21] R.S. Dobedoe, G.D. West, M.H. Lewis: Spark Plasma Sintering of Ceramics: Understanding Temperature Distribution Enables more Realistic Comparison with Conventional Processing. Adv. Appl. Ceram. **104** (3) p. 110-116, (2005)
[22] K.M. Liang, Evaluation by Indentation of Fracture Toughness of Ceramic Materials, *J. of Materials Science*, **25**, p.207-214, (1990).
[23] C.M. Phillippi, K.S.Mazdiyasni: Infrared and Raman spectra of zirconia polymorphs, Journ. Amer.Ceram. Soc,. **54,** (5), 254-256, (1971).

ACKNOWLEDGEMENTS

Part of this work was supported by the 2nd Romanian National Research Program, NCPM, in the frame of the 4th Program- Prioritary Domains Partnerships- the 8th Research Direction- Space and Security, Project No. 81-008/2007.

The authors thank to Prof. J.Groza and D.V.Quach from University of California, Davis, U.S.A and to Dr.Zorana Dohčević-Mitrović from Institute of Physics of Belgrade, Serbia for their technical support and helpful discussions.

FATIGUE CHARACTERIZATION OF A MELT-INFILTRATED WOVEN HI-NIC-S/BN/SIC CERAMIC MATRIX COMPOSITE (CMC) USING A UNIQUE COMBUSTION TEST FACILITY

Ted T. Kim[1,2], Shankar Mall[1] and Larry P. Zawada[2]

[1]Department of Aeronautics and Astronautics (AFIT/ENY)
Air Force Institute of Technology
[2]Materials and Manufacturing Directorate (AFRL/RXLN)
Air Force Research Laboratory
Wright-Patterson AFB, OH 45433, USA

ABSTRACT

A melt-infiltrated woven SiC/SiC Ceramic Matrix Composite (CMC) reinforced by BN inteface coated Hi-Nicalon type S SiC fiber (MI woven Hi-Nic-S/BN/SiC) was tested under tension-tension fatigue condition using the capabilities provided by a unique burner rig facility at Air Force Institute of Technology (AFIT) which simulated both the load and combustion condition of hot-section components of gas turbine engines such as turbine blades and vanes. A set of fatigue tests performed using the burner rig (stress ratio, R = 0.05 and frequency = 1 Hz) provided S-N data and fracture surfaces, which were analyzed for the role and effects of oxidation on the failure and damage mechanisms. These test results were then compared with those obtained from fatigue tests (R = 0.05 and 1 Hz) in a standard furnace under laboaratory air environement. Fatigue life in the combustion condition was lower by an order of magnitude in comparison to that under the laboratory condition across the range of applied stress. This suggested that the burner rig condition was more damaging to the woven MI Hi-Nic-S/BN/SiC CMC. The observed difference in fatigue strength is attributed to the thermal gradient stress and increased rate of oxidation due to a high moisture level in the burner rig test condition.

INTRODUCTION

Demands for next generation turbine engines for advanced aircraft with increased thrust-to-weight ratio and reliability have propelled on-going research efforts to develop light-weight materials capable of high-performance at elevated temperatures[1]. CMCs using silicon carbide (SiC) fiber reinforced in silicon carbide matrix composites with boron nitride (BN) interphase (SiC/BN/SiC) CMCs are the leading candidate materials that have received great attention due to their ability to retain strength at the temperature beyond the temperature limit of the nickel-based super alloys currently used in the gas turbine engine applications[2]. In the harsh combustion environment of gas turbine engines, however, oxidation can induce the deterioration of the structural integrity as well as load bearing capability and degradatation of the SiC/BN/SiC CMCs, and thus they may become inadequate for service over time. The majority of previous studies have focused on characterizing this deleterious effect of the oxidation[3,4,5] either by subjecting the material to a combustion condition for a controlled amount of time and then testing for the residual strength afterward or by applying controlled stress, either fatigue or creep, to the material in furnace from which either an S-N curve or creep-rupture data are obtained. These studies contributed to the understanding of the oxidation-induced damage mechanisms. However, the experimental approaches taken in these studies often did not involve a simulated gas turbine environment of a hot-section structural component, which undergoes mechanical and thermal loadings simultaneously in the presence of oxidizing chemical species that can only be simulated with a real combustion environment. To meet the needs for a test apparatus that provides a better link to a realistic service environment, a unique experimental facility was developed at the Air Force Institute of Technology (AFIT), in collaboration with the Air Force Research Laboratory (AFRL) Materials & Manufacturing Directorate that combined mechanical loading capability with a real and controllable combustion environment, so as to facilitate the exploration of a variety of test scenarios involving realistic settings. Additionally, a SiC/BN/SiC CMC was characterized under the tension-tension fatigue loading condition in a combustion environment using the AFIT/AFRL burner rig facility. Further, this CMC was also tested under the same fatigue loading and temperature conditions in a standard furnace involving laboaratory environment. This comparison will highlight the effect and role of oxidizing combustion environment on the fatigue response of the test material. These were the objectives of the present study.

EXPERIMENTAL PROCEDURES

AFIT/AFRL Burner Rig

The AFIT/AFRL burner rig, shown in Figure 1, is a one-of-a-kind mechanical test facility that provides a unique capability of characterizating a coupon-size specimen under various simulated combustion and mechanical loading conditions of gas turbine engine components. This experimental facility was named after the two organizations behind its development, namely AFIT and AFRL. The uniqueness of this facility lies in the integration of a true combustion environment to a mechanical test machine. An atmospheric pressure burner rig system mixes fuel and oxidizers to generate high temperature, high speed combustion flame, which is guided into the direction of a test specimen with the capability to vary the impingement angle in order to simulate an angled impingement for future studies. The specimen under the flame impingement simultaneously undergoes a controlled mechanical loading condition by an MTS servo-hydraulic material test system. Being an atmospheric burner rig, the test apparatus may not facilitate the exploration of the failure mechanism triggered by the surface recession, which is known to have an effect on mechanical degradation of material at high pressure[6].However, the burner rig is useful in characterizing the oxidation embrittlement as the failure mechanism under a variety of test scenarios[7] The burner rig system is capable of producing hot combustion gas which travels into the downstream at a sub-Mach speed, providing extra usefulness to simulate even a high-end thermal cycling condition of even the most advanced gas turbine engines. Thermal cycling is facilitated by a programmable mechanical actuation of the combustion rig in and out of the alignment with test specimen. The mechanical load of 25 kN can be applied either in a sustained manner to study a creep scenario or a cyclic loading to investigate fatigue behavior. With its programmed thermal cycling as well as mechanical loading capabilities in both sustained and cyclic forms, virtually any combination of thermal cycling and mechanical loading (involving either creep, fatigue or both) scenarios experienced by a component in gas turbine engine can be accommodated with the AFIT/AFRL burner rig. Table 1 shows a variety of mechanical loading and combustion conditions, which the burner rig is capable of simulating.

The current configuration of this facility, rendered for the task undertaken in this study, was to explore the potential of a SiC/BN/SiC CMC in a gas turbine engine by investigating the fatigue behavior in a realistic combustion condition. It was of particular interest to investigate the effects of thermal stress. Hence, the AFIT/AFRL burner rig was configured such that a non-symmetric thermal field was rendered on the test specimen by imposing the combustion condition on one side of the specimen only, while the other side was allowed to undergo natural convection with the ambient laboratory air. The thermal gradient induced by the combustion heating of such configuration accurately simulates conditions of air foils undergoing the impingement of hot gas stream on only one side during the operation of modern gas turbine engines. The stress induced by the temperature gradients across the specimen dimensions either add to or subtract from the applied peak stress such that a local stress state could be higher (or lower) than the peak stress applied globally. Specific objectives of this study were to characterize fatigue behavior of the woven MI Hi-Nic-S/BN/SiC in a prescribed combustion environment, and to compare the fatigue data and damge mechanisms with those obtained from a set of tests performed in a standard furnace at a similar temperature involving laboratory air environment.

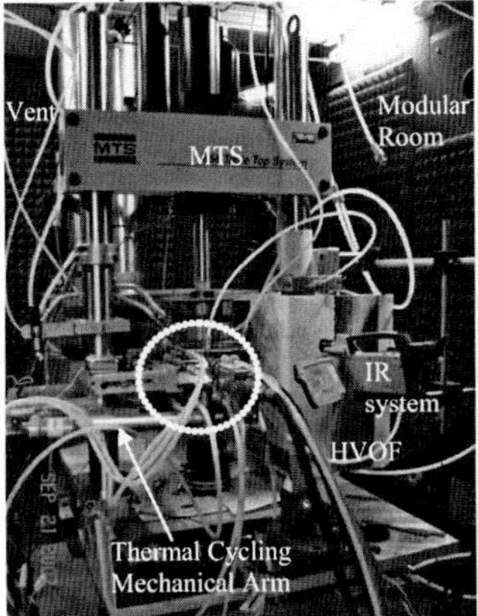

Figure 1: AFIT/AFRL Burner Rig; the right image represents the encircled area in the left.

Table 1: Capabilities of AFIT/AFRL Burner Rig

Mechanical Loading Type	*Thermal Loading Type*	*Test described as*	*Simulation of:*
Static	Constant (Elevated)	*Creep or Creep-rupture*	Turbine rotor blades under constant stress in radial direction
Cyclic	Constant (Elevated)	*Mechanical Fatigue Interacting with creep*	Hot-section components undergoing take-off and landing as well as maneuvering
Static	Cyclic	*Thermal Fatigue; Dwelling Effects*	Operating conditions of high-end engines
Cyclic	Cyclic	*In-phase or Out-of-phase Fatigue*	Worst case scenario for hot-section components

Material and Specimen

The woven MI Hi-Nic-S/BN/SiC was manufactured by GE Ceramic Composite Production, LLC. The fibers were Hi-Nicalon Type S, which have the density of 3.05 g/cm^3 and are non-stoichiometric SiC fibers having the chemical composition: 69 weight % (wt %) of silicon (Si), 31 wt % of carbon (C) and 0.2 wt % of oxygen (O)[8]. The fiber tows, each containing approximately 500 individual fibers, were woven in a five harness satin weave (5HS) into a cloth, eight layers of which were stacked using a warp aligned 0°/90° pattern before the preform was coated with CVI BN and CVI SiC. Final densification consisted of slurry casting with and then followed by Si melt-infiltration (MI) process for matrix densification. The fiber volume fraction was 35.7%[7] and the matrix porosity was approximately 15 ~ 20%. The specimens tested in this study were dog-bone shaped with the cross-sectional dimension of 2.31 ± 0.02 mm (thickness) by 10.15 ± 0.03 mm (width) and the total length of 150 mm. Out of ten specimens available for this study, one was tested under room temperature monotonic tension; six were fatigue tested using the AFIT/AFRL burner rig; and and three were fatigue-tested using the standard furnace.

Monotonic Tensile Testing

The test material was subjected to monotonic tension at room temperature and a stress-strain curve was established. The modulus of elasticity, proportional limit (PL) and ultimate tensile strength (UTS) as well as corresponding strains at PL and UTS obtained from this test provided a baseline to compare the residual strength data obtained from the runout specimens from the burner rig and furnace tests. Due to the limited availability of specimens, only one was tested under monotonic tension at room temperature.

AFIT/AFRL Burner Rig Testing

Prior to testing the woven MI Hi-Nic-S/BN/SiC specimens using the AFIT/AFRL burner rig, the combustion rig was calibrated for the test parameters as specified in Table 2. The combustion condition was to simulate an application environment for hot section components such as turbine vanes and blades in modern gas turbine engines. Gas temperature was measured using R-type thermocouple. Due to the harsh combustion environment that carries aggressive chemical species and accompanies physical force by means of the thrust, it was not possible to devise a conventional contact method of measuring surface temperature. Instead the temperature measurement was made using a non-contact method involving the FLIR ThermaCAM P640 infrared (IR) thermal imaging system, which was calibrated for the emissivity by means of R-type thermocouple such that surface temperature can be monitored and recorded in real time during the test. The calibration was performed for the nominal surface temperature of 1235 ± 50°C. Thermal images of each specimen were taken periodically during the burner rig test and later accessed to determine the temperature at the fractured cross-section on the front, side and back surfaces of specimen. With the current configuration of the burner rig system, the specimen was heated by the combustion flame on one side, while the other side was allowed to be heated through the natural convection with surrounding laboratory air. This mode of heating created a through-thickness thermal gradient, which involved a drop in surface temperature as much as 450°C across the thickness from the front to back surface. Gas composition in the combustion was assumed as predicted by a computer code, i.e. Chemical Equilibrium with Applications (CEA)[9] based on the calibration that showed a good match between the chemical contents predicted by the code and those experimentally collected using the TESTO 350 XL Gas Analyzer at lower temperature, i.e. less than 1000°C. The gas composition in the burner rig environment were estimated to be 26 vol % of CO_2, 14 vol % of O_2, 35 vol % of H_2O and approximately 25 vol % of other minor speices including flame radicals such as OH.

Table 2: Test Parameters

Test Parameter	Condition	Calibration Tool(s)
Surface Temperature	~1235°C	Furnace, R-type TC & IR
Gas Temperature	< 1800°C	R-type TC
Gas Velocity	~ Mach 0.5	XS-4 High Speed Camera
Equivalence Ratio	~ 0.9	HVOF™ Flow Controller
Gas Composition	H_2O, O_2, CO_2, CO, NOx	Testo XL 350 Gas Analyzer
Mechanical Loading	Fatigue (1 Hz & R = 0.05)	MTS
Test Duration	Up to 25 hours	N/A

Under the combustion environment, characterized by the parameters in Table 2, the woven MI Hi-Nic-S/BN/SiC was simultaneously fatigue loaded with a peak stress that was varied between each test, while the stress ratio (R) and cyclic frequency remained constant throughout the testing of all specimens at a stress ratio of R = 0.05 and frequency of 1 Hz. Each test was designed for the maximum number of cycles of 90,000 or 25 hours. If runout was reached, the test was terminated and the specimen was subsequently tested for residual strength at room temperature. The test data including applied peak stress, number of cycles at failure and the location of fracture were recorded from each fatigue test, while stress-strain curve was obtained from the residual strength test. After these tests, microscopic analysis using either Field Emission Scanning Electron Microscopy (FESEM) or Scanning Electron Microscopy (SEM) was subsequently carried out. The entire fracture cross-section was scanned thoroughly under FESEM or SEM to document the extent of oxidation.

Furnace Testing

A similar approach was taken for the furnace testing. The temparature of the furnace was calibrated prior to the actual testing of the woven MI Hi-Nic-S/BN/SiC specimens. The furnace was calibrated to maintain a specimen temperature of 1235 ± 5°C, which is equal to the surface temperature in the direct flame impingement area of specimen in the AFIT/AFRL burner rig. R-type thermocouple, the same type used during the calibration of the temperature in the burner rig test, was used to calibrate for the targeted surface temperature. Gas composition inside the furnace was assumed to be that of laboaratory air, i.e. 79 vol % of N_2 and 21 vol % of O_2, as the furnace was open to the air in the laboratory. The relative humidity in the room at the time of testing was 55 ± 5%, which translates approximately to less than 2 vol % of H_2O.

The rest of the test procedure in furnace testing was identical to that of the burner rig testing. The same stress ratio and cyclic frequency (R = 0.05 and 1 Hz) were used. Peak stress was varied with each test such that the correlation between the applied peak stress and corresponding number of cycles at failure could be established. The test runout condition was again set at 90,000, or 25 hours, and a specimen that survived this duration was tested for residual strength at room temperature. Microscopic study of fracture surface was carried out to document the damage and failure mechanisms as well as to determine the extent of oxidation.

RESULTS AND DISCUSSION

A room temperature monotonic tension test provided modulus of elasticity (E), ultimate tensile strength (UTS) and proportional limit (PL) of the woven MI Hi-Nic-S/BN/SiC, each of which served as the basis for interpreting the fatigue data from the two test environments. The test resulted in E of 195 GPa, PL of 121 MPa, UTS of 334 MPa and failure strain of 0.54%. The proportional limit (PL), which is often used as a reference in gauging the onset of matrix cracking, is the highest stress which a CMC can sustain without deviating from linearity of the stress-strain curve and was determined here using a 0.00005 mm/mm offset strain criteria[4,7,10].

For both the burner rig and furnace tests, the applied peak stress from each test was plotted against the corresponding number of cycles at failure to obtain the S-N curves as shown in Figure 2. The fatigue life increases with decreasing peak stress, indicating the rate of damage development is related to the applied peak stress. It is obvious from the comparison of the two curves that the fatigue strengths of the test material in the burner rig were lower than those obtained from the furnace test. Another intresting note is that the two curves are almost parallel, maintaining nearly constant gap between them across the span of stress values considered in this study. In the following sections, first, the test results obtained in each test condition is discussed briefly, secondly, the data and microscopic observations between the two test conditions are compared, and then the microscopic analysis that could potentially account for the difference in fatigue strength between the two test conditions is presented.

AFIT/AFRL Burner Rig Test Results

Six of the woven MI Hi-Nic-S/BN/SiC specimens were tested under cyclic fatigue at different applied peak stresses in the combustion condition as described earlier to develop the S-N relationship, shown in Figure 2.

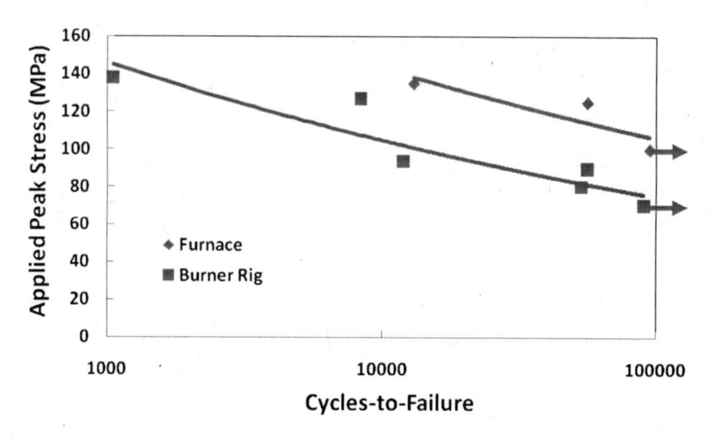

Figure 2: S-N Data of woven MI Hi-Nic-S/BN/SiC obtained from burner rig and furnace testing

The microscopic data from the fracture surfaces of all specimens fatigue tested are summarized and presented next. Part of the fracture surface on which the oxide overlayer was found to have formed was regarded as the oxidized region. In conjunction with the oxide layer, a planar fracture surface with no pullout or very short fiber pullout is considered to be an outcome of oxidation embrittlement and is termed here as "oxidized" region. The regions containing oxide overlayers that do not show planar fracture surface are regarded as the areas of secondary effects of oxidation. Both types of oxidized regions are denoted by "Ox" on the cross-sectional image. The rest of the fracture surface showing fiber pullout with no microscopic evidence of oxidation is denoted as "Non-embrittled" region. This region is from the failure of the material due to the inability of the remaining intact fibers in this area to sustain the applied stress, i.e. due to overload. The oxidized region is separated from the non-embrittled region by dotted lines as shown in Figure 3. This figure shows a typical fracture surface from tests conducted under the burner rig environment. For the sake of brevity, only one case is shown and difference between different tests was only in the extent of oxided area.

Figure 3: Fracture surface at the applied peak stresses of 127 MPa or 38% of UTS in burner rig

The specimen fatigue loaded with the applied peak stress of 80 MPa survived 53,034 cycles, more than 6 times as long as the number of cycles to failure for the specimen loaded with the peak stress of 127 MPa. The difference in fatigue life was observed as the difference in the extent of oxidation on the fracture surfaces. The fracture surface tested with the applied peak stress of 127 MPa is shown in Figure 3. The area of the oxidized region, compared to that on the fracture surface obtained in a test with the applied peak stress of 80 MPa, was relatively small indicating that the extent of oxidation is inversely proportional to the applied peak stress. An examination of the fracture surface also reveals that the oxidation front has advanced from the edge or corners on the left. This is in part due to the inherent susceptibility of the machined edges of 2D woven MI CMCs to oxidation through BN interphase and porosity open to the environment. The asymmetric development of the oxidized area from the left edge or corner could be due to a non-symmetric method of introducing the combustion flame to the specimen, which involved swinging of the burner rig into the position of alignment with specimen from the left of specimen at low temperature, i.e. 600~700°C.

Detailed microscopic features of the oxidized zone typically involve oxide overlayer on the fracture surface formed as a result of the capillary action of liquid borosilicate produced from the oxidation of BN interphase, as shown in Figure 4(a) and (b)[3,5]. Some parts of the oxide overlayer was seen to have bubbled up, as indicated by the arrow, suggesting that there may have been gasous oxidation products trapped underneath. Many fibers in the oxidized region appeared to have adhered to the matrix by means of the oxidation product such as SiO_2., as depicted in Figure 4(b). In the non-embrittled area, the fiber pullout was predominant. One observation of the fibers in the non-embrittled area was that the pullout length was inversely related to the fatigue life, as evident from the comparison of the two images in Figure 5 showing the fiber pullout lengths of two specimens fatigue loaded with different applied peak stresses.

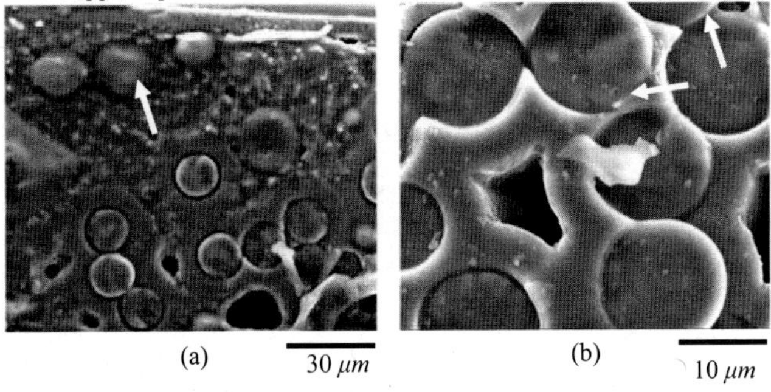

Figure 4: Typical microscopic features in oxidized region

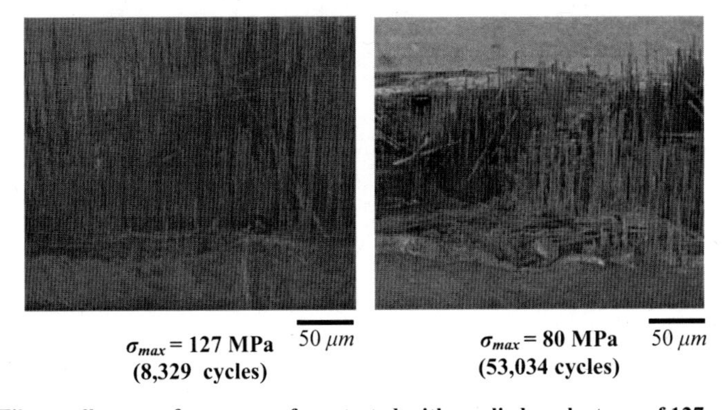

Figure 5: Fiber pullouts on fracture surfaces tested with applied peak stress of 127 and 80 MPa

The specimen that ran out the burner rig test under the applied peak stress of 70 MPa was found to have retained 186 MPa, 56% of UTS which was obtained from the subsequent residual strength test performed at room temperature. The stress-strain curve obtained from this test is shown in Figure 6, along with the two other stress-strain curves; one for a pristine woven MI Hi-Nic-S/BN/SiC specimen and the other from the residual strength test of a runout specimen that survived the applied peak stress of 100 MPa in the furnace test condition. The stress-strain behavior of a pristine sample in this study at room temperature represented in Figure 6 was very similar to that of the similar woven MI Hi-Nic/BN/SiC reported by Corman et al[10].

A comparison of the stress-strain behaviors of the specimen before and after fatigue test revealed that the fracture toughness of a runout specimen was greatly reduced, which was evident from the decrease in failure strain in the burner rig condition that involved the combined combustion environment and the fatigue loading with the applied peak stress at 70 MPa. This specimen, however, retained higher strength and strain than the one fatigued under the peak stress of 100 MPa in the furnace environment. Fractures ocurred outside the specimen gauge length of 12.7 mm; fractured at approximately 25 mm and 38 mm away from its geometric center for the specimen fatigued in burner rig and the one fatigued in furnace, respectively. Both the fracture surfaces, shown in Table 3, involved oxidation linked to the machined edge surfaces, and picture frame oxidation was not evident. The fraction of the oxidized area was only slightly larger for the burner rig tested specimen.

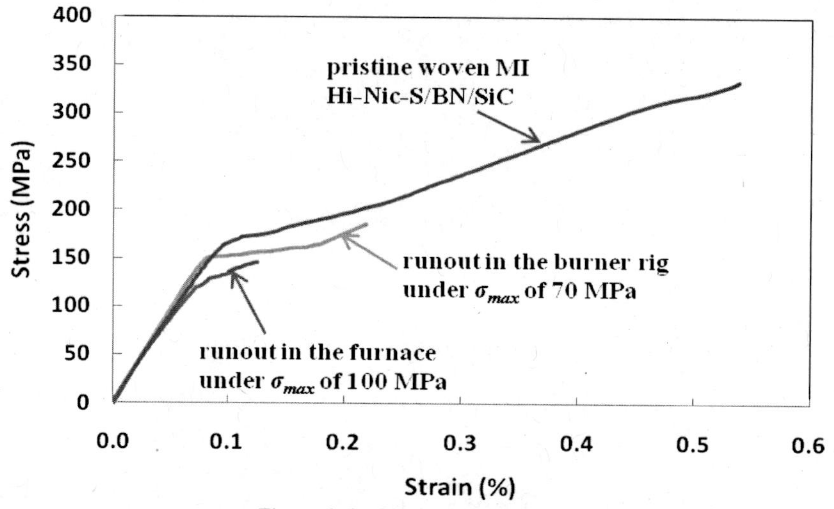

Figure 6: Residual strength test results

Furnace Test Results

Two of the three woven MI Hi-Nic-S/BN/SiC specimens fatigued in the furnace fractured during the test, and the resulting surfaces were analyzed to discern oxidized from non-embrittled region, which is indicated by dotted line that separates one region from the other as shown in Figure 7. Microscopic observations were very similar to those of the burner rig tested specimens. The oxide overlayer resulted from BN oxidation was observed on a relatively planar fracture surface of the oxidized area, along with fiber pullout in the intact non-embrittled regions, whose length was inversely

Table 3: Residual fracture surfaces

Applied Peak Stress (MPa)	Fatigued in	Residual Fracture Surface	Residual Strength (MPa)
70	Burner rig	Ox Non-embrittled	186
100	Furnace	Ox Non-embrittled	146

2 *mm*

proportional to the fatigue life of the specimen. Cracking phenomena that led to oxidation of the exposed fibers appeared to have occurred more uniformly around the machined edges of the specimen, unlike in the burner rig tested specimens. The decreasing trend, similar to that observed from the specimens tested in the burner rig, was found between the applied peak stress and fatigue life. The fatigue life was increased from 12,943 to 58,838 with the decrease in the applied peak stress from 135 MPa to 125 MPa. The difference in the extent of oxidation was subtle due to a relatively small difference in peak stress, i.e. 125 MPa vs. 135 MPa. Relatively large difference in fatigue life obtained from the two specimens that showed similar extents of oxidation on their fracture surfaces suggests that the specimen subjected to 125 MPa experienced higher rate of crack growth. Oxidation responsible for the observed microscopic features may occur immediately after the exposure due to the development of matrix cracks in relatively short time in comparison to the number of cycles being discussed here.

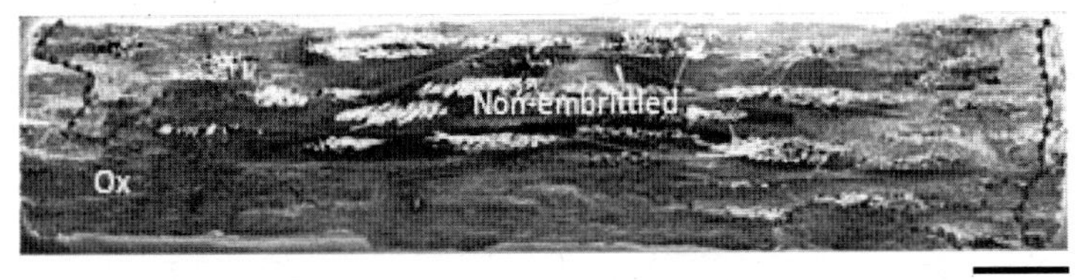

125 MPa (37 %*UTS*)

1 *mm*

Figure 7: Fracture surface at applied peak stress of 125 MPa or 37% of UTS in the furnace

The third specimen survived the furnace test without fracture at the end of 94,627 cycles under the applied peak stress of 100 MPa, and it was subsequently tested for residual strength. The stress-strain curve obtained from the test is shown in Figure 6, along with the runout specimen from the burner rig test to facilitate comparison. The test resulted in the retained strength of 146 MPa, 44% of UTS, compared to 186 MPa or 56% of UTS retained by the specimen that survived in the burner rig under a lower applied peak stress of 70 MPa. Thus, the reduction in fracture toughness was experienced in both test conditions. One could conjecture based on this result that the applied stress, which determines the rate of damage development in the matrix, may have a primary effect on failure while the severity of the oxidizing environment had a secondary role in degradation process.

The fracture surfaces shown in Figure 3 and 7, both of which were obtained under similar applied loading condition, i.e. 126 ± 1 MPa, 1 Hz and R = 0.05, except in different environments, are juxtaposed in Table 4 to facilitate comparison between specimens tested in the two test conditions. While the extents of oxidation in the two surfaces appear to be similar in size, the specimen failed in the furnace outlasted the other by nearly 7 times in number of cycles. This suggests that the specimen tested in the burner rig might have been subjected to a faster crack growth due to the thermal gradient stress that was added to the applied stress to increase the overall stress level as well as a more severe oxidation environment with significantly higher moisture content in the combustion environment of the burner rig.

DISCUSSION

Evidence is clear from the S-N data in Figure 2 that the combustion test condition was more detrimental to the fatigue strength and fatigue life of the woven MI Hi-Nic-S/BN/SiC than the same loading condition in the furnace environment. At a given applied peak stress common to both tests, the corresponding fatigue life of a burner rig tested specimen was approximately an order of magnitude smaller than that of a furnace tested specimen. The factors that may be accountable for the difference in fatigue behavior is discussed below. Each factor is diagnosed for their potential effect on failure.

Fatigue strength

From the comparison of the S-N results from the two test conditions, it is observed that the fatigue strength of the woven MI Hi-Nic-S/BN/SiC is different under different environments. The difference in fatigue strength between the two curves appeared to be constant throughout the range of stress values in this study, which suggests a systematic difference between the two test conditions. There are two potential factors which could explain the constant difference between the two curves, which are oxidation and thermal gradient induced stresses.

Table 4: Fracture surfaces of specimens tested in two environments at peak stress of 125 MPa

Applied Peak Stress (MPa)	Fatigued in	Fracture Surface	Cycles-to-failure
125	Burner Rig	 Ox Non-embrittled	8,329
	Furnace	 Ox Non-embrittled	58,838

2 *mm*

Oxidation is a chemical phenomenon that develops progressively with time. Certain chemical species such as moisture have been reported to accelerate the oxidation process and increase the rate of the oxidation-induced mechanical degradation of SiC/BN/SiC CMCs[3,5]. Realizing the relative abundance of water vapor in the combustion environment of the burner rig as compared to the relatively dry furnace environment, the propensity of an exposed BN interphase to oxidation will be higher in the burner rig leading to a faster oxidation embrittlement of the exposed fibers for a specimen in the burner rig than that in the furnace. This is one of the factors which account for the difference in fatigue strengths between the two test conditions.

However, oxidation alone may not translate into mechanical degradation of CMC. Pervasive oxidation, known to occur in the form of a picture frame[5], has relatively little effect on overall strength of CMCs since the fibers carrying the load are not affected by it. Thus, oxidation in the absence of matrix cracks that serve as conduit of oxidants may not have an appreciable effect on the strength of CMC. The applied peak stress level has a direct implication on the matrix crack density of a specimen. Hence, the comparison of the S-N data from the two tests at the same applied peak stress may seem warranted at first. However, the current configuration of the AFIT/AFRL burner rig involves directional heating that induces thermal gradients across the specimen dimensions, with the largest gradient rendered through the thickness. The temperature gradient induces thermal gradient stress, which applies compressive stress near the heated front surface while the plies on the back face would be under tension to balance the compression on the heated side. This imposes on the specimen a non-uniform stress field and, thus, the stress state higher than that applied can locally be experienced due to the non-uniform thermally induced stress field. This is the second factor which may contribute to the knockdown in the fatigue strength observed in Figure 2.

Thermal stress and oxidation

While the burner rig test imposed a harsh combustion environment with relative abundance of the chemical species that are known to promote oxidation, the furnace test condition subjected a relatively mildly oxidizing environment. Still, the microscopic features found in the oxidized region of the specimen fractured in the furnace were very similar to those observed from the burner rig test. This could be an indication that it might not take much oxygen to cause oxidation on the fracture surface[11]. Rather, it could be the difference in the overall peak stress including the the thermal gradient stress, which is related to the level of matrix cracking and subsequently the fatigue strength and life. While oxidation kinetics governing the rate at which embrittlement occurs, may vary with physical and chemical conditions, the embrittlement driven by oxidation will eventually occur as long as oxidants are available. The difference that exists in oxidation rate between the two test environments, though they may seem largely different with substantial difference in the content of moisture (i.e. 2 vol % vs. 35 vol %) and other oxidants present in the burner rig environment, could be a secondary factor in comparison to the difference in the overall stress applied due to the thermal gradient stress.

A fatigue study performed by Sharma[12] using both Sylramic fiber and Sylramic-iBN fiber reinforced Polymer Impregnation Pyrolosis (PIP) SiNC matrix in both air and full steam environments at 1300°C using 1 Hz and R = 0.05 revealed that there were relatively small decrease in the fatigue life at a given peak stress caused by the full steam environment. Under the peak stress of 120 MPa, the Sylramic fiber reinforced PIP SiC CMC failed after 14,688 cycles in the steam environment, which was about half as many as the number of cycles attained in the air, i.e. 28,515 cycles. Similar observation was made from the Sylramic-iBN reinforced PIP SiC CMC that failed after 5,764 cycles under 180 MPa in the steam, whereas the fatigue life in the air at the same peak stress was 22,807. Considering that the PIP SiC CMCs have matrix shrinkage cracks that would increase the matrix porosity which facilitates oxidation, the effect of oxidation in the MI and CVI matrices of the material of the current study would be most likely less. This, along with the relatively smaller difference in moisture content (i.e. 35 vol % vs. 2 vol % of H_2O) of the current study than in the referenced study for the larger observed difference in fatigue life, leads to a suspicion that the demonstrated difference in fatigue life may be affected by the thermal gradient stress that varied the stress state and caused localized acceleration of crack growth, which could have grown into a self-similar crack involving the fibers bridging the crack. The exposed fibers undergo oxidation embrittlement, initiated by the oxidation of BN interphase, which could occur fast enough to keep up with the advancement of cracks even in a mildly oxidizing environment, such that further increase in the oxidation rate by increased concentration of oxidants may not have as significant an effect on fatigue life as increased matrix cracking effect brought about by the thermal gradient stress in the burner rig test condition. Further in-depth studies incorporating the effect of moisture and other oxidants on chemical kinetics of BN oxidation are, therefore, needed to validate this conjecture in the future.

Residual strength

Keeping in mind that the two test conditions featured different peak stresses that resulted in runout, no definitive conclusion can be deduced from the comparison between the two residual strengths shown in Figure 6. However, the higher strength and strain retained by the burner rig tested specimen, in spite of relatively similar extent of oxidation, points to the possibility that the non-embrittled areas of the two runout cases may have undergone different levels of degradation, perhaps in the form of either frictional wear in the interphase under different peak stresses to affect the fatigue strength. The frictional wear could have been accompanied by the interphase oxidation that may have occurred by consuming residual oxygen content. The resulting embrittled interphase could have led to an accelerated erosion of the interphase by friction under cyclic loading. The comparison becomes more convoluted with the consideration of the thermal stress and oxidation. In particular, the non-uniform thermal stress was induced by the directional heating such that a stress level higher than that accounted for by the applied load was imposed on the plies adjacent to the unheated back surface of the specimen, whereas compressive thermal stress acting on the plies near the heated front surface relieved some of the applied stress to reduce the overall stress borne locally. Thus, the gradient nature of the overall stress condition leads to a conjecture that non-embrittled fibers in the burner rig test condition may have been subjected to the rates of degradation, such as in the sliding resistance, that varied in gradient along the thickness direction. The discrepancy in residual strength may also be related to the location of the fracture. The furnace tested specimen failed near the grip whereas the burner rig tested one failed within the reduced section. In general, the failure outside the reduced cross-section occurs at a lower stress. More in-depth studies in this direction are also needed to further investigate the phenomenon.

CONCLUSIONS

From the S-N curves and microstructures of the woven MI Hi-Nic-S/BN/SiC obtained under the burner rig and the furnace environments, the observations which apply to both the burner rig and furnace tests are: (1) that the fatigue life decreased with increasing applied peak stress, (2) that the evidence of oxidation was clear from the observation of the fibers with interphase adhered to adjacent fibers and the oxide overlayer in the oxidized zone, typically near the edges or corners, (3) that the extent of oxidation was directly related to fatigue life, and (4) that the strength and fracture toughness of the CMC was greatly reduced from both test conditions.

The fatigue life was greater in the furnace approximately by an order of magnitude than in the burner rig at the same applied peak stress across the stress range common in both tests. The fatigue life under the same applied peak stress of 125 MPa, 103% of PL, was significantly higher in the furnace than in the burner rig, i.e. 58,838 vs. 8,329 cycles. From this observation, it could be conjectured that the failure of the woven MI Hi-Nic-S/BN/SiC CMC depends not only on the severity of the oxidation environment, but also on the rate of crack propagation, which is determined by the state of matrix cracking that directly relates to the overall stress state. A thermal gradient stress field induced on a specimen by the burner rig heating elevated the stress state locally to cause the initiation and development of cracks in the matrix under the applied peak stress level that does not induce such matrix cracking for itself.

ACKNOWLEDGEMENT

We thank Dr. Ronald J. Kerans of Air Force Research Laboratory (AFRL) for providing vision and opportunity to put forth the development efforts for the AFIT/AFRL burner rig facility and also thank Dr. Michael J. Verrilli of General Electrics (GE) for providing specimens as well as invaluable insights of CMCs. We would also like to express sincere gratitude to Dr. Elizabeth Downie and the Dayton Area Graduate Studies Institute (DAGSI) for generous support and funding for this research.

REFERENCE

[1]D. Anson and D.W. Richerson, "The Benefits and Challenges of the Use of Ceramics in Gas Turbines," In Ceramic Gas Turbine Component Development and Characterization, eds M. van. Roode, M.K. Ferber, and D.W. Richerson, ASME press, New York, 2002.
[2]J.A. DiCarlo and M. van Roode, "Ceramic Matrix Composite Development for Gas Turbine Engine Hot Section Components," *Proceedings of GT2005 Turbo Expo 2005: Power for Land,Sea and Air* May 8-11, 2006, Barcelona, Spain.
[3]K.J. LaRochelle, "Tensile Stress Rupture Behavior of a Woven Ceramic Matrix Composite in Humid Environments at Intermediate Temperature," Ph.D. Dissertation, Air Force Institute of Technology (March 2005): Advisor: S. Mall.
[4]J.J. Brennan, "Interfacial Characterization of a Slurry-Cast Melt-Infiltrated SiC/SiC Ceramic-Matrix Composite," *Acta Materialia*, 48 4619-28 (2000).
[5]G. N. Morscher, "Intermediate Temperature Stress Rupture of Woven SiC Fiber, BN Interphase, SiC Matrix Composites in Air," Ph.D. Dissertation, Case Western Reserve University (January 2000).
[6] N.S. Jacobson, J.L. Smialek, and D.S. Fox; E.J. Opila, "Durability of Silica-Protected Ceramics in Combustion Atmospheres," *Proceedings of the 1995 2nd International Conference on High-Temperature Ceramic Matrix Composites*. Part 1, Santa Barbara, CA, August 21-24, 1995.
[7]M.J. Verrilli, Senior Research Engineer, GE Aviation Cincinnati, *OH*, Personal Correspondence, 18 September 08.
[8]J.A. DiCarlo and H.M. Yun, "Non-oxide (Silicon Carbide) Fibers," In Handbook of Ceramic Composites, ed N. P. Bansal, Kluwer Academic Publishers, Boston, 2005.
B.J. McBride and S. Gordon, Computer Program for Calculation of Complex Chemical Equilibrium Compositions and Application, NASA RP-1311, June 1996.
[10]G.S. Corman and K.L. Luthra, "Silicon Melt Infiltrated Ceramic Composite (HiPerCompTM)," In Handbook of Ceramic Composites, ed N. P. Bansal, Kluwer Academic Publishers, Boston, 2005.
[11]G.N. Morscher, G. Ojard, R. Miller, Y. Gowayed, U. Santhosh, J. Ahmed and R. John, "Tensile Creep and Fatigue of Sylramic-*iBN* Melt-Infiltrated SiC Matrix Composites: Retained Properties, Damage Development, and Failure Mechanisms," in press, *Composites Science and Technology*, 2008.
[12]V.Sharma, "Effects of Temperature and Steam Environment on Fatigue Behavior of Three SiC/SiC Ceramic Matrix Composites," Master Thesis, Air Force Institute of Technology (September 2008): Advisor: M. Ruggles-Wrenn.

EFFECT of SiC CONTENT and THIRD PHASE METAL ADDITIONS on THERMAL and MECHANICAL PROPERTIES OF Si/SiC CERAMICS

A. L. Marshall, P. Karandikar, A. L. McCormick, M. K. Aghajanian
M Cubed Technologies, Inc.
1 Tralee Industrial Park
Newark, DE 19711

ABSTRACT

Composites of silicon carbide (SiC) and silicon (Si) are fabricated by the reactive infiltration of molten Si into preforms of SiC particles and carbon. This product is often referred to as reaction bonded silicon carbide (RBSC). SiC materials are used in many applications due to their favorable properties including high hardness, high thermal conductivity, low thermal expansion and high stiffness. This paper demonstrates the manipulation of thermal and mechanical properties through the additions of third phase metals (e.g. Al and/or Ti) to the infiltration alloy and through the additions of ceramic-forming, reactive materials to the preform. The effects of these additions on microstructural, physical, mechanical, and thermal properties are presented and discussed.

INTRODUCTION

Reaction Bonded SiC (RBSC) is utilized in many applications for its high thermal conductivity, low coefficient of thermal expansion (CTE) and high specific stiffness. A large interest in this material for thermal management and precision applications is thus generated. The low density and high hardness of this material also leads to a large array of armor applications.

RBSC is typically created by reactively infiltrating a molten alloy, in this case Si, into a preform of SiC powder and a carbon binder. This is performed in a vacuum furnace generally around a processing temperature of 1500°C; whereby, the molten Si wicks into the preform and reacts with the carbon binder to form SiC. Unfortunately, it is not possible to react all of the Si with carbon to form a fully dense 100% SiC part. This process will "can off" when a certain threshold of carbon is met resulting in incomplete infiltration [1-3]. Thus, RBSC is a composite of SiC, reaction formed SiC, and residual Si.

Tailoring properties such as CTE and thermal conductivity becomes possible with changes in SiC content and third phase metal additions. The SiC content can be controlled through particle packing and through reactive material content, such as the varying amounts of carbon binder in the preform. Additions of alloying elements have the ability to lower the melting temperature of the alloy, allowing for lower processing temperatures. This presents the possibility of other material additions to the preform that would normally degrade or react at higher temperatures.

EXPERIMENTAL PROCEDURES

The samples generated for this paper were all produced as stated in the introduction with some minor variations. A systematic set of samples were made by varying the carbon of the preform, and the infiltration alloy composition. Lower infiltration temperatures were used with the Al additions. The SiC powders used were primarily the 6H polytype, and consisted of a bi-modal distribution of 45 and 10 μm particles.

Physical, mechanical, and thermal properties were measured with the methods shown in Table I.

Table I: Test Methods

Data Desired	Experimental Method
Density	ASTM B 311
Young's Modulus	ASTM E 494-05
Poisson's Ratio	ASTM E 494-05
Flexural Strength	ASTM C 1161
Fracture Toughness	ASTM C 1421 (chevron notch - K_{Ivb})
CTE	ASTM E 228-06
Thermal Diffusivity	ASTM E 1461-01
Specific Heat	Differential Scanning Calorimetry with a Perkin-Elmer DSC 7

Density, CTE, thermal diffusivity, and specific heat were measured only once for each composition. Young's modulus was measured in three different locations on each sample. Flexural strength and fracture toughness were each measured using ten samples.

Compositions were determined with quantitative x-ray powder diffraction (XRD) techniques on a Philips PW1800 diffractometer using Cu radiation at 40KV/30mA. Scans were done with a step size of 0.02° over a range of 10° to 70° with a 10 hour counting time. The "Powder Diffraction File" of the International Centre for Diffraction Data was used to identify the crystalline phases. The "Rietveld Refinement Method" was used to determine the quantitative phase data [4,5].

Microstructures of the samples were evaluated using a Leica D 2500 M optical microscope and the Clemex Vision PE imaging software.

Thermal conductivity was calculated from Equation 1 below:

$$\lambda = \kappa C_p \rho \qquad (1)$$

where λ is the thermal conductivity, κ is the diffusivity, C_p is the specific heat, and ρ is the density.

RESULTS AND DISCUSSION

A handbook property table for SiC, Si, Al, and Ti is provided in Table II. These data are used as inputs for modeling calculations throughout the paper. All models shown are only calculated for the SiC:Si systems. These models are used as guidelines for the cases where alloying elements are added to the Si. The data in Table II is provided for high purity materials. The materials incorporated into these composites are not necessarily at the same purity level, which in some cases may affect properties.

Table II: Property Data for Si and SiC

Property	Si [Ref.]	SiC [Ref.]	Al [Ref.]	Ti [Ref.]	Units
Density	2.33[6]	3.21[7]	2.70[6]	4.51[6]	g/cc
Young's Modulus	113[6]	475[7]	70[6]	120[6]	GPa
Poisson's Ratio	0.2[8]	0.27[7]	0.345[6]	0.361[6]	
Thermal Conductivity	84[6]	490 (300K) (α)[9]	222[6]	17[6]	W/m K
Coefficient of Thermal Expansion	2.6 (300K)[10]	2.3 (293K)*	23.6[6]	8.41[6]	10^-6/K

*Measured by M Cubed Technologies, Inc.

Microstructure and Composition

The compositions of the materials of interest consisted of SiC, Si, Al, and Ti in some combination. SiC:Si composites were made initially with varying amounts of carbon present in the perform. Specifically, four carbon levels were used, nominally 2, 5, 7, and 8 wt.% of the preform. Secondly, SiC:Si-Ti samples were made with 8, 11, and 18 volume % Ti in the alloy. One sample was generated for each separate combination. These numbers were chosen based on the phase diagram of Si:Ti and the ability to make fully infiltrated parts with limited solidification porosity. SiC:Si-Al samples were made with volume percents of 8, 18, and 37 Al in the alloy. In addition a SiC:Si-Al-Ti sample was created. This may allow for additional reaction formed SiC in the SiC:Si-Al samples as the Al reacts with C to form Al_4C_3 which is an unwanted phase. Example microstructures are provided in Figure I.

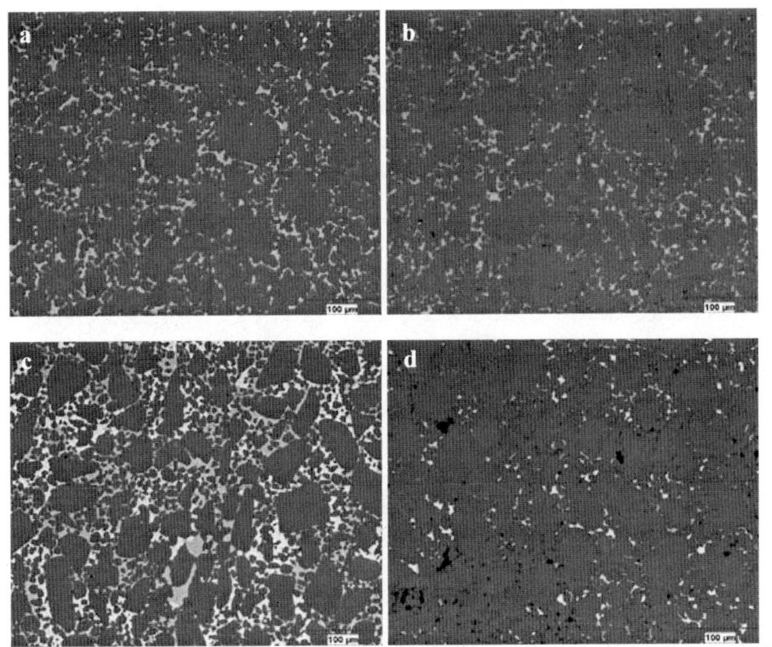

Figure I: Microstructures (a) SiC:Si, 2%C (b) SiC:Si, 8%C (c) SiC:Si-30Ti, 2%C and (d) SiC:Si-30Ti, 8%C

Comparing Figures I.a and I.b shows the effect of carbon additions, with greater carbon content in the perform leading to a more significant Si + C → SiC reaction during infiltration. This leads to a greater connectivity of the SiC phase (dark) and less residual Si phase (bright). Comparing Figures I.a and I.c demonstrates the effect of a Ti addition to the Si infiltration alloy. The metallic second phase now consists of two materials, namely Si (gray) and $TiSi_2$ (bright). As with Figures I.a and I.b, comparing Figures I.c and I.d confirms the formation of additional SiC due to a higher carbon content in the perform. Moreover, some solidification porosity is seen in Figure I.d. This is due to the Si-30Ti alloy shrinking upon solidification, as opposed to pure Si which expands.

Physical and Mechanical Properties

Data for density, Young's modulus, Poisson's ratio, flexural strength and fracture toughness are presented in this section. These are all room temperature measurements, with results shown in Table III. In each case, the "%C" refers to the weight percent of carbon added to the perform. Later figures represent the data versus volume fraction of SiC. Volume fraction in the SiC:Si system was calculated using the "rule of mixtures" based on density. These calculated numbers were later used for the alloyed systems of matching percent carbon additions as the SiC volume fraction should not vary significantly.

Table III: Physical and Mechanical Properties

Label	Density (g/cc)	Young's Modulus (Gpa)	Poisson's Ratio	Flexural Strength (MPa)	Fracture Toughness (MPa √m)
SiC:Si, 2%	3.02	360	0.20	243	3.71
SiC:Si, 5%	3.06	378	0.19	232	3.88
SiC:Si, 7%	3.11	405	0.18	242	3.55
SiC:Si, 8%	3.13	416	0.18	300	3.54
SiC:Si-9Al, 2%	3.02	359	0.19	245	4.65
SiC:Si-18Al, 2%	3.02	353	0.19	254	5.30
SiC:Si-40Al, 2%	3.03	337	0.20	271	4.60
SiC:Si-15Ti, 2%	3.11	371	0.19	190	3.98
SiC:Si-15Ti, 5%	3.14	389	0.19	236	4.09
SiC:Si-15Ti, 7%	3.17	404	0.18	240	3.66
SiC:Si-15Ti, 8%	3.17	410	0.18	223	3.44
SiC:Si-20Ti, 2%	3.15	383	0.20	197	3.97
SiC:Si-20Ti, 5%	3.18	397	0.19	248	4.14
SiC:Si-20Ti, 7%	3.18	411	0.19	256	3.70
SiC:Si-20Ti, 8%	3.19	417	0.18	254	3.69
SiC:Si-30Ti, 2%	3.18	375	0.19	204	4.22
SiC:Si-30Ti, 5%	3.21	389	0.18	254	3.65
SiC:Si-30Ti, 7%*	3.17	384	0.18	218	4.41
SiC:Si-30Ti, 8%*	3.18	388	0.18	241	3.30
SiC:Si-37Al-14Ti, 2%*	3.07	322	0.19	- -	5.32

*This sample demonstrated some porosity, thus the low Young's modulus

Several conclusions are apparent when evaluating the data provided in Table III. The first observation being Young's modulus increases with increasing amounts of reactive carbon present in the preform. In the base SiC:Si samples, an increase in carbon content from 2 to 8% in the preform results in a 56 GPa increase in Young's modulus from the larger amounts of reaction formed SiC. This is the equivalent of a 10 vol. % increase in SiC content. Samples with Ti added to the infiltration alloy have a higher initial Young's modulus at similar SiC volume percents without carbon added. These all approach a maximum allowable modulus as the SiC content approaches 100 %. The varying Ti additions seemed to have little difference in the attendant properties. This may stem from the additions being small in volume or from porosity occurring due to the simultaneous increases in carbon and Ti. The Al samples Young's modulus decreases, as expected due to the low stiffness of Al.

The flexural strength increases with Al additions to the base where Ti additions have little to no effect. A similar result is presented for the fracture toughness where the Al additions yield higher results (including the Si-37Al-14Ti, 2% sample).

Young's modulus data is plotted below with the following models also in evidence: Voigt (Rule of Mixtures), Ress, and Hashin-Shtrikman. The Voigt bound assumes proportional stiffness contributions where as the Ress and Hashin-Shtrikman models account for stress contributions [11,12]. These bounds are provided below respectively:

$$E_c = E_m V_m + E_p V_p \tag{2}$$

$$E_c = \left(\frac{V_m}{E_m} + \frac{V_p}{E_p} \right)^{-1} \tag{3}$$

$$E_c = \frac{E_m \left(E_m V_m + E_p \left(V_p + 1 \right) \right)}{\left(E_p V_m + E_m \left(V_p + 1 \right) \right)} \tag{4}$$

where E is the Young's modulus, and in the case of Equation 4 the subscripts just need to be changed to get the other bound. These models were created for the base system of SiC:Si and used as guidelines for the subsequent composite systems. Property inputs can be found in Table II.

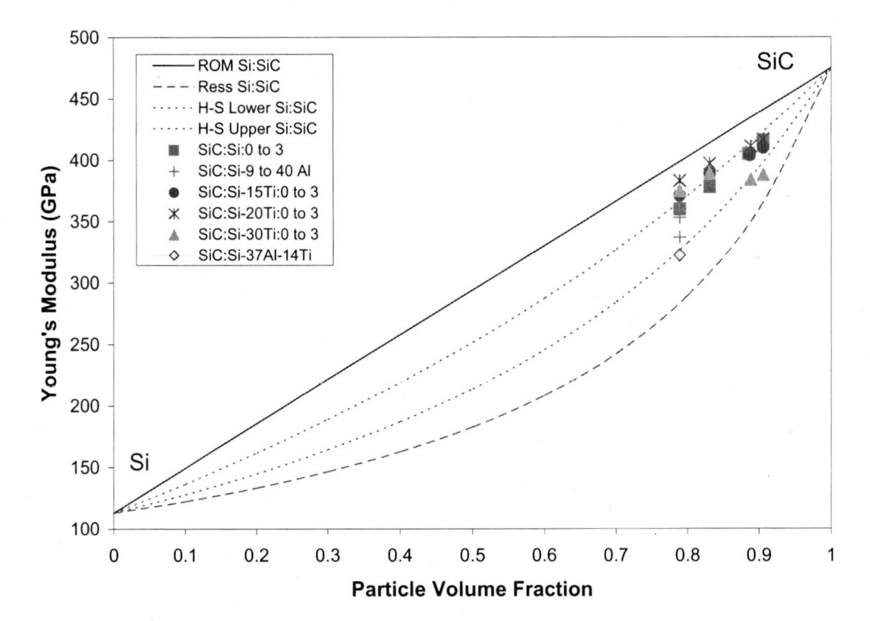

Figure II: Young's Modulus Data with Bounds

These bounds were constructed in regard to the SiC:Si material and are merely a guide. The particle volume percents were calculated using a rule of mixtures with regard to the densities of the SiC:Si samples. The Ti and Al containing samples were estimated based on this. Expectedly, the base material follows the upper Hashin-Shtrikman bound at these volume fractions. The other materials fall within range of these bounds for the most part. They are not expected to necessarily follow these particular bounds as the other additions were not taken into account. Evidence of porosity appears in the SiC:Si-

30Ti, 7 and 8% samples along with the SiC:Si-37Al-14Ti, 2% sample; This is noticeable from the deviation in behavior with respect to the lower Ti containing data sets.

Thermal Properties

Thermal property evaluation took place with respect to thermal expansion, diffusivity, specific heat, and ultimately thermal conductivity. Instantaneous CTE data at 20°C is presented in Figure III. The thermal expansion of these materials is an important consideration to take account of, as many applications require matching CTE's to help reduce thermal mismatch stresses during cycling. Results are plotted with a rule of mixture model as well as the Turner and Kerner models for CTE. These are shown in Equations 5, 6, and 7 respectively. These predictions were also based on the SiC:Si system, with the property inputs provided in Table II.

$$\alpha_c = \alpha_m V_m + \alpha_p V_p \tag{5}$$

$$\alpha_c = \frac{\alpha_m V_m K_m + \alpha_p V_p K_p}{V_m K_m + V_p K_p} \tag{6}$$

$$\alpha_c = \alpha_m - (\alpha_m - \alpha_p) * \frac{K_p(3K_m + 4G_m)V_p}{K_m(3K_p + 4G_m)V_p + 4(K_p - K_m)G_m V_p} \tag{7}$$

where α is the CTE, K is the bulk modulus, G is the shear modulus, V is the volume fraction, and c, m, and p represent the composite, matrix, and particulate respectively. The shear and bulk modulus were calculated from the data provided in Table II for the individual constiuents. The Turner model (Equation 6) incorporates the idea that the components of the composite undergo a homogeneous static stress. The Kerner model (Equation 7) assumes the component phases are under both isostatic and shear stresses [12,13].

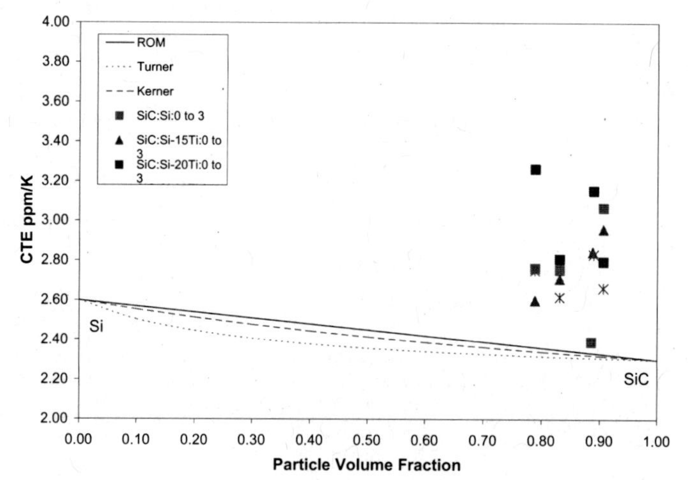

Figure III: Instantaneous CTE Data at 20°C

Measurements were taken by first cooling the sample to -40°C and then heating to 500°C. The data presented do not correlate with the literature predictions. Based on these results, it appears as if residual stress due to a CTE mismatch of the component materials is being relieved during the measurement. Further analysis will need to be conducted by cycling the measurement multiple times to discover any hysteresis taking place. This is the most likely explanation as it is similar to results found by Shu and Tu with regards to SiC_p/Cu composites for the initial heating and cooling cycle [13].

Thermal diffusivity data at room temperature and 200°C is presented below with specific heat data. Thermal conductivity data is also presented. Table IV demonstrates how the thermal conductivity increases with increasing reactive carbon content. This is expected as the SiC formed has a much higher thermal conductivity than that of Si. That is, the reaction formed SiC decreases the presence of Si as depicted in Figures I.a and I.b.

Table IV: Thermal Property Data

Label	25°C			200°C		
	κ (mm^2/s)	C_p (J/g K)	λ (W/m*K)	κ (mm^2/s)	C_p (J/g K)	λ (W/m*K)
SiC:Si, 2%	92.6	0.678	190	40.3	0.926	113
SiC:Si, 5%	102.9	0.679	214	40.2	0.940	116
SiC:Si, 7%	101.5	0.675	213	40.7	0.932	118
SiC:Si, 8%	105.3	0.674	222	43.0	0.931	125
SiC:Si-9Al, 2%	99.8	0.677	204	38.2	0.928	107
SiC:Si-18Al, 2%	98.4	0.677	201	38.2	0.934	108
SiC:Si-40Al, 2%	100.6	0.681	208	39.1	0.941	112
SiC:Si-15Ti, 2%	98.4	0.664	203	37.9	0.910	107
SiC:Si-15Ti, 5%	100.3	0.661	208	39.8	0.912	114
SiC:Si-15Ti, 7%	101.4	0.665	214	40.6	0.922	119
SiC:Si-15Ti, 8%	103.6	0.668	219	42.3	0.916	123
SiC:Si-20Ti, 2%	88.7	0.668	187	37.8	0.920	110
SiC:Si-20Ti, 5%	97.8	0.667	207	38.6	0.917	113
SiC:Si-20Ti, 7%	90.4	0.659	190	38.2	0.910	111
SiC:Si-20Ti, 8%	93.0	0.666	198	38.8	0.913	113
SiC:Si-30Ti, 2%	90.7	0.663	191	38.9	0.900	111
SiC:Si-30Ti, 5%	88.5	0.663	189	38.8	0.906	113
SiC:Si-30Ti, 7%*	89.2	0.664	188	39.0	0.912	113
SiC:Si-30Ti, 8%*	83.1	0.663	175	38.4	0.913	111

*This sample demonstrated some porosity, thus the low Young's modulus

The specific heats of SiC, Si, Al, and Ti are respectively 0.667, 0.709, 0.897, and 0.527 J/g K [14].As expected the specific heat decreases with increasing SiC content, decreases when Ti is added, and increases slightly with the additions of Al.

High thermal conductivity in these materials provides an additional benefit to these light-weight, high specific stiffness composites. A number of factors can influence the thermal conductivity of particulate based composites such as the conductivity of the individual constituents, the size and volume of particles present, and the interfacial thermal barrier between two materials. Models for predicting the thermal conductivity of materials have been developed and are provided in the literature such as the Hasselman-Johnson model of the form [12,15]:

$$\lambda_c = \lambda_m \left(\frac{2(1 - V_p) + (\lambda_p/\lambda_m)(1 + 2V_p) + 4(\lambda_p/Dh)(1 - V_p)}{(2 + V_p) + (\lambda_p/\lambda_m)(1 - V_p) + 2(\lambda_p/Dh)(2 + V_p)} \right) \tag{8}$$

Here, D represents the reinforcement particle diameter, and h is the thermal boundary conductance. The thermal boundary conductance is dependant on the contact between the faces along with the CTE mismatch. When the particle diameter becomes large (D→∞) Equation 8 reduces to:

$$\lambda_c = \lambda_m \left(\frac{2(1 - V_p) + (\lambda_p/\lambda_m)(1 + 2V_p)}{(2 + V_p) + (\lambda_p/\lambda_m)(1 - V_p)} \right) \tag{9}$$

which effectively negates the existence of a thermal barrier between components and provides the Maxwell model [11]. Thermal conductivity data are presented in Figure IV. Curves for a particle diameter of 10 and 45 μm are also presented in Figure IV along with the Maxwell model. From previously published data [16], the thermal barrier conductance was estimated by fitting results from a monodisperse SiC:Si samples with a nominal diameter of 10 μm., which gives $h = 0.55 * 10^8$ W/m^2 K. This number was used to predict the curve with a diameter of 45 μm as well.

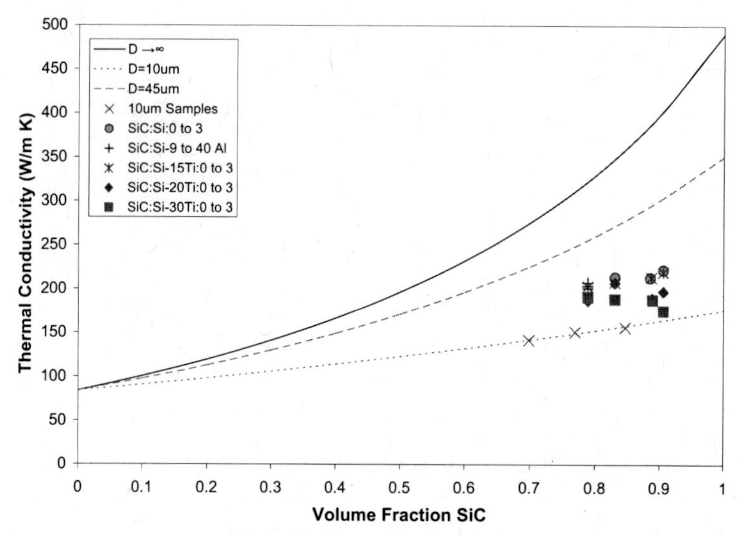

Figure IV: Thermal Conductivity Data

The SiC:Si samples fall between the curves generated from the diameter estimates of 10 and 45 μm, which is expected as this was the nominal binomial distribution used to generate bulk parts. The models in Figure IV were generated using the SiC and Si properties from Table II. This analysis demonstrates the need to reduce the number of fine SiC particles in relation to increasing the thermal conductivity. Providing reactively formed SiC in these composites is another way to reduce the thermal barrier effect as the surfaces will be in a more intimate contact. Small additions of Ti to the alloy have a positive effect on increasing the thermal conductivity. This may occur as $TiSi_2$ can be used to strip the SiO_2 layer on the surface of SiC [17].

SUMMARY

Dense RBSC ceramic composites were formed and characterized with varying carbon and alloy contents. Carbon addition lead to significant increases in Young's modulus provided there is not a detrimental reaction with third phase alloy additions. Al additions to the matrix increased the fracture toughness slightly while Ti additions do not have an effect either way. CTE results demonstrated the onset of stress relief as the samples were cooled and subsequently heated. Thermal conductivity increased with carbon, Al, and Ti additions to a point. Finally, a ternary Si-Al-Ti alloy was successfully used to infiltrate a preform with only minor porosity (modifications will be made going forward).

REFERENCES

1. K.M. Taylor, "Cold Molded Dense Silicon Carbide Articles and Methods of Making the Same," U.S. Pat. No. 3 205 043, Sept. 7, 1965.
2. P.P. Popper, "Production of Dense Bodies of Silicon Carbide," U.S. Pat. No. 3 275 722, Sept. 27, 1966.
3. C.W. Forrest, "Manufacture of Dense Bodies of Silicon Carbide," U.S. Patent No. 3 495 939, Feb. 17, 1970.
4. H.M. Rietveld, "The Crystal Structure of some Alkaline Earth Metal Uranates of the Type M_3UO_6*," *Acta Cryst.* **20** 508-513 (1966).
5. H.M. Rietveld, "A Profile Refinement Method for Nuclear and Magnetic Structures," *J. Appl. Cryst.* **2** 65-71 (1969).
6. *Metals Handbook: Desk Edition* (ASM International, Metals Park, OH, 1985).
7. *Engineered Materials Handbook, Vol. 4, Ceramics and Glasses* (ASM International, Metals Park, OH, 1991).
8. J. J. Wortman, R. A. Evans, "Young's Modulus, Shear Modulus, and Poisson's Ratio in Silicon and Germanium," *J. Appl. Phys.*, **36** [1] 153-156 (1965).
9. G.A. Slack, "Nonmetallic Crystals with High Thermal Conductivity," *J. Phys. Chem. Solids*, **34** 321-335 (1973)
10. *ASM Handbook, Volume 2 Properties and Selection: Nonferrous Alloys and Special-Purpose Materials* (ASM International, USA, 1990).
11. Z. Hashin, S. Shtrikman, "A Variational Approach to the Theory of the Elastic Behaviour of Multiphase Materials," *J. Mech. Phys. Solids*, **11** 127-140 (1963).
12. L. Zhang, et. Al., "Thermo-physical and mechanical properties of high volume fraction SiC_p/Cu composites prepared by pressureless infiltration," *Mat. Sci. & Eng.* **A 489** 285-293 (2008).
13. Kuen-Ming Shu, G.C. Tu, "The microstructure and thermal expansion characteristics of Cu/SiC_p composites," *Mat Sci & Eng* **A329** 236-247 (2003).
14. http://webbook.nist.gov (calculated)
15. A.G. Every, et. al., "The Effect of Particle Size on the Thermal Conductivity of ZnS/Diamond Composites," Acta metal. Mater. Vol 40, No 1. 123-129 (1992).
16. A. L. Marshall, et. al., "The Effects of Si Content and SiC Polytype on the Microstructure and Properties of RBSC," *Proceedings of the 32nd International Conference on Advanced Ceramics and Composites*, 115 (2008).
17. S. P. Murarka, "Silicide thin films and their applications in microelectronics," *Intermetallics 3*, 173-186 (1995)

COMPRESSIVE STRENGTH DEGRADATION IN ZrB_2-SiC AND ZrB_2-SiC-C ULTRA HIGH TEMPERATURE COMPOSITES

J. Ramírez-Rico, M. A. Bautista, J. Martínez-Fernández
Dpto. Física de la Materia Condensada-ICMSE
Universidad de Sevilla-CSIC
Avda. Reina Mercedes, s/n, 41012 Sevilla, Spain

M. Singh
Ohio Aerospace Institute
MS 106-5, Ceramics Branch
NASA Glenn Research Center
Cleveland, OH 44135-3191

ABSTRACT

The high melting point of refractory metal diborides makes them promising materials for ultra high temperature applications. In this work, we study the compressive strength of two ZrB_2-SiC and ZrB_2-SiC-C composites. Samples have been studied in compression at room temperature, 1400°C and 1550°C, in atmospheric air. The degradation of the mechanical properties as a result of atmospheric air exposure at high temperatures has also been studied as a function of exposure time. The presence of C is detrimental to the compressive strength, as carbon burns out at high temperatures in air. After exposure to air at high temperatures an external SiO_2 layer is formed, below which ZrB_2 oxidizes to ZrO_2. A reduction of 30% in room temperature strength occurs after 16-24h of exposure to air at 1400°C for the ZrB_2-SiC material, while for the ZrB_2-SiC-C composition this reduction is observed after less than 6h. The thickness of the oxide layer has been measured and the oxidation process is discussed in terms of the existing models.

INTRODUCTION

The high melting point of refractory metal diborides coupled with their ability to form refractory oxide scales give these materials the capacity to withstand temperatures in the 1900-2500°C range. These Ultra-High Temperature Ceramics (UHTC) were developed in the 1960s[1]. Fenter[2] provides a comprehensive review of the work accomplished in the 1960s and early 1970s. Additions of Silicon Carbide are used to enhance oxidation resistance and limit diboride grain growth[3-6]. Carbon is also sometimes used as an additive to enhance thermal stress resistance[7, 8].

These materials offer a good combination of properties that make them candidates for airframe leading edges on sharp bodied reentry vehicles[9]. UHTCs have some potential to perform well in such applications' environment, i.e. air at low pressure. However, for hypersonic flight in the upper atmosphere one must recognize that stagnation pressures can be greater than one atmosphere. Some interest has also been shown in these materials for single use propulsion applications[10].

Major improvements in the manufacturing and characterization of ZrB_2 materials and composites have been put forward in recent years, and now several important aspects of their properties and processing are well understood[4, 11-19]. However, the study of high temperature properties has been mostly limited to oxidation behavior, area which is also well understood[8, 20-26]. To the best of our knowledge, very few studies of high-temperature mechanical properties exist[7, 10, 27].

In this work, we study the mechanical strength of ZrB_2-SiC and ZrB_2-SiC-C composites. Samples were studied in compression at room temperature, 1400°C and 1550°C, in atmospheric air. The degradation of the mechanical properties as a result of atmospheric air exposure at high temperatures has also been studied as a function of exposure time. The microstructure and composition of the as-fabricated and tested materials has been studied by means of Scanning Electron Microscopy (SEM) and Energy Dispersive Spectroscopy (EDS).

MATERIALS AND METHODS

Samples of ZrB_2-SiC composites were fabricated by uniaxial hot-pressing by Materials and Machines, Inc., Tucson, AZ, using a procedure previously described[7]. Two polycrystalline composites were studied, containing variable amounts of ZrB_2 and SiC; one of them also contained C. In both cases the ZrB_2/SiC volume ratio was 4. The nominal compositions of the materials investigated are summarized in Table 1. ZrB_2 (d_{50} = 3-5 μm) and α-SiC (d_{50} = 1.4 μm) powders were obtained from H. C. Starck, and C powders were obtained from Asbury Graphite Mills. Densities obtained were 5.57 g/cm^3 (99.9% theoretical density) for ZS and 4.50 g/cm^3 (99.0% theoretical density).

Compression tests were carried out on an electromechanical universal testing machine with a furnace attached to its frame, at constant cross-head displacement rate. Load was applied using alumina rods with SiC pads. Samples were cut into parallelepiped shape using a low speed diamond saw. Nominal sample dimensions were 3x3x5 mm, and the load was applied to the longest dimension. High temperature mechanical tests were conducted at room temperature, 1400°C and 1550°C. Several samples were exposed to oxidation by annealing at 1400°C in atmospheric air in a tube furnace, and exposure times ranged from 6h to 48h. The room temperature strength was measured after oxidation, to study the degradation of the mechanical properties after exposure to an oxidizing environment. At least three samples were studied at each temperature or exposure time at 1400 °C. Error bars throughout represent one standard deviation.

Microstructural studies were carried out using SEM and EDS techniques, in both the as-fabricated and tested specimens. Samples were prepared using conventional metallographic techniques which involved cutting, grinding and lapping. A conductive coating of either carbon or gold was applied to the specimens prior to observation.

Table 1. Nominal composition and designation of the two composites studied in this work

Acronym	ZrB_2 (vol. %)	SiC (vol. %)	C (vol. %)
ZS	80	20	0
ZSC	56	14	30

RESULTS AND DISCUSSION

Microstructure of as-received specimens

Figure 1 shows the as fabricated microstructure of the two compositions studied. Our observations match those reported in Ref. 7. The ZS composite appears to be fully dense, while the ZSC suffered from significant grain pullout during polishing. This is attributed to the weak C bonding to the ZrB_2 and SiC, phases, which results in removal of the C phase during polishing. In ZS, ZrB_2 grains are equiaxed with reported grain size in the 6-12 μm range, while the SiC grains are elongated with sizes of approximately 1.5-3 μm thick by 3-11 μm wide/long[7]. In ZSC, the grain pullout during polishing made the estimation of grain size difficult, although it can be seen from Figure 1 that ZrB_2 grain size is smaller in ZSC than in ZS. It should be noted that, at least for the ZS composite, the grain size is close to the critical grain size for microcracking due to the anisotropic thermal expansion coefficient of ZrB_2, which has been reported to be around 15 μm[28].

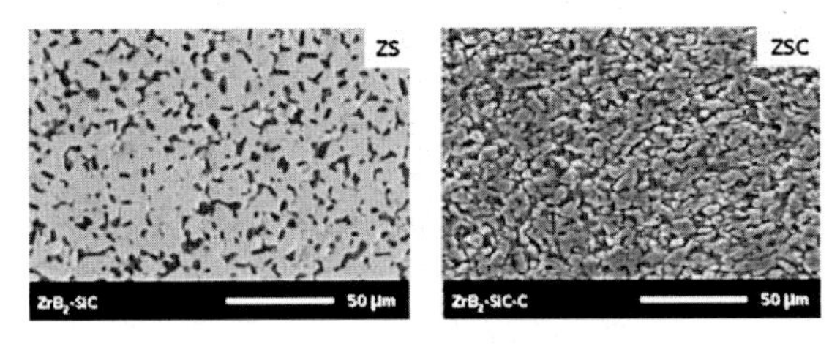

Figure 1. Microstructure of the as-received materials. (ZS) ZrB_2 – 20% vol. SiC. (ZSC) ZrB_2 – 30% vol. C – 14% vol. SiC. In the ZSC material, grain pullout and C removal during the metallographic preparation is evident.

High Temperature Strength

The high temperature compressive strength was measured at 1400°C and 1550°C in atmospheric air for both the ZS and ZSC compositions. Figure 2 shows representative stress-displacement curves for the compositions and temperatures studied, while Figure 3 shows the average compressive strength as a function of temperature. The ZS composite shows higher compressive strength than the carbon containing ZSC. This is attributed to both the weak carbon bonding to the other phases, evidenced as grain pullout in the SEM observations, and the burnout of carbon at high temperature in air. This creates porosity and also produces channels through which air can enter, oxidizing the ZrB_2 phase not only on the surface but also inside the sample. At 1550°C, creep of the SiC pads used as a protection to the alumina rods could be observed at high stresses for the ZS material. For the ZSC material this effect was not observed due the much lower maximum applied stress.

The relatively low values for the compressive strength in ZS at room temperature could be attributed to the large grain size, which probably induces some microcracking in the ZrB_2 matrix. This can also explain why at room temperature the composite containing carbon is stronger, as it has a smaller grain size due to the presence of grain-growth inhibiting carbon. Additionally, the weak bonding of C to ZrB_2 evidenced as grain pullout in Figure 1 can contribute to the relaxation of microcrack-inducing residual stresses developed upon cooling. At high temperatures however, carbon burn-out is responsible for the lower strength observed in ZSC when compared to ZS.

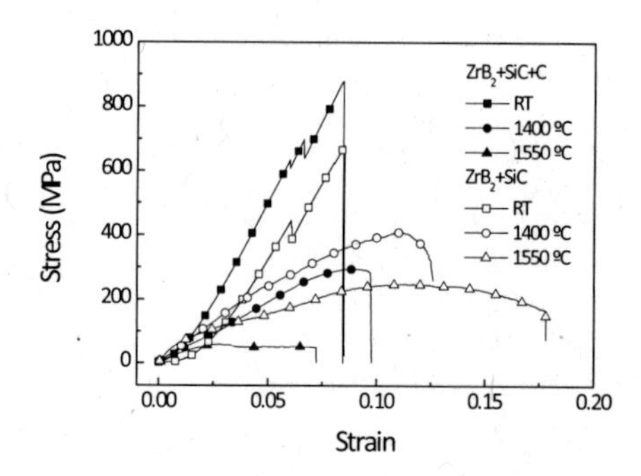

Figure 2. Typical Stress-Displacement curves for ZS and ZSC at room temperature, 1400 °C and 1550°C.

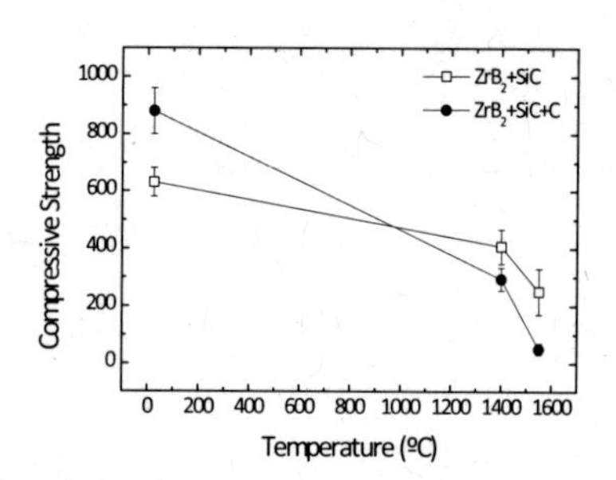

Figure 3. Compressive strengths of ZS and ZSC as a function of temperature.

Room Temperature Strength and Degradation

The degradation in the mechanical properties was studied as a function of exposure time to atmospheric air at 1400°C. Figure 4 shows some selected stress-strain curves for ZS (left) and ZSC (right), for exposure times ranging from 0 to 48h at 1400°C in atmospheric air. These results are summarized in figure 5, where the room temperature strength of the materials is plotted as a function of exposure time. For ZS, a reduction of 50% in strength occurred between 16h and 24h, while for ZSC less than 6 hours were needed. This can be explained considering that C burns in the oxidizing atmosphere, creating pores and channels through which air can permeate.

It can be seen in Figure 5 that the strength of ZS improves after 1h exposure to atmospheric air at 1400 °C. This effect can probably be explained as a blunting of defects and anisotropy-induced microcracks by the silica glass layer that is formed during oxidation, as will be described in the next section.

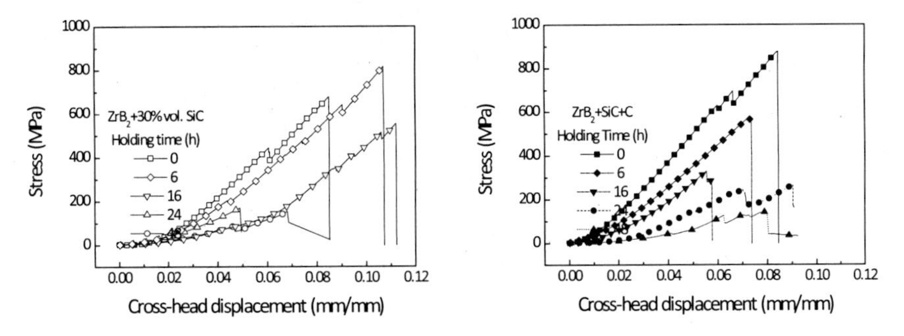

Figure 4. Typical stress-displacement behavior for ZS (left) and ZSC (right) samples oxidized in atmospheric air at 1400 °C for different holding times.

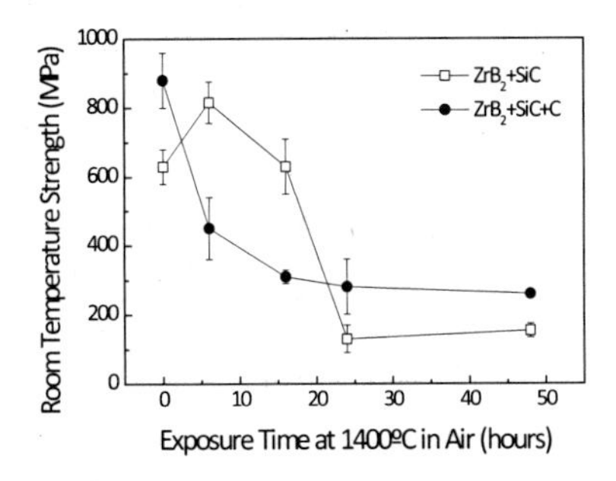

Figure 5. Compressive strength degradation after oxidation in atmospheric air at 1400 °C. For ZS samples, 50% strength is lost after 16-24h of exposure time, while for ZSC samples 50% of the strength is lost during the first 0-6h of oxidation.

Oxide layer formation

The microstructure and thickness of the oxide layers formed after exposure to atmospheric air at 1400°C was studied for both ZS and ZSC samples, which were cut and polished for observation in the SEM. Figure 6 shows a micrograph and compositional maps for a ZS sample annealed for 1 hour. The outer, Si and O rich layer can be concluded to be SiO_2, while the intermediate layer, which is O rich and B poor, is ZrO_2. Similar conclusions can be drawn from Figure 7, which represents compositional maps for a ZSC sample annealed for 24h. Again, the outer layer is mainly composed of SiO_2, and an intermediate layer of ZrO_2 separates the former layer and the bulk interior of the sample.

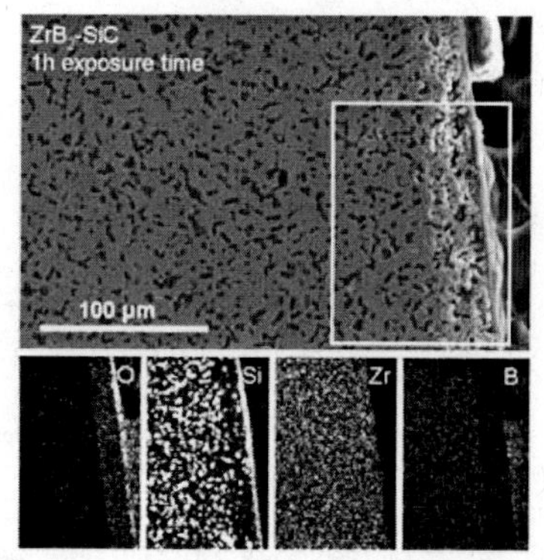

Figure 6. Typical microstructure of ZS samples after oxidation for 1h at 1400 °C. (Top) BSE micrograph of a section. (Bottom) O, Si, Zr and B EDS maps. An outer, glassy SiO_2 layer is followed by a porous ZrO_2 layer.

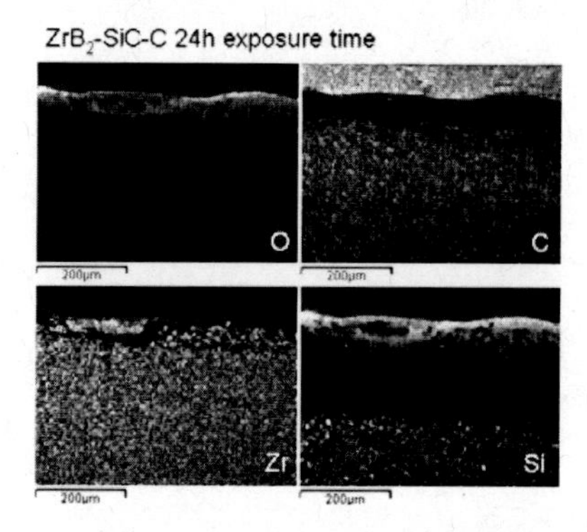

Figure 7. Characteristic EDS composition maps of ZSC after oxidation for 24h at 1400 °C.

These observations were confirmed by quantitative analysis, of which an example is presented. Figure 8 is a cross section of a ZSC sample annealed for 24h at 1400°C in air, and contains several of the features previously observed. The microstructure of the oxide layer can be divided into four different zones or regions. The first, outer one is composed of ZrO_2 grains embedded in a glassy SiO_2 matrix, as can be deduced from the elemental composition. The zone labeled as number 2 is composed only of SiO_2, while the zone labeled as 3 is composed almost exclusively of ZrO_2 and contains no Si. The fourth zone corresponds to the composition of the bulk, as-fabricated material. For ease of comparison, raw spectra obtained from all four different zones are depicted.

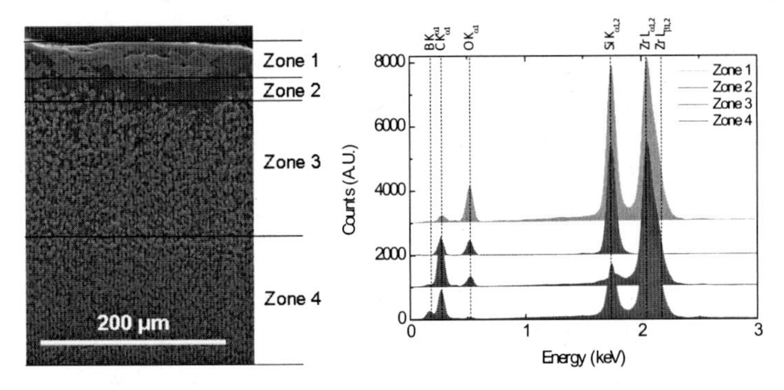

Figure 8. (Left) Characteristic microstructure of ZSC samples after 24h exposure to oxidizing atmosphere at 1400°C. Several zones can be distinguished. (Right) Relevant EDS spectra obtained from the four zones depicted.

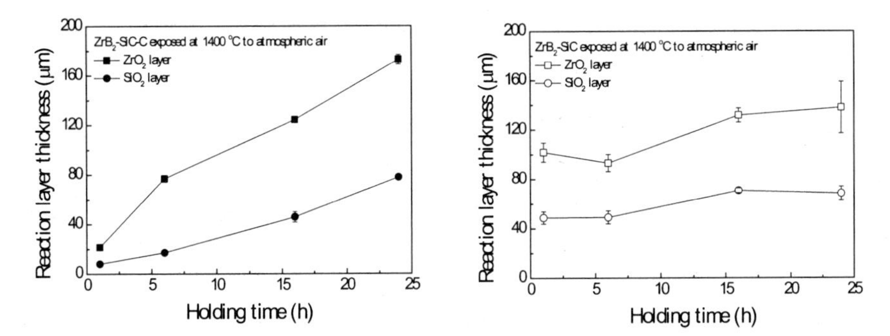

Figure 9. Reaction layer thickness as a function of holding time for ZSC (left) and ZS (right) exposed to atmospheric air at 1400 °C.

These observations confirm the oxidation process already outlined in previous references, such as Refs. [7, 8, 23, 24, 26]. Both ZrB_2 and SiC are oxidized, producing B_2O_3 that evaporates at high temperatures. The SiO_2 formed, which is liquid at the studied temperatures, is expelled towards the surface of the sample by capillary forces, and acts as a protective layer. The intermediate layer is thus composed mostly of ZrO_2 and pores that allow for oxygen permeation. It is thus expected that the ZSC samples, containing carbon that burns out at high temperature, will oxidize at a faster rate because of the porosity produced during carbon combustion.

To ascertain this effect, the thickness of both the SiO_2 and ZrO_2 oxide layers were measured as a function of annealing time, for both ZS and ZSC samples. These results are presented in Figure 9. It can be seen that the reaction rate is faster for samples containing C, and that the thickness of the oxide layers is smaller for ZS samples.

CONCLUSIONS

It has been shown that the addition of C to a ZrB_2-SiC composite is detrimental to the high temperature mechanical properties, since the degradation in strength and oxidation resistance observed counters any possible improvement to the thermal shock resistance. The observed strengths are low when compared to other materials in the literature, especially for finer grained materials, which can be attributed to the grain size of ZrB_2 being close to the critical size for microcracking. The ZS composition shows the best performance of the two compositions studied, both in terms of strength and oxidation resistance. The ZS material can withstand exposures up to 24h in air at 1400°C before its compressive strength falls below 50% when compared to the as-fabricated material, but the ZSC composition is probably not suitable for application if no protective coating is applied.

ACKNOWLEDGEMENTS

This material is based upon work supported by the European Office of Aerospace Research and Development, Air Force Office of Scientific Research, Air Force Research Laboratory, under Grant No. FA8655-07-1-3087. Any opinions, findings and conclusions or recommendations expressed in this material are those of the authors and do not necessarily reflect the views of the European Office of Aerospace Research and Development, Air Force Office of Scientific Research or Air Force Research Laboratory. SEM observations were done at the CITIUS of the Universidad de Sevilla. J.R-R. was funded by a Junta de Andalucía FPDI grant.

REFERENCES

1. E.V. Clougherty, R.L. Pober and L. Kaufman: Synthesis of oxidation resistant metal diboride composites. *Trans. Met. Soc. AIME* **242**, 1077-1082 (1968).
2. J.R. Fenter: Refractory diborides as engineering materials. *SAMPE Quarterly* **2**, 1-15 (1971).
3. G.J. Zhang, Z.Y. Deng, N. Kondo, J.F. Yang and T. Ohji: Reactive hot pressing of ZrB2-SiC composites. *Journal of the American Ceramic Society* **83**, 2330-2332 (2000).
4. F. Monteverde: Beneficial effects of an ultra-fine a-SiC incorporation on the sinterability and mechanical properties of ZrB2. *Applied Physics A: Materials Science and Processing* **82**, 329-337 (2006).
5. Y. Yan, Z. Huang, S. Dong and D. Jiang: Pressureless sintering of high-density ZrB2-SiC ceramic composites. *Journal of the American Ceramic Society* **89**, 3589-3592 (2006).
6. S. Zhu, W.G. Fahrenholtz and G.E. Hilmas: Influence of silicon carbide particle size on the microstructure and mechanical properties of zirconium diboride-silicon carbide ceramics. *Journal of the European Ceramic Society* **27**, 2077-2083 (2007).
7. S.R. Levine, E.J. Opila, M.C. Halbig, J.D. Kiser, M. Singh and J.A. Salem: Evaluation of ultra-high temperature ceramics for aeropropulsion use. *Journal of the European Ceramic Society* **22**, 2757-2767 (2002).
8. F. Monteverde and L. Scatteia: Resistance to thermal shock and to oxidation of metal diborides-SiC ceramics for aerospace application. *Journal of the American Ceramic Society* **90**, 1130-1138 (2007).
9. J.D. Bull, D.J. Rasky and K. C.C. *Stability Characterization of diboride composites under high velocity atmospheric flight conditions*. in *24th International SAMPE Technical Conference*. 1992.

10. M.M. Opeka, I.G. Talmy, E.J. Wuchina, J.A. Zaykoski and S.J. Causey: Mechanical, Thermal, and Oxidation Properties of Refractory Hafnium and zirconium Compounds. *Journal of the European Ceramic Society* **19**, 2405-2414 (1999).
11. F. Monteverde, A. Bellosi and S. Guicciardi: Processing and properties of zirconium diboride-based composites. *Journal of the European Ceramic Society* **22**, 279-288 (2002).
12. F. Monteverde, S. Guicciardi and A. Bellosi: Advances in microstructure and mechanical properties of zirconium diboride based ceramics. *Materials Science and Engineering A* **346**, 310-319 (2003).
13. A.L. Chamberlain, W.G. Fahrenholtz, G.E. Hilmas and D.T. Ellerby: High-strength zirconium diboride-based ceramics. *Journal of the American Ceramic Society* **87**, 1170-1172 (2004).
14. W.G. Fahrenholtz, G.E. Hilmas, A.L. Chamberlain, J.W. Zimmermann and B. Fahrenholtz: Processing and characterization of ZrB2-based ultra-high temperature monolithic and fibrous monolithic ceramics. *Journal of Materials Science* **39**, 5951-5957 (2004).
15. A. Bellosi, F. Monteverde and D. Sciti: Fast densification of ultra-high-temperature ceramics by spark plasma sintering. *International Journal of Applied Ceramic Technology* **3**, 32-40 (2006).
16. A.L. Chamberlain, W.G. Fahrenholtz and G.E. Hilmas: Pressureless sintering of zirconium diboride. *Journal of the American Ceramic Society* **89**, 450-456 (2006).
17. A. Rezaie, W.G. Fahrenholtz and G.E. Hilmas: Effect of hot pressing time and temperature on the microstructure and mechanical properties of ZrB2-SiC. *Journal of Materials Science* **42**, 2735-2744 (2007).
18. J.W. Zimmermann, G.E. Hilmas, W.G. Fahrenholtz, F. Monteverde and A. Bellosi: Fabrication and properties of reactively hot pressed ZrB2-SiC ceramics. *Journal of the European Ceramic Society* **27**, 2729-2736 (2007).
19. W.G. Fahrenholtz, G.E. Hilmas, S.C. Zhang and S. Zhu: Pressureless sintering of zirconium diboride: Particle size and additive effects. *Journal of the American Ceramic Society* **91**, 1398-1404 (2008).
20. F. Monteverde and A. Bellosi: Oxidation of ZrB2-based ceramics in dry air. *Journal of the Electrochemical Society* **150**, B552-B559 (2003).
21. M.M. Opeka, I.G. Talmy and J.A. Zaykoski: Oxidation-based materials selection for 2000 °C + hypersonic aerosurfaces: Theoretical considerations and historical experience. *Journal of Materials Science* **39**, 5887-5904 (2004).
22. W.G. Fahrenholtz: The ZrB2 volatility diagram. *Journal of the American Ceramic Society* **88**, 3509-3512 (2005).
23. F. Monteverde: The thermal stability in air of hot-pressed diboride matrix composites for uses at ultra-high temperatures. *Corrosion Science* **47**, 2020-2033 (2005).
24. W.G. Fahrenholtz: Thermodynamic analysis of ZrB2-SiC oxidation: Formation of a SiC-depleted region. *Journal of the American Ceramic Society* **90**, 143-148 (2007).
25. F. Monteverde and R. Savino: Stability of ultra-high-temperature ZrB2-SiC ceramics under simulated atmospheric re-entry conditions. *Journal of the European Ceramic Society* **27**, 4797-4805 (2007).

26. A. Rezaie, W.G. Fahrenholtz and G.E. Hilmas: Evolution of structure during the oxidation of zirconium diboride-silicon carbide in air up to 1500 °C. *Journal of the European Ceramic Society* **27**, 2495-2501 (2007).
27. J.J. Meléndez-Martínez, A. Domínguez-Rodríguez, F. Monteverde, C. Melandri and G. De Portu: Characterisation and high temperature mechanical properties of zirconium boride-based materials. *Journal of the European Ceramic Society* **22**, 2543-2549 (2002).
28. W.G. Fahrenholtz, G.E. Hilmas, I.G. Talmy and J.A. Zaykoski: Refractory diborides of zirconium and hafnium. *Journal of the American Ceramic Society* **90**, 1347-1364 (2007).

SiC NANOMETER SIZING EFFECT ON SELF HEALING ABILITY OF STRUCTURAL CERAMICS

Wataru NAKAO, Shihomi ABE and Kotoji ANDO
Yokohama National University,
79-5, Tokiwadai, Hodogaya-ku,
Yokohama, 240-8501, Japan

ABSTRACT

Enhancement in self healing ability has been necessary to actualize the self healing ceramics. For the purpose, SiC nanometer sizing was attempted as self healing is driven by the oxidation of SiC. In the present study, the self-healing behavior of alumina composites containing nanometer sized SiC particles whose particle sizes are less than 10 nm and 20-50 nm were investigated at several temperatures.

Observations of the surface after healing suggested that the oxidation rate increases as the particle size of SiC decreases. In alumina composite containing SiC particles with particle size = 20- 50 nm, complete strength recovery could be attained by heat treatment above low temperature at which alumina containing commercial SiC particles (particle size = 270 nm) recovered scarcely the cracked strength. However, alumina composite containing SiC particles whose particle size is less than 10 nm cannot recover completely the cracked strength under every condition. This might result from the space between crack walls cannot be filled with the formed oxide due to little volume of SiC on the crack walls. Thus, it was found that there is that optimal SiC particle size for endowing self healing ability.

INTRODUCTION

Self-healing of surface cracks is effective to improve the reliabilities of structural ceramics. Surface cracks cause critical strength degradation to structural ceramics, because the general ceramic has high sensitivity to flaws. Thereby, for using the traditional ceramics component, it is necessary to change the cracked one for new one immediately, because the applied stress during the service is possible to exceed the strength decreased by cracking. On the other hand, ceramics possessing self-healing ability can automatically recover the cracked strength. Thus, the ceramics not only keep high structural integrity but also are elongated their life time.

The present authors[1] succeeded to endow the structural ceramics with the self-healing ability driven by the high temperature oxidation of dispersed SiC particles. The schematic is shown in Fig. 1. The SiC particles located on the crack walls are allowed to react first with the oxygen by cracking.

$$SiC + 3/2\ O_2 = SiO_2 + CO \quad (1)$$

The oxidation (1) includes 80% volume expansion of the condensed phases. As a result, the space between crack walls is filled with the formed oxide. Moreover, the reaction (1) exhibits huge exothermic heat. The heat makes the matrix and the formed oxide melt at once, and to form the strong bond. For this phenomenon, the healed cracks gains higher strength than the base materials. This implies that the crack healing can give rise to complete recovery to the cracked ceramics.

Enhancement of the healing rate is important to actualize the function. The self-healing is invalid at low temperature, because the healing rate is strongly affected by the temperature change. Thus, the function can be valid above the correspond temperature. Actually alumina containing 15 vol.% commercially available SiC particles[2], which have means particle size of 270 nm, can heal the indentation semi-elliptical crack having depth of 0.05 mm by heat-treatment above 1000 °C.

In the present study, it was attempted to enhance the reaction rate which induces the self-healing by nanometer-sizing the dispersed silicon carbide particles. Nanometer-sized silicon carbide particles include many dangling bonds at the surface and exhibit high energy state. Thereby, self-healing would progress actively at low temperature, because the apparent activation energy of the reaction (1) becomes low. Therefore, few silicon carbide nanometer sized particles containing alumina composites were prepared. The healing behaviors were investigated. From the obtained results, the silicon carbide nanometer sizing effect on self-healing was discussed.

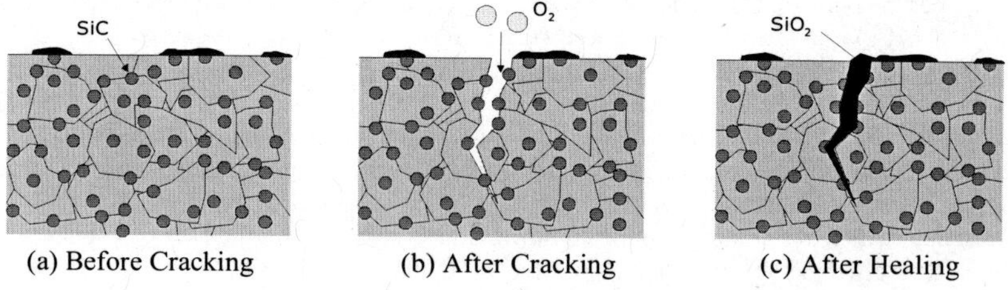

(a) Before Cracking (b) After Cracking (c) After Healing

Figure 1 Schematic of self crack-healing behavior

PREPARATION OF ALUMINA SiC NANOMETER SIZED PARTICLES COMPOSITE

In the present study, the following self-propagating high-temperature synthesis was adapted to produce alumina/ SiC nanometer-sized particles composite,

$$3(3Al_2O_32SiO_2) + 8Al + 6C = 13Al_2O_3 + 6SiC \quad (2).$$

By this method, Zhang et al.[3] succeeded fabricating alumina/ SiC nanometer-sized particles composite in which SiC particle has a means particle size of 50 nm. When reaction (2) progressed below 1400 °C, the formed SiC particles became almost same shape and size as the raw carbon powder, because the diffusion of carbon can be neglect. Moreover, the volume fraction of SiC is 18 vol.% in the composites.

The used mullite (KM101, Kyoritsu Materials Ltd, Japan), aluminum (600F, Minaruko Ltd., Japan) and carbon black fine powders (#4000B, Mitsubishi Chemical Ltd., Japan) have means of the particles size of 780 nm, 5400 nm and 20 nm, respectively. Carbon powder was dispersed well in ethanol using ultra sonic vibration, before the ball-milling. Mullite and aluminum powders were added to carbon ultra-fine powder dispersed ethanol and the powders were mixed well via a Teflon pot and alumina balls for 24 h. To finish reaction (2), the dried mixed powder was heated at 1400 °C for 5 h in a carbon crucible and Ar atmosphere. After that, the formed alumina/ SiC powder was sintered via hot-press at 1800 °C for 1 h in Ar. The sintered plate was cut into rectangular test specimens (Width = 4 mm, Length = 22 mm, Height = 3 mm). The alumina/ SiC composite is abbreviated as AS18NP in this paper.

Moreover, similar alumina/ SiC composite was also prepared by substituting the raw carbon powder from the carbon black fine powder to mixed fullerene powder (nanomix, Frontier Carbon, Ltd., Japan). The composite was expected to include smaller SiC nanometer sized particles than AS18NP. Also the alumina/ SiC composite is abbreviated as AS18NPF in this paper.

MICROSTRUCTURE

Figure 2 shows the microstructure of AS18NP. The grains size of the alumina matrix ranges from 1 μm to 10 μm. The formed nanometer sized SiC particles mainly entrapped inside alumina grains, as pointed by black arrow in Fig. 2. Alternatively, the submicron-sized SiC particles are located at the grain boundaries of alumina grains, as pointed by white arrow in Fig. 2. The grains size of the intra SiC particles ranges from 10 nm to 30 nm, corresponding to one tenth that of commercially available SiC particles. The nanometer sized SiC particles are beneficial for improving crack-healing ability.

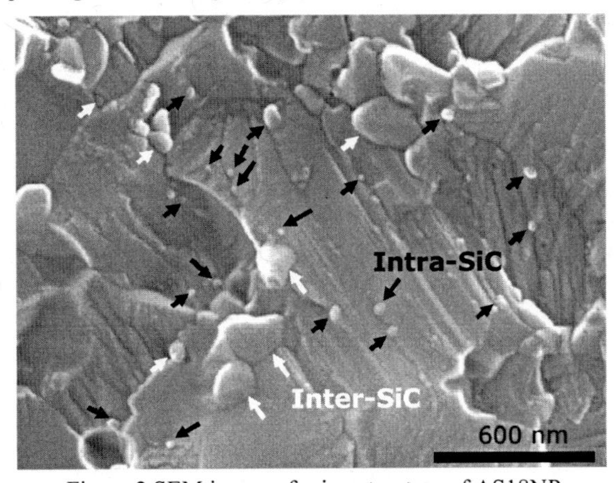

Figure 2 SEM image of microstructure of AS18NP

Figure 3 shows the microstructure of AS18NPF. As same as AS18NP, the formed nanometer sized SiC particles also entrapped inside alumina grains, as pointed by black arrow in Fig. 3. The SiC nanometer sized particles formed in AS18NPF had particles size of less than 10 nm and was smaller than that in AS18NP. Thus, AS18NPF was expected to exhibit rapid self-healing at lower temperature than AS18NP.

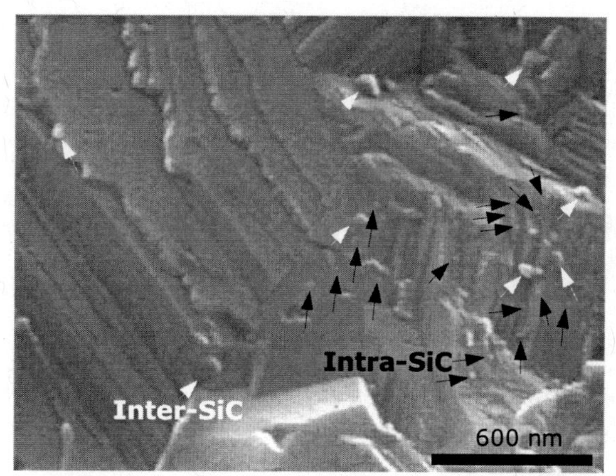

Figure 3 SEM image of microstructure of AS18NPF

SELF-HEALING BEHAVIOR

Figure 4 shows the healed strengths of AS18NP and AS18NPF as a function of the healing temperature, where the strengths were measured by three point bending with rectangular bar specimen. All the healing time is 10 h. For comparison, that of alumina containing 18 vol.% commercially available SiC particles (means particle size = 270 nm) composite, abbreviated as AS18P, is also shown in Fig. 4. Before the healing treatment, a semi-elliptical surface pre-crack of 0.1 mm in surface length was introduced at the center of the specimen in order to clear the strength recovery behavior at the healed pre-crack. If the healing is incomplete, the fracture occurred from the pre-crack, thereby the specimen described as open symbols in Fig. 4. Alternatively, closed symbols were indicted as the specimen, of whom fracture did not occur from the pre-crack, i.e., the pre-crack can be completely healed.

AS18NP can heal completely the pre-crack by the heat treatment above 1100 °C for 10 h in air. The threshold temperature is 200 °C less than that of AS18P. Moreover, AS18NP healed at 950 °C and 1000 °C exhibited same strength as completely healed AS18NP. It is confirmed that the strength recovery is satisfied by the heat treatment at 950 °C and 1000 °C. The temperature to be able to obtain the adequate strength recovery is 350 °C less than that of AS18P. However, AS18NPF cannot heal completely the pre-crack, although the healed strength of AS18NPF increases significantly with increase in the healing temperature. Alternatively, observations of the surface after healing suggested that the oxidation rate increases as the particle size of SiC decreases. This might result from the space between crack walls cannot be filled with the formed oxide due to little volume of SiC on the crack walls. Thus, it was found that there is that optimal SiC particle size for endowing self-healing ability.

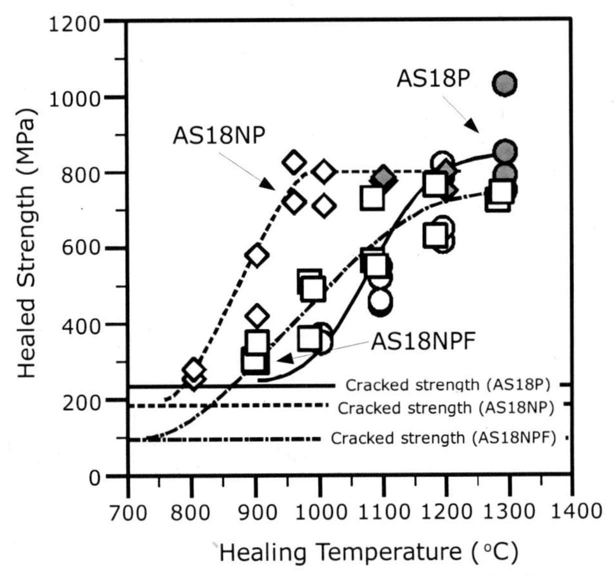

Figure 4 Healed strength of AS18NP and AS18NPF as a function of healing temperature, with that of AS18P.

CONCLUSIONS

Enhancement in self-healing ability has been necessary to actualize the self-healing ceramics. For the purpose, SiC nanometer sizing was attempted as self-healing is driven by the oxidation of SiC.

In the present study, the alumina containing nanometer-sized SiC particles composites were sintered by the reaction synthesis,

$$3(3Al_2O_32SiO_2) + 8Al + 6C = 13Al_2O_3 + 6SiC$$

Using nanometer-sized carbon black and mixed fullerene as raw carbon powder, the present authors succeeded to fabricate alumina composites containing nanometer-sized SiC particles whose particle sizes are 20-50 nm and less than 10 nm, respectively. The self-healing behaviors of the prepared composites were investigated at several temperatures.

Observations of the surface after healing suggested that the oxidation rate increases as the particle size of SiC decreases. In alumina composite containing SiC particles with particle size = 20- 50 nm, complete strength recoveries could be attained by heat treatment above low temperature at which alumina containing commercial SiC particles (particle size = 270 nm) recovered scarcely the cracked strength. However, alumina composite containing SiC particles whose particle size is less than 10 nm cannot recover completely the cracked strength under every condition. This might result from the space between crack walls cannot be filled with the formed oxide due to little volume of SiC on the crack walls. Thus, it was found that there is that optimal SiC particle size for endowing self healing ability.

REFERENCES

[1]W. Nakao, K. Takahashi and K. Ando, Self-healing of Surface Cracks in Structural Ceramics. Self-healing Materials: Fundamentals, Design, Strategies, and Applications. Edited by Swapan Kumar Ghosh, WILEY-VCH Verlag GmbH & Co. KgaA, Weinheim, 2009. pp. 183-217.

[2] K. Ando, B.S. Kim, M.C. Chu, S. Saito and K. Takahashi, Crack-healing and Mechanical Behaviour of Al_2O_3/SiC Composites at Elevated Temperature. Fatigue and fracture of Engineering Materials and structures. 2004, 27, 533-541

[3]G. J. Zhang, J. F. Yang, M. Ando and T. Ohji. Reactive Hot Pressing of Alumina-Silicon Carbide Nanocomposite. Journal of the American Ceramics Society. 2004, 87(2), 299-301

CREEP AND FATIGUE BEHAVIOR OF MI SIC/SIC COMPOSITES AT TEMPERATURE

Ojard, G.[2], Gowayed, Y.[3], Morscher, G.[4], Santhosh, U.[5], Ahmad, J.[5], Miller, R.[2] and John, R.[1]

[1] Air Force Research Laboratory, AFRL/RXLMN, Wright-Patterson AFB, OH
[2] Pratt & Whitney, East Hartford, CT
[3] Auburn University, Auburn, AL
[4] Ohio Aerospace Institute, Cleveland, OH
[5] Research Applications, Inc., San Diego, CA

ABSTRACT

As interest grows in considering the use of ceramic matrix composites for critical components, the response of the ceramic matrix composites to anticipated service conditions needs to be understood. To this end, this work presents an attempt to understand the fatigue behavior and its impact on the life of SiC/SiC composites. The effect of cyclic loading rate of 1 Hz and 30 Hz was evaluated at room temperature and 1204°C. Multiple maximum stress levels were used at a stress ratio of 0.05. Past work by the authors documented the response of the material under creep and long term dwell fatigue (2 hour hold) loading. During this investigation, fatigue tests were carried out at similar maximum peak cyclic stress levels and temperatures.

INTRODUCTION

As increasing interest is shown in Ceramic Matrix Composites (CMCs) for potential applications such as in gas turbine engines, interest has been expressed in the durability of the material. This has been shown by the authors by looking at long term "creep" type testing and residual properties [1-4]. This work highlighted the long term testing and behavior. This is usually considered to be important for long term applications that could be seen in such applications such as industrial gas turbines [5, 6]. As potential applications mature and additional insight is gained, expansion of durability testing into additional areas such as fatigue is done. The results of this effort for a Melt Infiltrated SiC/SiC (MI SiC/SiC) composite will be discussed.

The MI SiC/SiC system was chosen as the material for this effort based on the past work [1-4] as well as the vast data that exists on the material system [5]. This material system was developed out of the Enabling Propulsion Materials Program run out of NASA-Glenn Research Center. NASA-GRC is still evaluating this material as a candidate for advanced applications.

PROCEDURE

Material Description

The Melt Infiltrated SiC/SiC CMC system chosen for this study was the one which was initially developed under the Enabling Propulsion Materials Program (EPM) and is still under further refinement at NASA-Glenn Research Center (GRC). This material system has been systematically studied at various development periods and the most promising was the 01/01 Melt Infiltrated Sylramic-iBN SiC/SiC (01/01 is indicative of the month and year that development was frozen) [7]. There is a wide set of data from NASA for this system as well as a broad historic database from the material development [8]. This allowed a testing system to be put into place to look for key development properties which would be needed from a modeling effort and would hence leverage existing data generated by NASA-GRC.

The Sylramic® fiber was fabricated by Dow Corning Company as a 10 μm diameter stochiometric SiC fiber and bundled into tows of 800 fibers each. The sizing applied was polyvinyl alcohol (PVA). For this study, the four lots of fibers, which were used, were wound on 19 different spools. The tow spools were then woven into a 5HS balanced weave at 20 EPI (36% V_f for all panels in this study). An in-situ Boron Nitride (iBN) treatment was performed on the weave (at NASA-GRC), which created a fine layer of BN on every fiber. The fabric was then laid in graphite tooling to correspond to the final part design (flat plates for this experimental program). All the panels were manufactured from a symmetric cross ply laminate using a total of 8 plies. The graphite tooling has holes to allow the CVI deposition to occur. At this stage, another BN layer was applied. This BN coating was doped with Si to provide better environmental protection of the interface. This was followed by SiC vapor deposition around the tows. Typically, densification is done to about 30% open porosity. SiC particulates are then slurry cast into the material followed by melt infiltration of a Si alloy to arrive at a nearly full density material. The material at this time has less than 2% open porosity. Through this process, 15 panels were fabricated in 3 lots of material. Typical cross sections of this material are shown in Figure 1 showing the material phases.

After fabrication, all the panels were interrogated by pulse echo ultrasound (10 MHz) and film X-ray. There was no indication of any delamination and no large-scale porosity was noted in the panels. In addition, each panel had 2 tensile bars extracted for witness testing at room temperature. All samples tested failed above a 0.3% strain to failure requirement. Hence, all panels were accepted into the testing effort.

Testing

Testing for this effort ranged from standard tensile testing per ASTM C-1275 (room temperature) and ASTM C-1349 (elevated temperature) to durability testing. The creep and dwell fatigue testing has been reported in past papers [1-4]. The fatigue testing was done in a load controlled manner per ASTM C-1360 at elevated temperatures. Testing was done at 1 Hz or 30 Hz using a sine wave with an R ratio of 0.05. During the fatigue testing, strain was recorded.

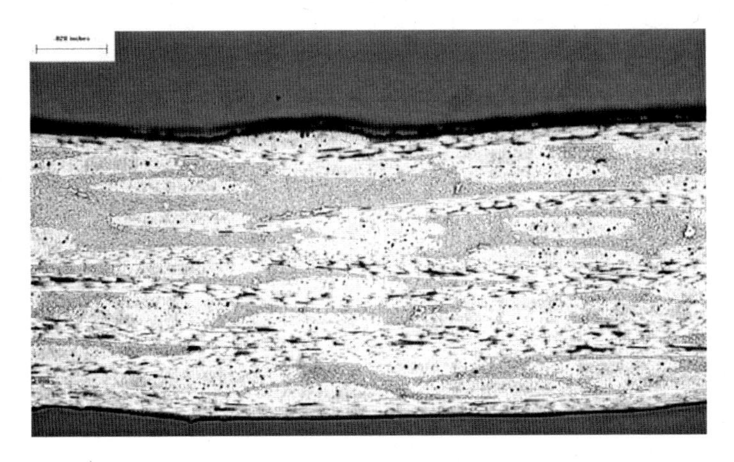

a) cross section showing fiber tows and matrix rich regions

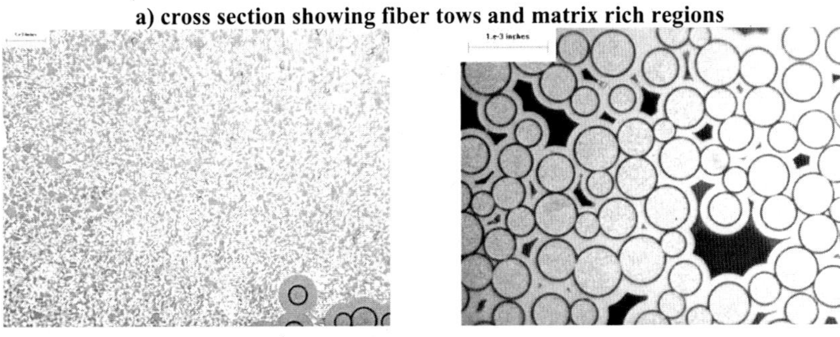

b) SiC particulate with Si **c) Interface coating (BN)**

Figure 1. Micro-structure of MI SiC/SiC

RESULTS

Modulus

One of the key properties determined from the tensile test is the modulus of the material. Early in the testing, it was decided that the modulus would be analyzed by a linear fit of the data between 14 and 56 MPa. This range was determined based strictly on limited 1204°C tensile tests (2 samples) and applied to all subsequent tensile testing. Since this property is a key in determining other stress-strain properties, a review of the curves was conducted to assure that this range was acceptable. It was decided to determine the instantaneous modulus of the stress-strain curves versus strain to assure that there was no deviation of linearity within the range picked. An example of this is shown Figure 2. Figure 2c clearly shows that the modulus range chosen for the program remained in the linear region of the stress-strain curve.

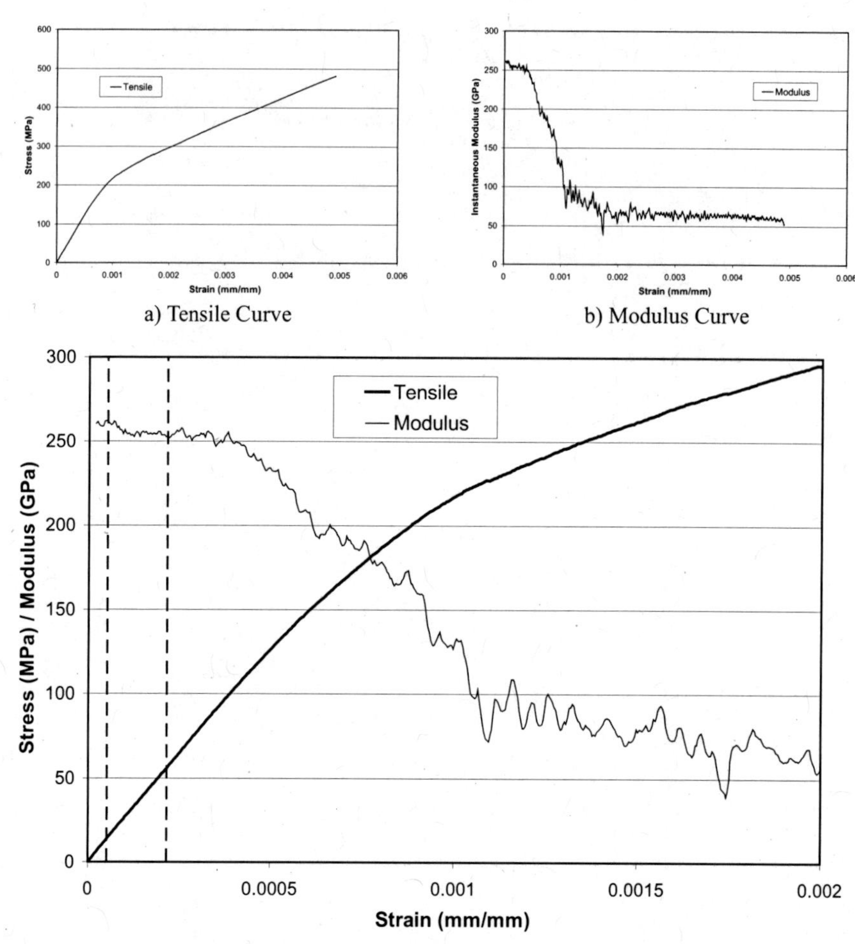

a) Tensile Curve

b) Modulus Curve

c) combined curves showing lines at the strains equivalent to 14 and 56 MPa

Figure 2. Determination of Modulus from Stress-Strain Data

The resultant tensile modulus values for the RT testing done are shown in Table I. Even though the material was tested in 3 lots, lot effects were not reviewed since fatigue effects were known to have larger influences than lot effects [9].

Table I. Room Temperature Modulus Results

	Average (GPa)	St. Deviation (GPa)	# of Tests
Modulus	268.5	13.09	36

Tensile Results

In considering the durability testing, a review was undertaken of the tensile curves: modulus, ultimate tensile strength and strain to failure. This is shown in Table II. Table II shows that overall the data is somewhat constant between 24°C and 815°C with some drop at 1204°C.

Table II. Summary of Tensile Properties

Temperature	Modulus		Ultimate Tensile Strength		Failure Strain	
	Average	St.Deviation	Average	St.Deviation	Average	St.Deviation
(°C)	(GPa)		(MPa)		(mm/mm)	
24	268.5	13.09	466.4	40.51	0.0042	0.00080
815*	247.0		461.1		0.0045	
1204**	227.4	30.50	377.4	8.78	0.0038	0.00020

* = there was only a single acceptable test performed at 815°C that allowed analysis to be performed
* * = there were two test performed at 1204°C

The review was then extended against the key damage points of the stress strain curve as shown schematically in Figure 3 [10]. These points show: σ_{mc}, $\sigma_{proportional\ limit}$ and $\sigma_{saturation}$. σ_{mc} is where the matrix starts cracking. $\sigma_{proportional\ limit}$ is an offset method (0.005% strain [11]) to eliminate the subjective nature of determining the location of the initial damage (σ_{mc}). $\sigma_{saturation}$ is where the matrix is fully cracked and all the load is on the fibers. This review was done at all temperatures where stress-strain data was taken: room temperature, 815°C and 1204°C. These values are shown in Table III. As can be seen, the values for σ_{mc}, and $\sigma_{proportional\ limit}$ require a consistently determined modulus value as noted above.

The proportional limit was determined graphically by offsetting a line (0.005% offset) that was parallel to the initial linear region (modulus) and graphically extrapolating that line to see where it intersected the stress-strain curve. The micro-cracking and saturation stresses were determined by fitting linear lines of the three regions of the tangent modulus curve for the two linear regions (Figure 2c) and then the transition region where the proportional limit exists. This then resulted in two equations with two unknowns. This was then solved to determine the strain and then that value was reviewed against the stress strain curve to determine the resulting stress. This was done for both stress values: micro-cracking and saturation.

Table III. Damage Points from the Stress-Strain Curve

Temperature	Micro Cracking Stress		Proportional Limit Stress		Saturation Stress	
	Average	St.Deviation	Average	St.Deviation	Average	St.Deviation
(°C)	(MPa)		(MPa)		(MPa)	
24	92.5	23.13	170.4	12.15	258.2	15.77
815*	69.5		165.6		254.1	
1204	74.5	24.40	158.1	36.46	280.5	0.49

* = there was only a single acceptable test performed at 815°C that allowed analysis to be performed

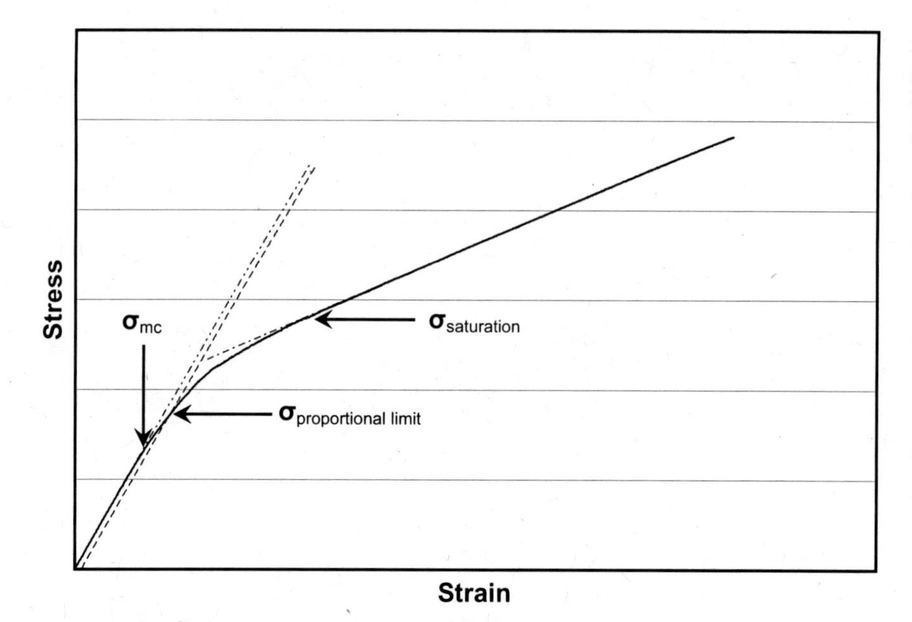

Figure 3. Schematic of CMC Stress-strain Curve

The data generated in the previous tables can be plotted against temperature (Stress only in Figure 4). It would be expected that a lot of these values should be relatively constant with temperature since the MI SiC/SiC system has been demonstrated to have capability up to 1315°C and all the testing done under this effort was up to 1204°C. Figure 4 shows that the values appear to be relatively constant with temperature except for the ultimate tensile strength that drops at 1204°C.

Creep and Dwell Fatigue Results

A series of creep and dwell fatigue tests were done and discussed in past papers by the authors [1, 2]. These series of durability testing consisted of standard creep tests as well as a dwell fatigue test where the sample is tested in what can essentially be considered a creep test. The dwell fatigue test is unloaded at every 2 hours of hold time (R = 0.05). This was intended to break up any oxides that could have been forming in cracks that may be protective in order to make the test a more severe oxidative test. The results obtained at 1204°C are shown in Figure 5. Some tests were stopped so that residual properties could be considered [1].

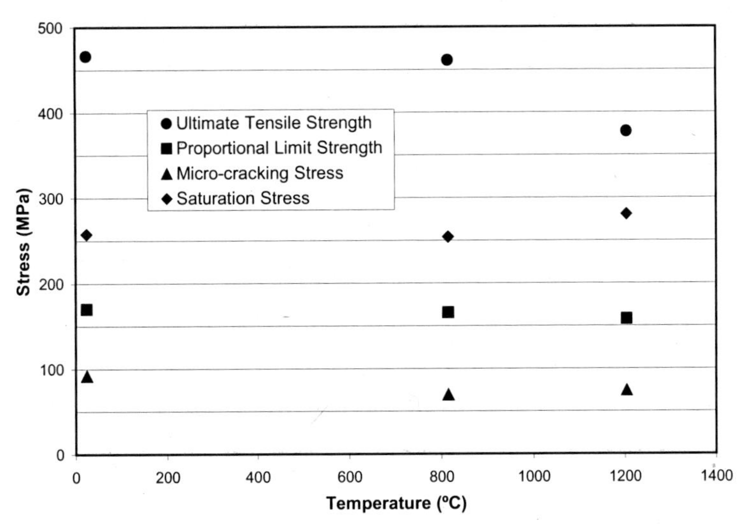

Figure 4. Effect of Temperature on Tensile Properties (from Tables II and III)

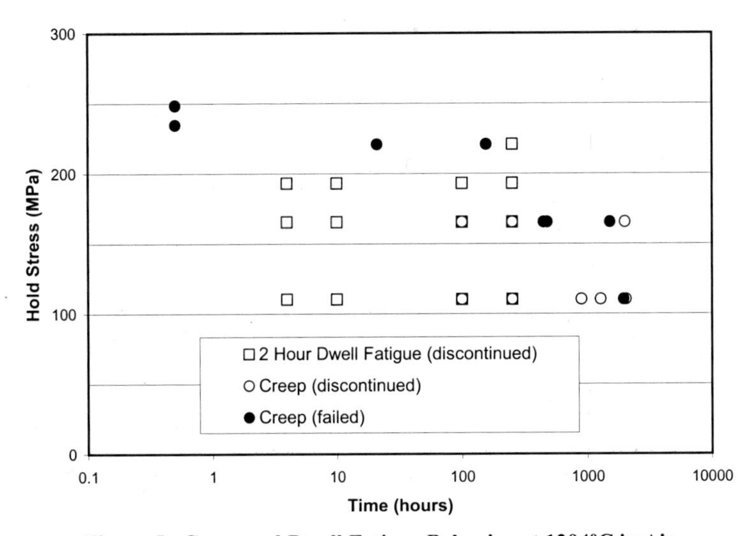

Figure 5. Creep and Dwell Fatigue Behavior at 1204°C in Air

Fatigue Results

There were a series of fatigue tests that were run at 1204°C in both 1 Hz and 30 Hz conditions (R = 0.05). The resulting data from that effort is shown in Figure 6 in terms of applied cycles. In order to better compare the data, the data are re-plotted against time in Figure 7. Figures 6 and 7 show that there is not a large effect for the frequencies tested.

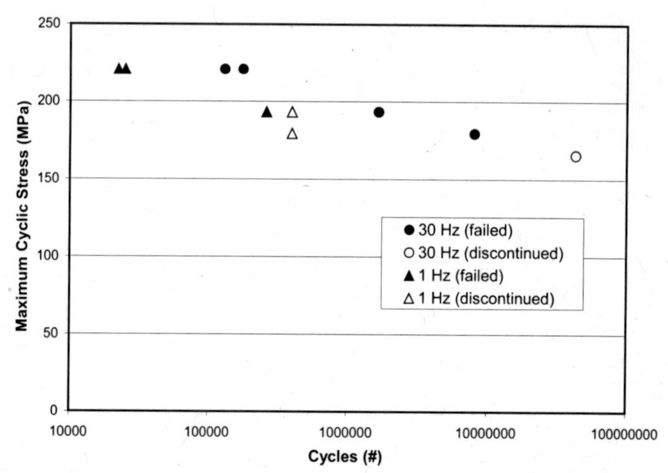

Figure 6. Fatigue behavior as a function of applied stress at 1204°C

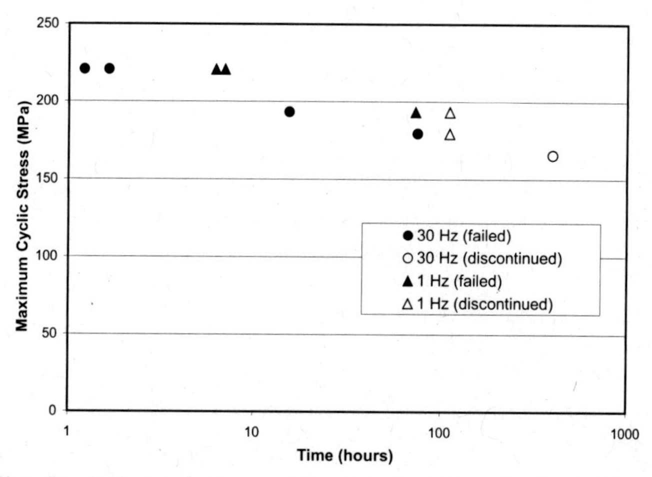

Figure 7. Fatigue behavior as a function of test duration (from Figure 6.)

DISCUSSION

The creep and fatigue data can be compared by looking at the stress versus time to failure as shown in Figure 8. This figure has been simplified in that only the failed data points are shown. This figure shows that there is very little differentiation between the fatigue testing and creep testing in regards to time to failure. This is consistent with the main durability being affected by time at temperature and not dependent on the loading rate.

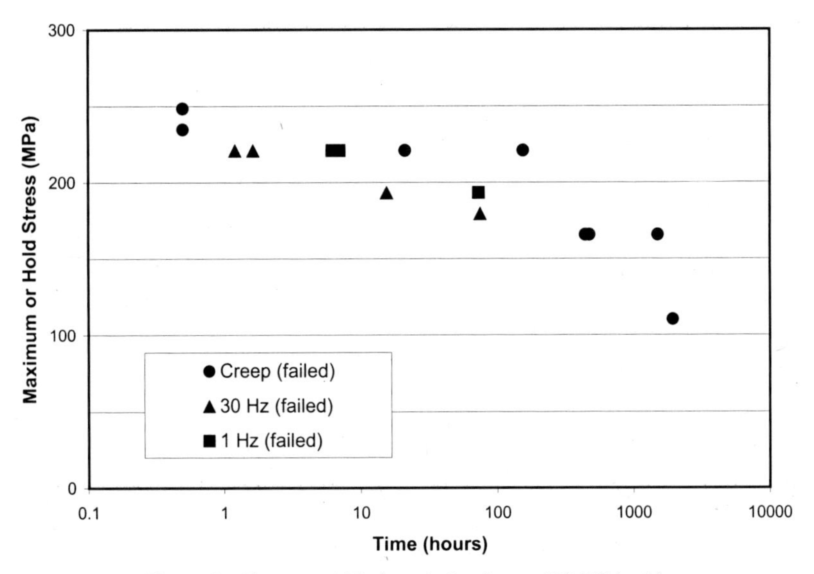

Figure 8. Creep and Fatigue behavior at 1204°C in Air
(Only failed data are included in this plot)

Throughout this study, most testing was done around or slightly above the proportional limit. This is highlighted in Figure 9 where the micro-cracking, proportional limit and saturation stresses are added to the plot shown in Figure 8. There was very limited testing done near the micro-cracking stress and this confirms the reason for no run-out durability stress being noted for this CMC Material system. Figure 9 also shows that the failed data points extrapolate towards the saturation stress at very short times.

Since it is not intuitive that the behavior between tests should result in similar overall lifetime behavior, the strain response was reviewed between testing types. This was done between several stress levels and temperatures where it was found that the strain evolved during testing was independent of the test being run. This is shown in Figure 10 for a 1Hz test and a dwell fatigue test run at 193.2 MPa and 1204°C. As can be seen, there is no difference in the strain evolved during these two very different tests.

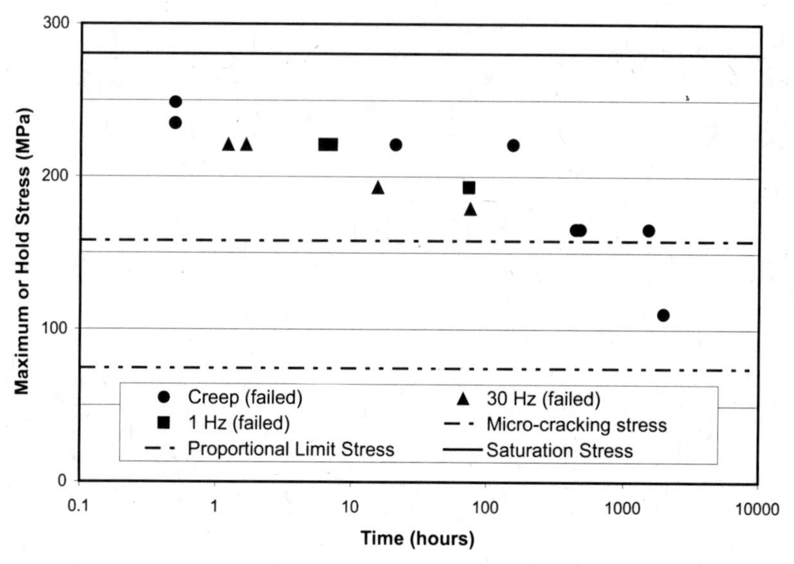

Figure 9. Figure 8 with key points from the stress-strain curve (in terms of stress)

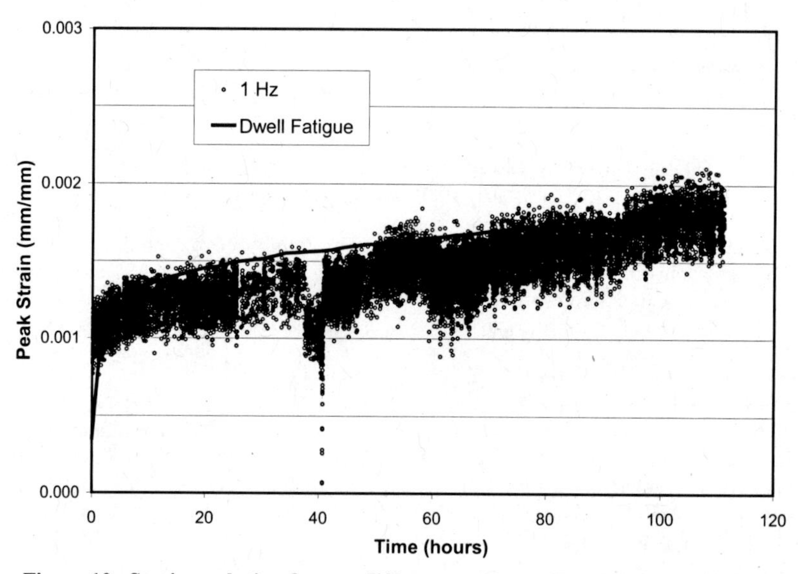

Figure 10. Strain evolution for two different test types (Dwell Fatigue and 1 Hz)

If the interface coating is well protected by the matrix or part of the matrix, then the environmental ingress would control failure. It can be envisaged that in a fully dense CMC systems such as MI SiC/SiC that this would be the case. This is one possible explanation at this time for the lack of differentiation between the failure times seen between the different durability tests done. There may also be an effect that there was not a large enough difference between the tests to see a frequency affect on the loading. This is an area for ongoing research.

CONCLUSION

The results of this study shows that time at temperature is the key to understanding failure in ceramic matrix composites. There does not appear to be any affect on the loading frequency under fatigue loading between 1 and 30 Hz. There was no change seen in the overall lifetimes for the various durability testing done and this was confirmed in seeing the strain response between different tests within the large scatter generated during the creep testing. The failed durability data extrapolates towards the saturation stress as well as the micro-cracking stress (Figure 9). Future investigations are required at stresses near or below the micro-cracking stress to see if a run-out stress can be found for long durations (~ 1000 hours) on this material system.

It is apparent that the effect of an environment is affecting the lifetimes (and not loading method) since samples are clearly failing between the micro-cracking and saturation stresses. An area for future study would be to conduct testing at, near or above the saturation stress to determine the effect the environment has on the lifetime of the material.

ACKNOWLEDGMENTS

The Air Force Research Laboratory, Materials and Manufacturing Directorate (AFRL/RX) sponsored this work under contracts F33615-01-C-5234 and F33615-03-D-2354-D04.

REFERENCES

1. Ojard, G., Calomino, A., Morscher, G., Gowayed, Y., Santhosh, U., Ahmad J., Miller, R. and John, R., "Post Creep/Dwell Fatigue Testing of MI SiC/SiC Composites", American Ceramic Society, Ceramic Engineering and Science Proceedings, pp. 135-143, 2008.
2. Ojard, G., Gowayed, Y., Chen, J., Santhosh, U., Ahmad J., Miller, R., and John, R., "Time-Dependent Response of MI SiC/SiC Composites Part 1: Standard Samples", American Ceramic Society, Ceramic Engineering and Science Proceedings, pp. 145-153, 2008.
3. Y. Gowayed, G. Ojard, J. Chen, R. Miller, U. Santhosh, J. Ahmad and R. John. "Time-Dependent Response of MI SiC/SiC Composites Part 2: Samples with Holes", American Ceramic Society, Ceramic Engineering and Science Proceedings, pp. 155-162, 2008.
4. Ojard, G., Gowayed, Y., Santhosh, U., Ahmad J., Miller, R., and John, R., "The Effect of Holes on the Residual Strength of SiC/SiC Ceramic Composites", In Press, American Ceramic Society, Proceedings of Ceramic Engineering and Science, 2008.
5. Brewer, D., Ojard, G. and Gibler, M., "Ceramic Matrix Composite Combustor Liner Rig Test:, ASME Turbo Expo 2000, Munich, Germany, May 8-11, 2000, ASME Paper 2000-GT-670.
6. Wedell, James K. and Ahluwalia, K.S., "Development of CVI SiC/SiC CFCCs for Industrial Applications" 39th International SAMPE Symposium April 11- 14, 1994, Anaheim California, Volume 2, pg. 2326.
7. Calomino, A., NASA-Glenn Research Center, personal communication.

8. J.A. DiCarlo, H-M. Yun, G.N. Morscher, and R.T. Bhatt, "SiC/SiC Composites for 1200°C and Above" Handbook of Ceramic Composites, Chapter 4; pp. 77-98 (Kluwer Academic; NY, NY: 2005)
9. Robert J. Miller, Pratt & Whitney, personal communication.
10. Evans, A.G., Zok, F.W. and Mackin, T.J., "The Structural Performance of Ceramic Matrix Composites", in High Temperature Mechanical Behavior of Ceramic Composites, Nair, S.V. and Jakus, K. Editors, Butterworth-Heinemann, Newton, MA, 1995.
11. HSR/EPM D-001-93 Consensus Standard, "Monotonic Tensile Testing of Ceramic Matrix, Intermetallic Matrix and Metal Matrix Composites Materials", NASA-Glenn Research Center, Cleveland, OH.

SELF-CRACK HEALING BEHAVIOR UNDER COMBUSTION GAS ATMOSPHERE

Self-crack-healing behavior under combustion gas atmosphere
Toshio Osada
Post-graduate student, Department of Energy and Safety Engineering, Yokohama National University, 79-5 Tokiwadai, Hodogaya-ku, Yokohama, 240-8501, Japan.

Wataru Nakao
Interdisciplinary Research Center, Yokohama National University,
79-5 Tokiwadai, Hodogaya-ku, Yokohama, 240-8501, Japan.

Koji Takahashi and Kotoji Ando
Department of Energy and Safety Engineering, Yokohama National University,
79-5 Tokiwadai, Hodogaya-ku, Yokohama, 240-8501, Japan.

ABSTRACT

Self-crack-healing behavior under combustion gas atmosphere with low oxygen partial pressure, P_{O2}, is important for actualizing ceramic gas turbines, but to date only self-crack-healing behavior in air has been investigated. In this study, we investigated crack-healing behaviors at 1273 K – 1773 K under several levels of P_{O2}. Crack-healing in atmospheres with $P_{O2} \geq 50.0$ Pa gave rise to the complete strength recovery of cracked specimens, resulting from passive oxidation. Based on the obtained results, the kinetics for strength recovery by self-crack-healing was expressed as a function of healing temperature, T_H (K) and P_{O2} (Pa). The strength recovery rate for complete crack-healing, v_H (s^{-1}), could be expressed as

$$v_H = 6.95 \times 10^5 \exp\left(\frac{-4.65\times10^4}{T_H}\right) P_{O_2}{}^{0.835}$$

Using this rate equation, one can evaluate the healing time for complete strength recovery under combustion gas in a gas turbine.

INTRODUCTION

Self-healing of surface cracks in structural ceramics is a valuable technique for ensuring structural integrity[1-4]. Ceramics are brittle materials that are sensitive to flaws. Furthermore, the ease with which cracks are introduced by machining prior to service as well as by contact during service cause the reliability of ceramics parts to be decreased significantly. Therefore, flaws, especially surface cracks, directly lead to low structural reliability. Among the many types of crack-healing[1-21], that driven by the oxidation of SiC is the only one that educes a "complete" strength recovery. The SiC oxidation concept was proposed by Ando *et al.*[12-14]. Since then, there have been many reports [13-21] on crack-healing behavior due to the oxidation

$$\mathrm{SiC(s)} + \tfrac{3}{2}\mathrm{O_2(g)} = \mathrm{SiO_2(s)} + \mathrm{CO(g)} \quad (1)$$

for alumina/SiC[14, 17-19], mullite/SiC[4, 12, 19], and Si_3N_4/SiC[13, 20-21] composites. The oxidation includes exothermic heat and approximately 80% volume expansion of the condensing phase. These two phenomena result in (1) material that exhibits the same or a higher level of strength as that of the base material, (2) healing material that is bonded strongly to the base material, and (3) spaces between the crack walls that are filled with the healing material. To satisfy situation (3), it is also necessary to contain more than 10 vol% SiC particles or whiskers.

Ando *et al.* reported that crack-healing was useful even under tensile stress.[18-19, 21] If the surface cracks initiated due to contact events and thermal shock can be healed, the structural components, e.g., advanced heat engines and micro ceramics gas turbine blades, can offer improved reliability. In the case of gas turbine blades, the oxidation environment differs substantially between the application and the conditions examined in previous studies. The structural component, i.e., the gas turbine, generally operates under combustion gas with lower oxygen partial pressure, P_{O2} than that in air. For example, P_{O2} in the 1500 °C-class gas turbine is approximately 8×10^3 Pa - 1×10^4 Pa. The oxidation of SiC has a transition P_{O2} from active to passive oxidation[22-25]. Kim *et al.*[26] investigated the oxidation effect on the strength of SiC-whisker reinforced alumina and reported that a critical point in determining the strength of the polished sample was the P_{O2} transition from active to passive oxidation of monolithic SiC at 1673 K. Also, Jung *et al.*[27] investigated the P_{O2} dependence of healing behavior for indentation cracks in silicon nitride/SiC composite and reported that strength recovery is also attained by crack-healing even when P_{O2} is 50.0 Pa, corresponding to 0.24% of air. As mentioned above, there are many valuable qualitative reports about this process, but the kinetics of this behavior including the P_{O2} effect has not been explored.

In the present study, our goal was to clarify the kinetics of strength recovery by self-crack-healing as a function of P_{O2}. To accomplish this, we investigated the crack-healing behavior of alumina/15 vol. % SiC particle composites under several P_{O2} atmospheres.

EXPERIMENTAL PROCEDURES

The alumina powder (AKP-20, Sumitomo Chemical Co., Ltd., Tokyo, Japan) used has an average particle size of 0.5 μm. The SiC powder (ultrafine grade, Ibiden Co., Ltd., Japan) used has an average particle size of 0.27 μm. The mixture of alumina powders and 15 vol. % SiC powders was well blended in alcohol using alumina balls and an alumina mill pot. Rectangular plates of 90 × 90 × 6 mm were hot-pressed in N_2 at 1973 K for 2 h under 35 MPa. The sintered alumina/silicon carbide composite had an average grain size of approximately 1 μm and relative density of 99%. The SiC particles were distributed uniformly in the grain boundaries; however, several nanometer-size SiC particles were distributed in the alumina grain as reported previous study[14]. The hot-pressed plates were cut into 3× 4× 22 mm rectangular bar specimens, as shown in figure 1. The specimens were polished to a mirror finish on one face according to the Japanese Industrial Standard (JIS) [28]. The edges of the specimens were beveled 45° to prevent fracturing from edge cracks. In this paper these specimens are called "as-polished specimens".

A semi-elliptical surface crack of 100 μm in surface length was made at the center of the polished surface of the as-polished specimen with a Vickers indenter, using a load of 19.8 N. The indentation crack, as shown in figure 2, is called a "pre-crack" in this paper. The ratio of the depth (*d*) to half the surface length (*c*) of the crack (aspect ratio *d*/*c*) was 0.9. We call these specimens the "as-cracked specimens".

Crack-healing treatments were performed in a horizontal alumina tube furnace, as shown in figure 3. The specimens containing the pre-crack were subjected to crack-healing treatment at 1273 K - 1773 K for 6.00×10^2 - 1.80×10^5 s. The oxygen partial pressure of 50.0 Pa and 5.00×10^3 Pa was controlled by passing N_2 (impurity O_2 of 500 ppm) and N_2-5.00% O_2 (mixture of high purity N_2 and high purity O_2) gases through the reaction tube with a flow rate of 0.1 l/min, respectively. Moreover, by passing N_2 gas deoxidized by reacting with heated graphite, the specimens were subjected to a reductive atmosphere. For comparison, the cracked specimens were also subjected to the healing treatment in air ($P_{O2} = 2.10 \times 10^4$ Pa). We call these specimens the "crack-healed specimens".

All specimen fracture tests were performed on a three-point loading system with a span of 16 mm at room temperature, as shown in figure 1. The cross-head speed in the monotonic test was 0.5 mm/min. The bending strength, σ_B of as-polished specimens (without pre-crack) was also investigated. It has been reported that small surface flaws caused by polishing decreased the bending strength. Thus, all polished specimens were heat-treated at 1573 K for 3.60×10^3 s in air before the bending test to heal preexisting surface flaws[14]. The specimens which were indented have not received this heat-treatment before indentation because the polishing flaws were much smaller than the indentation crack.

The specimen surfaces were analyzed using a scanning electron microscope (SEM). The X-ray diffraction (XRD) method was used to identify the surface oxidized layer. The radiation used in X-ray was Cu$K\alpha$ at 50-kV accelerated voltage and 60-mA electron current.

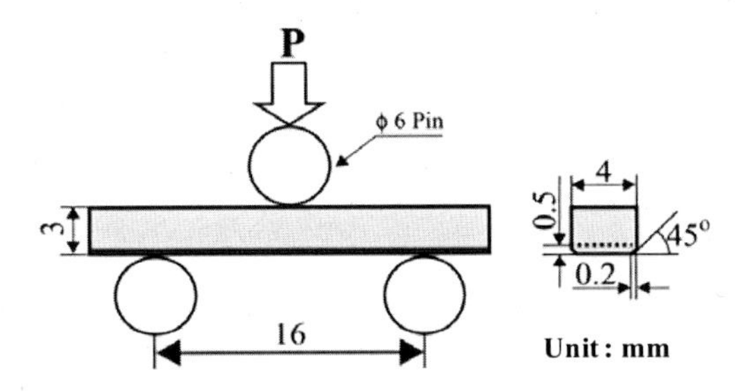

Figure 1. Schematic illustration of the three-point loading system and test specimen size.

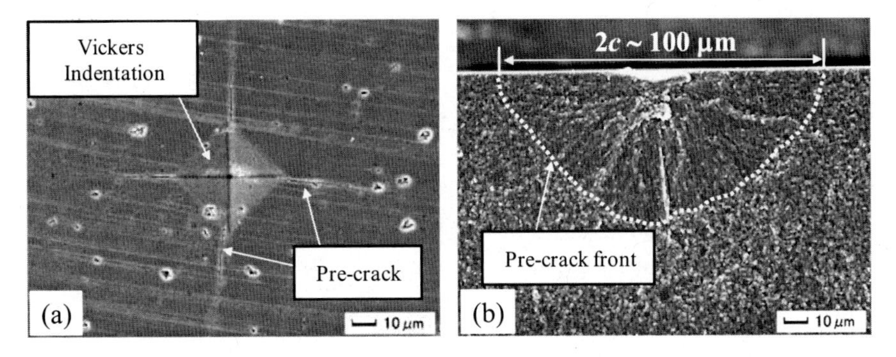

Figure 2. SEM images of (a) surface shapes and (b) cross-sectional shapes of the pre-crack and indentation

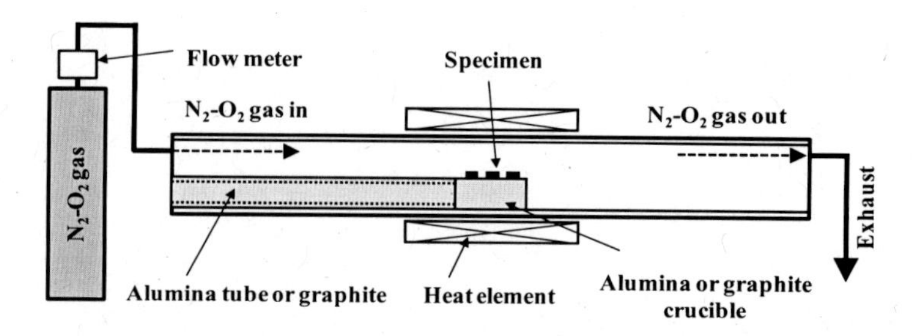

Figure 3. Schematic illustration of the experimental apparatus.

RESULTS AND DISCUSSIONS

Influence of oxygen partial pressure on crack-healing behavior

Before discussing the kinetics of crack-healing, we describe thermodynamically the experimental conditions associated with the temperature and partial oxygen pressure. Generally SiC oxidation includes active-to-passive transition, where passive oxidation is given by Eq. (1) and active oxidation is given by

$$SiC(s) + O_2(g) = SiO(g) + CO(g) \qquad (2).$$

The active-to-passive transition P_{O2} increases logarithmically with decreasing inverse of temperature. Under the present experimental conditions, the nearest condition to active-to-passive transition is a temperature of 1773 K and P_{O2} of 50.0 Pa. According to Hinze and Graham[23], the transition P_{O2} of monolithic SiC at 1773 K is approximately 50.0 Pa. Thus, all experimental conditions except the reductive atmosphere can be estimated to cause passive oxidation of SiC.

Figure 4 shows the strength recovery behavior by crack-healing for 3.60×10^4 s at several healing temperatures, T_H, in several P_{O2} atmospheres. Crack-healing in atmospheres with $P_{O2} \geq$ 50.0 Pa gave rise to large strength recovery of the cracked specimens. This resulted from the condensed phase formation due to passive oxidation. Also, SEM observation and XRD analyses indicated traces of passive oxidation. Figure 5 (a) and (b) shows a SEM image and XRD pattern representative of a specimen surface after crack-healing treatment at 1773 K in N_2 gas (P_{O2} = 50.0 Pa). In the crack-healed surface, many reaction products like "sweat" appear due to passive oxidation. Additionally, these reaction products were identified as mullite and β-cristobalite SiO_2.

The lowest oxygen partial pressure for educing crack-healing ability could be confirmed to be the P_{O2} transition from active to passive oxidation at a given temperature. When the as-cracked specimen was heat-treated under a reductive atmosphere, small strength recovery was attained. However, strength recovery is well known to be caused not by crack-healing but rather by the stress relaxation of the residual stress at the site of the Vickers indentation [17].

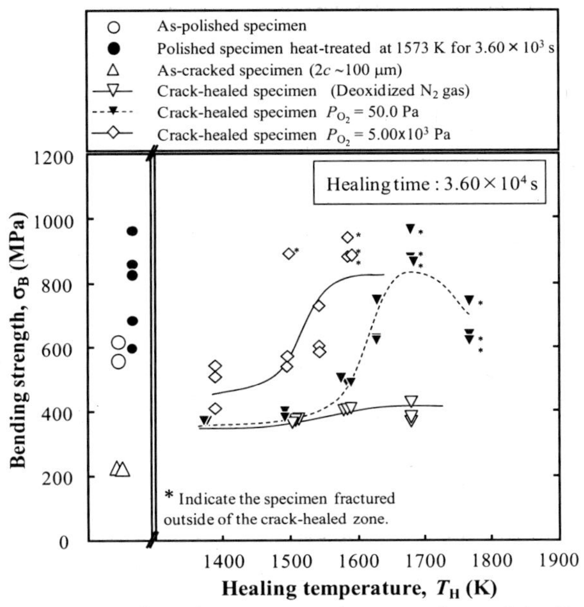

Figure 4. Healing temperature dependence on strength recovery by crack-healing for 3.60×10^4 s in several oxygen partial pressure atmospheres.

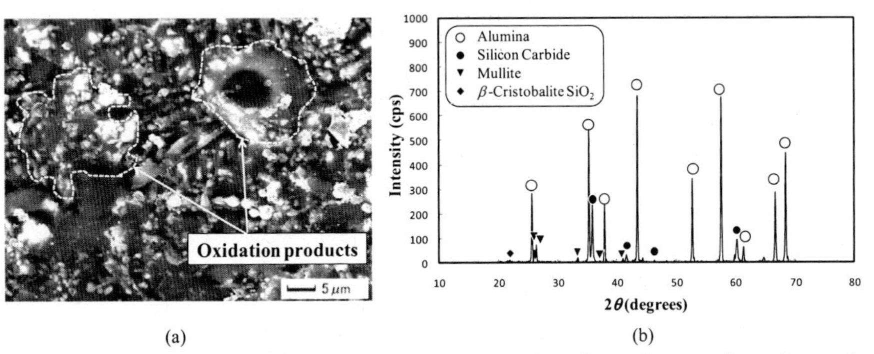

Figure 5 (a). SEM image and (b) XRD pattern representative of a specimen surface after crack-healing treatment at 1773 K for 10 h in flowing N_2 gas with $P_{O2} = 50.0$ Pa.

Dependence of oxygen partial pressure on the strength recovery rate

Figure 6 shows time dependence of the strength recovery by crack-healing in several P_{O2} atmospheres at 1623 K. The closed diamonds, open diamonds, and inversed closed triangles indicate the crack-healing behaviors in atmospheres with P_{O2} of 2.10×10^4, 5.00×10^3, and 50.0 Pa, respectively. In addition, the bending strength, σ_B of the polished specimen and the polished specimen heat-treated at 1573 K for 3.60×10^3 s and the as-cracked specimens are shown in the left column of this figure.

The atmosphere with P_{O2} of 5.00×10^3 Pa caused approximately 400 MPa strength recovery due to crack-healing for 1.20×10^3 s. However, fracturing still initiated from the pre-crack (indentation crack). Additional 200 MPa strength enhancements and changes in fracture initiation from the pre-crack to the other flaws (preexisting embedded flaws) were caused by crack-healing in the presence of oxygen under 5.00×10^3 Pa for 3.60×10^3 s. Moreover, the crack-healed specimen exhibited almost the same strength as the polished specimen heat-treated at 1573 K for 3.60×10^3 s. Thus, the pre-crack could be confirmed to be erased from the view of fracture mechanics. From the results, we determined that the minimum healing time for the complete strength recovery, t_{HMin}, under P_{O2} of 5.00×10^3 Pa was 3.60×10^3 s. Similar crack-healing behaviors occurred in the atmospheres with the other P_{O2}. Thereby, the present authors determined the t_{HMin} in atmospheres with P_{O2} of 2.10×10^4 and 50.0 Pa to be 1.20×10^3 s and 1.80×10^5 s, respectively. The obtained t_{HMin} increases significantly with decreasing P_{O2}. In other words, the strength recovery rate decreases as P_{O2} decreases.

Table I summarizes the t_{HMin} (s) obtained at several healing temperatures, T_H (K). The hashmarked value was obtained in a previous study[14]. Lower temperature and P_{O2} atmosphere are clearly understood to require longer healing time.

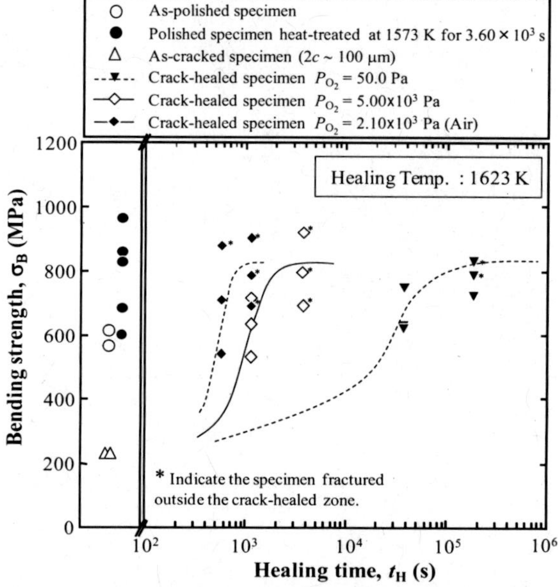

Figure 6. Healing time dependence on strength recovery by crack-healing at 1623 K in several oxygen partial pressure atmospheres.

Table I. Minimum crack-healing time for complete strength recovery, t_{HMin}, in several healing temperatures and oxygen partial pressures

Oxygen partial pressure, P_{O2} (Pa)	Minimum healing time for complete strength recovery, t_{HMin} (s)					
	Healing temperature, T_H (K)					
	1273	1473	1573	1623	1673	1723
50.0	-	-	-	1.80×10^5	3.60×10^4	1.08×10^4
5.00×10^3	-	1.44×10^5	1.08×10^4	3.60×10^3	-	-
2.10×10^4	$^{\#}1.08\times10^6$	$^{\#}3.60\times10^4$	3.60×10^3	1.20×10^3	-	-

Crack-healing kinetics

Crack-healing can be expressed as a function of oxygen partial pressure, P_{O2}, and healing temperature, T_H. For this purpose, two assumptions were employed. One is that the crack-healing rate, v_H, which can be assumed to be the exponential function of P_{O2}, and the exponential index, corresponding to the reaction order of oxygen, are independent of temperature. The other is that the rate constant, which is a proportional constant of crack-healing rate versus n-th powered P_{O2}, obeys Arrhenius' law.

The inverse of the minimum healing time, $1/t_{HMin}$, represents the crack-healing rate for complete strength recovery for the given temperature (v_H). The crack-healing rate can be expressed as the following exponential function of the P_{O2}:

$$v_H = \frac{1}{t_{HMin}} = k\left(\frac{P_{O_2}}{P^o_{O_2}}\right)^n \quad (3),$$

where n is the reaction order of O_2, k is the rate constant (s^{-1}), P_{O2} is an oxygen partial pressure, and P_{O2}^{o} is the standard pressure of 0.1 MPa. The reaction order, n, must be independent of temperature. Thus, the crack-healing rate should exhibit the same value of n for whole temperatures. If the natural logarithm of the crack-healing rate is a function of the natural logarithm of P_{O2}/ P_{O2}^{o} for each temperature, parallel lines should be written by obeying Eq. (4) as follows:

$$\mathrm{Ln}\left(\frac{1}{t_{HMin}}\right) = \mathrm{Ln}k + n\mathrm{Ln}\left(\frac{P_{O_2}}{P^o_{O_2}}\right) \quad (4).$$

Figure 7 shows the relationship between $1/t_{HMin}$ and P_{O2}/ P_{O2}^{o}. The closed diamonds, open diamonds, and inversed closed triangles indicate the natural logarithm of $1/t_{HMin}$ at the healing temperature of 1623 K, 1573K, and 1473K, respectively. As shown in the figure, the lines are almost parallel to each other. Thus, the temperature-independent constant n can be evaluated as follows:

$$n = 0.835 \pm 0.082 \quad (5).$$

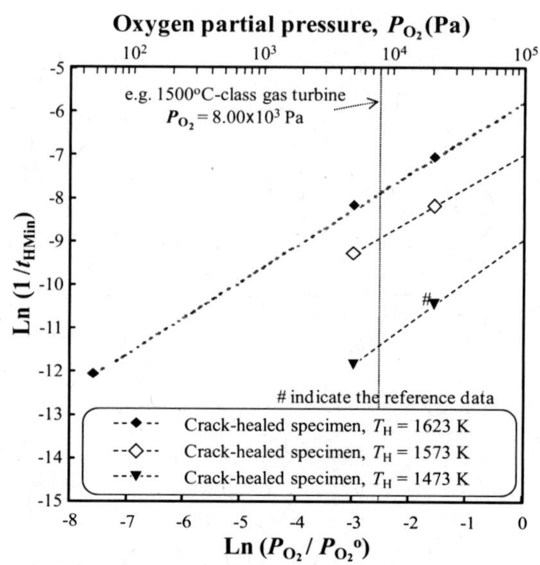

Figure 7. Relationship between the minimum healing time for complete strength recovery and oxygen activity at 1473 K, 1573 K, and 1623 K.

Using the value obtained in Eq. (5), one can evaluate the rate constant, k, for each temperature. k can be given by

$$k = A_H \exp\left(\frac{-Q_H}{RT_H}\right) \quad (6),$$

where A_H is a frequency factor (s^{-1}), Q_H is the activation energy for crack-healing (kJ/mol), R is the gas constant (J/mol K), and T_H is a healing temperature (K). If the natural logarithm of k is plotted as a function of the inverse of the healing temperature, one can obtain only one straight line. Furthermore, the slope and the intercept of this line should be the values Q_H and Ln A_H, respectively. The temperature dependence of the Ln k calculated from various levels of P_{O2} is plotted in figure 8. The closed diamonds, open diamonds, and inversed closed triangles indicate the Ln k calculated from the P_{O2} of 2.10×10^4 Pa, 5.00×10^3 Pa, and 50.0 Pa, respectively. The error bars represent a root-mean square deviation of Ln k calculated from that of n, corresponding to $\pm 8.2\times10^{-2}$. Despite crack-healing in the atmospheres with several P_{O2}, the data exhibit only one straight line. Thus, the results agreed with Arrhenius' law at least in the experimental P_{O2} and T_H range. The values of Q_H and A_H in the temperature range could be determined to be 387 ± 66 (kJ/mol) and $(1.04 \pm 0.00) \times10^{10}$ (s^{-1}), respectively, from the least-square fitting ($R^2 = 0.949$).

Narushima *et. al.* [29] reported that the oxidation of chemical vapor deposited (CVD)-SiC proceeded in two stages. The amorphous SiO_2 films formed in first stage was crystallized to β-cristobalite in second stage. The reported activation energy for the crystallization was 387kJ/mol.

Similarly, Hinze *et. al.* [30] reported that the activation energy of approximately 400 kJ/mol for the hot-pressed SiC oxidation. The value of Q_H obtained in the present study was close to the values

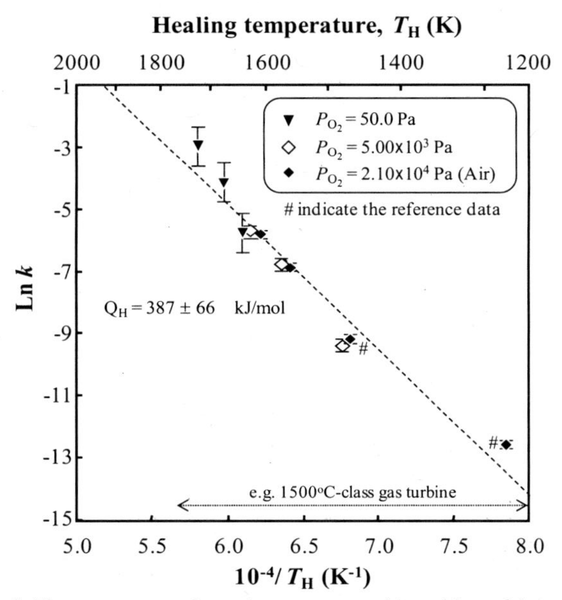

Figure 8. The temperature dependence on natural logarithm of the rate constant, k, calculated from various levels of P_{O2}.

for the crystallization to β-cristobalite in second stage. These results suggested that the strength would be completely recovered when the SiO_2, which completely filled the space between the crack walls, was completely crystallized. The crystallization process is rate-controlling step of the complete crack-healing. Additionally, the diffraction peaks of mullite in figure 5 are that formed as inter-compound at the interface between these oxidation products and alumina.

Substituting Eq. (6) into Eq. (3), the strength recovery rate for complete crack-healing, v_H, can be written as follows:

$$v_H = \frac{1}{t_{HMin}} = A_H \exp\left(\frac{-Q_H}{RT_H}\right)\left(\frac{P_{O_2}}{P^o_{O_2}}\right)^n \tag{7}$$

Thus, one can estimate the v_H, substituting the values of A_H, Q_H and n, as follows:

$$v_H = \frac{1}{t_{HMin}} = 6.95 \times 10^5 \exp\left(\frac{-4.65\times10^4}{T_H}\right) {P_{O_2}}^{0.835} \tag{8}$$

The equation allows the prediction of t_{HMin} (s) exclusively as the function of P_{O2} (Pa) and T_H (K). As an example, the crack-healing behavior in the atmosphere of a 1500 °C-class gas turbine will be estimated. In the situation, the turbine inlet - outlet temperature ranges from 1723 K to 1273 K, and the P_{O2} in the combustion gas is 8.00×10^3 Pa. From the estimation, t_{HMin} at 1373 K, 1573 K, and 1673 K could be estimated to be approximately $(4.1 \pm 0.85) \times 10^5$ s (110 ± 24 h), $(5.5 \pm 1.1) \times 10^3$ s (1.5 ± 0.3 h), and $(9.4 \pm 2.0) \times 10^2$ s (0.27 ± 0.06 h), respectively. Thus, we confirmed that crack-healing is valuable for ensuring structural integrity, even if the components are used in the combustion gas atmosphere produced by the gas turbine. Our study results suggest that surface cracks introduced in the blade can be completely healed during operation, enhancing the potential uses for ceramics turbine blades with self-crack-healing ability.

CONCLUSIONS

In this study we established the kinetics of strength recovery by self-crack-healing as a function of P_{O2}. For this purpose, we investigated the crack-healing behavior of alumina/15 vol. % SiC particles composites under several P_{O2} atmospheres. We derived the following conclusions based on our results:

(1) Crack-healing in atmospheres with $P_{O2} \geq 50.0$ Pa gave rise to the complete strength recovery of cracked specimens, resulting from passive oxidation.

(2) The strength recovery rate depended on T_H as well as P_{O2}, and the time required to completely heal the indentation crack increased with decreasing P_{O2}.

(3) Despite crack-healing in the atmospheres with several P_{O2}, the Ln k as a function of the $1/T_H$ exhibited only one straight line using $n = 0.835 \pm 0.082$. Thus, the results agreed with Arrhenius' law at least in the experimental P_{O2} and T_H range. The value of Q_H and A_H in the temperature range could be determined to be 387 ± 66 (kJ/mol) and $(1.04 \pm 0.00) \times 10^{10}$ (s^{-1}), respectively. Thus, the rate, v_H (s^{-1}) could be expressed as a function of P_{O2} (Pa) and T_H (K), and given by

$$v_H = \frac{1}{t_{HMin}} = 6.95 \times 10^5 \exp\left(\frac{-4.65\times10^4}{T_H}\right) P_{O_2}{}^{0.835}.$$

(4) Using the equation in (3), the strength recovery rate by crack-healing in an atmosphere with P_{O2} of 8.0×10^3 Pa, corresponding to P_{O2} in the combustion gas of an actual gas turbine, could be estimated.

REFERENCES

[1] K. Ando, K. Furusawa, K. Takahashi and S. Sato, Crack-healing Ability of Structural Ceramics and A New Methodology to Guarantee the Structural Integrity Using the Ability and Proof-Test, *J. Eur. Ceram Soc.*, **25**, 549-58 (2005).

[2] M. Ono, W. Nakao, K. Takahashi, M. Nakatani and K. Ando, A New Methodology to Guarantee the Structural Integrity of Al_2O_3/SiC Composite Using Crack Healing and A Proof Test, *Fatigue Fract. Engng. Mater. Struct.*, **30**, 599-607 (2007).

[3] T. Osada, W. Nakao, K. Takahashi, K. Ando and S. Saito, Strength Recovery Behavior of Machined Al_2O_3/SiC Nano-Composite Ceramics by Crack-Healing, *J. Eur. Ceram Soc.*, **27**, 3261-7 (2007).

[4] W. Nakao, S. Mori, J. Nakamura, K. Takahashi, K. Ando and M. Yokouchi, Self-Crack-Healing Behavior of Mullite/SiC Particle/SiC Whisker Multi-Composites and Potential Use for Ceramic Springs, *J. Am. Ceram. Soc.*, **79**, 1352-7 (2006).
[5] F. F. Lange and T. K. Gupta, Crack Healing by Heat Treatment, *J. Am. Ceram. Soc.*, **53**, 54-55 (1970).
[6] F. F. Lange and K. C. Radford, Healing of Surface Cracks in Polycrystalline Al_2O_3, *J. Am. Ceram. Soc.*, **53**, 420-21 (1970).
[7] T. K. Gupta, Crack Healing and Strengthening of Thermally Shocked Alumina, *J. Am. Ceram. Soc.*, **59**, 259-262 (1976).
[8] J. Zhao, L. C. Stearns, M. P. harmer, H. M. Chan., G. A. Miller and R. F. Cook, Mechanical Behavior of Al_2O_3-SiC 'Nanocomposite', *J. Am. Ceram. Soc.*, **76**, 503-10 (1993).
[9] I. A. Chu, H. M. Chan and M. P. Harmer, Effect of Annealing Environment on the Crack Healing and Mechanical Behavior of Silicon Carbide-Reinforced Alumina Nanocomposite, *J. Am. Ceram. Soc.*, **81**, 1203-208 (1998).
[10] A. M. Thompson, H. M. Chan and M. P. Harmer, Crack Healing and Surface Relaxation in Al_2O_3-SiC 'Nanocomposite', *J. Am. Ceram. Soc.*, **78**, 567-571 (1995).
[11] K. Niihara, New Design Concept of Structural Ceramics - Ceramic Nanocomposites, The Chemical Issue of the Ceramic Society of Japan, *J. Ceram. Soc. Jpn.*, **99**, 974-82 (1991).
[12] M. C. Chu, S. Sato, Y. Kobayashi and K. Ando, Damage Healing and Strengthening Behavior in Intelligent Mullite/ SiC Ceramics, *Fatigue Fract. Engng. Mater. Struct.*, **18**, 1019-29 (1995).
[13] K. Ando, T. Ikeda, S. Sato, F. Yao, and Y. Kobayasi, A Preliminary Study on Crack healing Behavior of Si_3N_4/SiC Composite Ceramics, *Fatigue Fract. Engng. Mater. Struct.*, **21**, 119-22 (1998).
[14] K. Ando, B. S. Kim, M. C. Chu, S. Saito and K. Takahashi, Crack-Healing and mechanical Behavior of Al_2O_3/SiC composites at elevated temperature, *Fatigue Fract. Engng. Mater. Struct.*, **27**, 533-41 (2004).
[15] H. Z. WU, S. G. Roberts and B. Derby, The strength of Al_2O_3/SiC Nanocomposite after Grinding and Annealing, *Acta Mater.*, **46**, 3839-48 (1998).
[16] I. A. Chou H. M. Chan, M. P. Harmer, Effect of Annealing Environment on the Crack Healing and Mechanical Behavior of Silicon Carbide-Reinforced Alumina Nanocomposites, *J. Am. Ceram. Soc.*, **81**, 1203-1208 (1998).
[17] K. Ando, M. Yokouchi, S. K. Lee K. Takahashi, W. Nakao and H. Suenaga, Crack-healing Behavior, High Temperature Strength and Fracture Toughness of Alumina Reinforced by SiC Whiskers, *J. Soc. Mat. Sci. Jpn.*, **53**, 599-606 (2004).
[18] W. Nakao, M. Ono, S. K. Lee, K. Takahashi and K. Ando, Critical Crack-healing Condition under Stress of SiC Whisker Reinforced Alumina, *J. Eur. Ceram. Soc.*, **25**, 3649-55 (2005).
[19] W. Nakao, K. Takahashi and K. Ando, Threshold Stress During Crack-Healing Treatment of Structural Ceramics Having the Crack-Healing Ability, Materials Letters, **61**, 2711-13 (2007).
[20] K. Ando, M. C. Chu, M. Matsushita and S. Sato, Effect of Crack-healing and Proof-testing procedures on fatigue strength and reliability of Si_3N_4/SiC Components, *J. Eur. Ceram. Soc.*, **23**, 977-84 (2003)
[21] K. Takahashi, K. Ando, H. Murase, S. Nakayama and S. Saito, Threshold Stress for Crack-Healing of Si_3N_4/SiC and Resultant Cyclic Fatigue Strength at the Healing Temperature, *J. Am. Ceram. Soc.*, **88**, 645-51 (2005).
[22] C. Wagner, Passivity during the oxidation of Silicon at Elevated Temperatures, *J. Appl. Phys.*, **29**, 1295 (1958).

[23] J. W. Hinze and H. C. Graham, The Active Oxidation of Si and SiC in the Viscous Gas-Flow Regime, *J. Electrochem.Soc.*, **123**, 1066-73 (1976).
[24] T. Narushima, T. Goto, T. Hirai and Y. Iguchi, High-Temperature Oxidation of Silicon Carbide and Silicon Nitride, *Mater. Trans.*, JIM, **38**, 821-835 (1997).
[25] N. S. Jacobson, Corrosion of Silicon-Based Ceramics in Combusion Enviroments, *J. Am. Ceram. Soc.*, **76**, 3-28 (1993).
[26] H. E. Kim, A. J. Moorhead, Oxidation Behavior and Effects of Oxidation on the Strength of SiC-Whisker Reinforced Alumina, *J. Mater. Sci.*, **29**, 1656-61 (1994).
[27] Y. S. Jung, W. Nakao, K. Takahashi, K. Ando and S. Saito, Crack-Healing Behavior of Si_3N_4/SiC Composite under Low Oxygen Partial Pressure, *J. Soc. Mat. Sci., Jpn.* (in press)
[28] Japan Industrial Standard R1601, Testing Method for Flexural Strength of High Performance Ceramics, Japan Standard Association, Tokyo (1993).
[29] T. Narusima, T. Goto and T. Hirai, High-Temperature Passive Oxidation of Chemically Vapor Deposited Silicon Carbide, *J. Am. Ceram. Soc.*, **72**, 1386-90 (1989).
[30] W. Hinze, W. C. Tripp and H. C. Graham, The High-Temperature Oxidation of Hot-Pressed Silicon Carbide, pp. 409-19 in Mass Transport Phenomena inCeramic. Edited by A. R. Cooper and A. H. Heuer. Plenum, New York (1975).

SELECTION OF A TOUGHENED MULLITE FOR A MINIATURE GAS TURBINE ENGINE

Barry A. Bender and Ming-Jen Pan
Naval Research Laboratory
Code 6351
Washington, DC 20375

ABSTRACT

A small 4 hp recuperated gas turbine engine is being developed at the Naval Research Laboratory. The high turbine inlet temperature of 1225°C on a small-sized engine along with a design life of 1000 hours places stringent conditions on what materials can be used. Mullite appears to be an excellent candidate due to its low density, low thermal conductivity, relatively low coefficient of thermal expansion, and decent thermal shock resistance. However, mullite has low fracture toughness. To improve the toughness six different zirconia-toughened mullite systems were tested using commercially available powders. Two pressureless sintering processes were explored: *in-situ* reaction sintering between zircon, alumina and various dopants and conventional solid state reaction sintering of mullite with various zirconia powders. Dopants included ceria, yttria, magnesia, and dysprosia. Fracture toughness was measured by an indentation method. From these measurements, X-ray diffraction results, and density measurements two systems were downselected for further study and optimization: mullite doped with 18 vol% yttria-doped ZrO_2 or doped with 18 vol% ceria-doped ZrO_2. Detailed microstructural and mechanical characterization of these two systems are presented along with the fracture toughness and X-ray diffraction characterization of the other unselected systems.

INTRODUCTION

A fundamental focus of the Navy's UAV (unmanned aerial vehicle) program is improved pervasive reconnaissance capabilities. The program would like to develop an UAV with increased range and endurance, which is capable of high-altitude flight with the ability to carry and operate a high power payload. One approach to achieve this goal is to develop a smaller UAV powered by a small (< 10 hp) fuel engine. However, this engine would have to work efficiently using Navy JP5/JP8 fuel which is safer than standard fuel due to its higher flash point. At the moment there are no commercial-off-the-shelf small engines that would meet the needed requirements. Small piston engines running on JP8 fuel are unreliable as they are hard to start, knock and show efficiencies <15%. There are small simple cycle gas turbines that run on heavy fuel but typically they have efficiencies of <5% and a short engine life.

What is needed is an engine with an engine life of 1000 hours that runs on heavy fuels with an efficiency approaching 30%. Researchers at NRL have developed a proprietary design for such an engine. It is a miniature 4 hp recuperated gas turbine engine designed to run on JP5/JP8 fuel. Its higher projected efficiency of 30% is a result of the engine being engineered to run at a temperature of 1225°C. To run at this high of temperature places stringent conditions on the material requirements of the engine material. The perfect candidate material for the heat exchanger and possibly the stator blades would require a material with a low creep rate, high tensile strength of 200 MPa at 1225°C, a toughness of 5 MPa-$m^{1/2}$, a low thermal conductivity, a low coefficient of thermal expansion of less than 7×10^{-6}/°C, excellent thermal shock resistance and excellent thermal stability. Ideally the material would be able to be fabricated to a high relative density (limits gas permeability problems) using inexpensive commercially-available starting materials.

One candidate material that meets most of these material requirements is mullite ($3Al_2O_3$-$2SiO_2$). Commercial mullites have thermal conductivities that are 5 to 6 times less than that of silicon nitrides which eliminates Si_3N_4 as a candidate material because low thermal conductivity is crucial to the design of the miniature turbine. Mullite also exhibits good high temperatures properties as it is refractory (melting point as high as 1905°C), and exhibits both good thermal shock and creep resistance.[1-3] It also has excellent intrinsic thermal stability under oxidizing conditions and shows little degradation of strength with temperatures as high as 1500°C.[2,4] Mullite is inexpensive and has a density of 3.17 g/cm^3 which is about one third the densities of super alloys.

However, there are two major drawbacks of mullite- its poor sinterability and weak mechanical properties. Commercial undoped stoichiometric mullite powders have to be sintered at high temperatures of 1650°C or higher.[1,2] While its room temperature strength of 200 MPa meets design requirements its fracture toughness of 2.2 MPa-m$^{1/2}$ is deficient.[5] Researchers have overcome these obstacles through various techniques such as sol-gel processing, hot-pressing, and additives such as SiC whiskers. However, the design requirements for fabrication of the miniature turbine require pressureless sintering of commercial powders without the complications of using SiC whiskers. Researchers have shown that dense mullite composites can be sintered at temperatures as low as 1450°C through the additions of doped-zirconia or via an *in-situ* reaction between zircon and alumina. Both pathways yield mullite-zirconia composites with improved toughness and strength.[6,7] The purpose of this research was to screen six different possible mullite-zirconia composite systems where previous researchers had shown the potential for the composite materials to meet the design requirements of using commercial starting powders and pressureless sintering. Screening was done via indentation fracture toughness testing, density measurements, and phase analysis using X-ray diffraction (XRD) and scanning electron microscopy (SEM). From this initial survey two systems were selected for further study which included sintering billets for test bars to be used in single-edge notch beam (SENB) fracture toughness testing and four point bend testing.

EXPERIMENTAL PROCEDURE

The test samples were prepared using two different techniques- conventional solid state reaction processing of mullite and zirconia powders or *in-situ* reaction processing of zircon and alumina powder to form mullite dispersed with zirconia particles. Six different systems were tested initially (see Results and Discussion section for synthesis details for each system). For each system the starting powders were mixed into a purified water solution containing a dispersant (Tamol 901) and a surfactant (Triton CF-10). To improve the reactivity of the powders the slurries were attrition-milled (1 h) instead of being ball-milled[8] follow by drying the slurries at 90°C. A 2% PVA binder solution was mixed with the powders and they were sieved to eliminate large agglomerates. The dried powder was uniaxially pressed into discs typically 13 mm in diameter and 1 mm in thickness. The discs were then placed on porous mullite foam setter plates and sintered between 1450 to 1615°C for times varying from 0.5 to 6 h.

A fine particle size (0.7 μm) stoichiometric (71.83% alumina) mullite powder (KM101, KCM Corp., Nagoya, Japan) was used for conventional sintering along with doped-zirconia powders. The two zirconia powders were zirconia partially-stabilized with 3 mol% yttria (3YSZ-TZ-3Y-E, 99.7%, Tosoh USA Inc., Grove City, OH) or 10 mol% ceria (Zircar Inc., Florida, New York). A fine particle size (0.4 μm) alumina powder (A16 SG, 99.8%, Almatis, New Milford, CT) and zircon powder (<2 μm, Alfa Aesar, Ward Hill, MA) powder were the main constituents used for *in-situ* reaction processing along with various dopants. These dopants were MgO (99.5%, Cerac Inc., Milwaukee, WI), CeO_2 (99.9%, Alfa Aesar, Ward Hill, MA) and Dy_2O_3 (99.9%, Alfa Aesar, Ward Hill, MA).

Material characterization was done on the discs after processing. X-ray diffraction scans were made from both as-fired and polished surfaces using monochromated CuKα radiation. The densities of the fired-specimens were measured by the Archimedes method using water. Fracture toughness was determined using an indentation microcrack technique[9] assuming an elastic modulus of 200 GPa which has been measured for a mullite dispersed with 20 vol% zirconia.[5] Microstructures were observed by scanning electron microscopy (SEM) on fracture surfaces, polished surfaces, and thermally-etched (1450°C– 5 h) polished surfaces.

Results of the initial testing led to a selection of two systems for further mechanical characterization. Test specimens were machined from sintered billets that were 50 x 50 x 5 mm in size. Flexural strength was done using test bars that were nominally 3.4 mm x 1.7 mm in cross section and 35 mm in length. The bars were ground flat with a 20 μm grinding wheel and beveled. Four to five bars were broken for each system using an outer span of 19 mm and an inner span of 9.5 mm with a crosshead speed of 0.25 mm/min. Toughness was measured using the single-edge notch-beam (SENB) test with bars that were nominally 6.5 mm by 2.8 mm in cross section. Three bars were broken for each composite using the same loading configuration. The notch depth was a constant percentage (28%) of the specimen's thickness.

RESULTS AND DISCUSSION

Initial Screening

Claussen and Jahn[10] were the first researchers to show that the mechanical properties of mullite could be improved by the dispersion of zirconia particles. Fracture toughness improved from 2.0 to 4.5 MPa-m½ and the strength doubled to 400 MPa. They were able to fabricate the dense (>98% theoretical density (T.D.)) mullite-zirconia composite at relatively low sintering temperatures of 1575-1600 °C by the *in-situ* reaction of alumina and zircon to form mullite with approximately 25 vol% zirconia:

$$2\ ZrSiO_4 + 3\ Al_2O_3 \rightarrow 3Al_2O_3 \cdot 2SiO_2\ (\text{mullite}) + 2\ ZrO_2 \qquad (1)$$

They found that the key to obtaining dense mullite at a relatively low sintering temperature was to hold their compacts during their sintering runs at a temperature between 1400 and 1440°C for 2 h before increasing the temperature to 1575 to 1600°C for 2 h. During the 1400°C hold a dense compact of alumina and zircon grains is formed. With increasing temperature a subsolidus reaction occurs between the alumina and zircon to form an amorphous mullite with zirconia particles dispersed throughout and at higher temperatures the amorphous mullite transforms to crystalline mullite.

Due to the above excellent results, similar processing was done using the commercial powders that we had obtained. The cold-pressed samples (AZ) of alumina and zircon were held at 1420°C for 2 hours followed by final sintering at 1600°C for 1 h. The results were not encouraging (see Table I). XRD indicated that the reaction was incomplete as a large amount of zircon was detected. Also the compact didn't sinter well showing an estimated relative density below 90%. Increasing the final sintering conditions to 1615°C with a hold of 4 h did not lead to any significant improvements.

It was thought that perhaps the reactivity of the commercial zircon powder was poor due to its particle size of around 2 microns. So nano-sized zircon powder (particle size <0.1 μm) was tried using the same sintering schedules. The amount of pellet shrinkage of these samples (AZN) improved by 30% (see Table I) but XRD indicated that the reaction was incomplete and in this case alumina was detected along with zircon. Due to the presence of these unwanted secondary phases no toughness testing were done on these samples and the AZ and AZN samples were not downselected for testing.

To improve the reactivity of zircon and alumina reaction-sintering additives have been used. Pena *et al.*[11] found that the presence of MgO decreases the temperature of dissociation of zircon leading to an increase in the reaction of formation at a lower temperature. Orange *et al.*[4] used 1.8 w% MgO as an additive in their reaction-sintering mullite work. This allowed them to fabricate fine-grained mullite-zirconia samples at a lower temperature of 1500°C (0.5 h hold). The mullite composites showed desirable mechanical properties with a strength of 270 MPa and a fracture toughness of 4.6 MPa-m½. Our MgO-doped samples (1.8% MgO- AZM) were fabricated under similar conditions. XRD (see table I) indicated again that the reaction was incomplete as the presence of zircon was detected. This was verified by SEM that showed (see Fig. 1a) zircon surrounding grains of zirconia. However, the amount of zircon is much less as compared to the AZ and AZN samples. Also the sample shrunk 15% more than the AZN disc. Due to the presence of zircon no indentation fracture toughness was done and the AZM system was not selected for further study.

Table I. Screening Data Results

Starting Materials	Processing Parameters	XRD Results	K_{ic}/Sintering (sintered disc diameter)
Alumina + Zircon: (AZM)	1600°C- 1 h 1615°C- 4 h	Zircon present	Sintered very poorly (11.7 mm)
Alumina + Zircon (nano-sized): (AZN)	1600°C- 1 h 1615°C- 4 h	Zircon and alumina present	Sintered poorly (11.2 mm)
Alumina + Zircon + 1.8% MgO: (AZM)	1500°C- 0.5 h	Zircon present	Pockets of porosity (10.9 mm)
Alumina + Zircon + 2.5% Dy_2O_3: (AZD)	1550°C- 2 h	Only mullite and zirconia	Sintered well- 3.2 MPa-m½
Alumina + Zircon + 8% CeO_2: (AZC)	1500°C- 4 h	Only mullite and zirconia	Sintered well but had liquid phase formation
Mullite + 18% Zirconia (3% Y_2O_3): M3YSZ	1550°C- 6 h	Only mullite and zirconia	Sintered well- 4.0 MPa-m½
Mullite + 18% Zirconia (10% CeO_2): M10CeSZ	1500°C- 4 h	Only mullite and zirconia	Sintered well- 4.8 MPa-m½

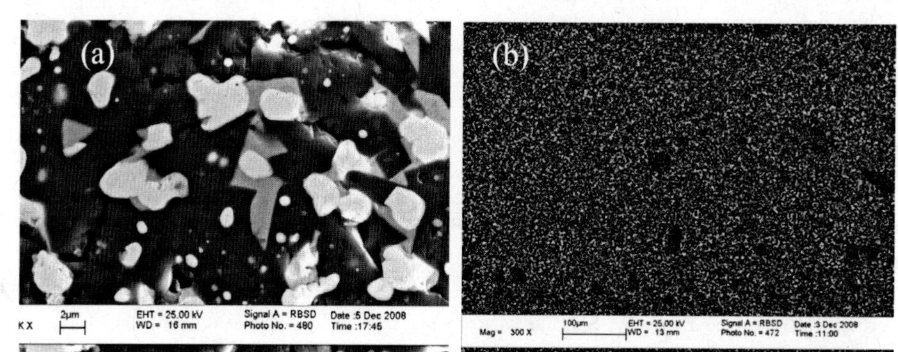

Figure 1. SEM micrographs of polished surfaces of (a) AZM highlighting the presence of zircon (light gray phase) and (b) AZD showing the excellent dispersion of zirconia throughout the mullite.

Reaction-sintering additives can also be used to increase reaction sintering via the formation of a liquid phase. One such additive is dysprosia which has been shown to react with zircon and alumina to form a liquid phase at a temperature of 1375°C which enhances the *in-situ* reaction process.[12] Das and Banderjee[12] found that by adding 2.5 mol% Dy_2O_3 to alumina and zircon that they could fabricate fine-grained dense mullite-zirconia samples at a temperature of 1550°C (2 h hold). The resultant composites had excellent mechanical properties with a flexural strength of 328 MPa and a toughness of 5.03 MPa-m½. Interestingly, the thermal expansion coefficient dropped from 4.87 for pure mullite to 3.83 for the mullite-zirconia ceramic.[12] These are excellent properties so the sintering conditions were duplicated for our AZD samples doped with 2.5 mol% Dy_2O_3. The XRD results (Table I) showed that complete reaction was occurring as only mullite and zirconia were detected. Density measurements indicated that densities approaching 99% T.D. were obtained which was confirmed by electron microscopy which showed the presence of very little porosity (see Fig. 1b). Fig.1b also shows that the dysprosia-doped *in-situ* reaction leads to a fine dispersion of the zirconia particles (white phase) in the mullite matrix. Indentation fracture toughness was undertaken and showed that AZD had a fracture toughness of 3.2 MPa-m½. This was lower than expected and further evaluation was not undertaken.

Reaction-sintering additives can also be used in the alumina-zircon system to enhance the reaction of zircon and alumina to directly form mullite and zirconia. In this case Wu and Lin[7] showed that with increasing amounts of ceria that it slowed down the decomposition of zircon and increased the direct *in-situ* reaction rate of alumina and zircon to form mullite and zirconia. They also showed that some liquid phase formation was occurring and that the ceria was also going into solid solution with the dispersed zirconia particles. This increased the amount of retained tetragonal zirconia which can increase the fracture toughness of the ceramic via stress-induced transformation toughening. They found that by doping with 8 mol% CeO_2 that they could fabricate fully dense fine-grain mullite ceramics via sintering for 4h at 1500°C. Though the strength of the mullite composites was weak (125 MPa) they had a desirable fracture toughness of 5.5 MPa-m½ or higher. AZC samples doped with 8 mol% CeO_2 were sintered using similar parameters. XRD showed that the samples fully reacted and showed no signs of a secondary phase (see Table I). Density measurements showed that samples were almost fully dense (99% T.D.) which was confirmed by SEM characterization. However, small pools of a solidified liquid phase were observed on the surfaces of the samples. Sintering at 1450°C with smaller amounts of ceria did not solve the problem. As a result, fracture toughness testing was not carried out on the AZC samples. However, due to the complete reaction of the AZC system, its high density, and the high fracture toughness values measured by Wu and Lin[7] the ceria-zirconia-mullite system was selected for further evaluation.

An alternative fabrication route for pressureless sintering of mullite-zirconia composites is conventional solid state sintering of mullite and zirconia. One advantage of this process is that the volume fraction of zirconia can be varied which could be important in optimizing both the mechanical and thermal properties of the composites. Prochazka *et al.*[8] showed that the presence of zirconia improved the rate of densification of the mullite, which was confirmed by Osendi.[13] Ishitsuka *et al.*[6] then used partially stabilized zirconia to increase the toughness of the dispersed zirconia powders. Using 20 vol% 3YSZ powder they reacted it with mullite at 1450°C for 10 h to sinter a composite with a relative density of 98%. The sample had excellent mechanical properties with a toughness of 5 MPa-m½ and a strength of 225 MPa. Mullite composites with 18 vol% zirconia (M3YSZ) were fabricated under similar conditions. However, very little densification occurred as the samples shrunk less than 60% of the amount fully-densified samples shrunk. The sintering temperature was increased to 1500°C and the sintering time reduced to 6 h. This resulted in samples that were 97% of T.D. XRD confirmed the presence of only zirconia and mullite. Fracture indentation testing was done and the M3YSZ (see Table I) samples had a toughness of 4.1 MPa-m½. Increasing the sintering temperature to 1550°C improved the T.D. of the M3YSZ samples to 99%. Fracture toughness of the 1550°C sample was measured to be 4.0 MPa-m½. As a result of these good properties the M3YSZ system was downselected for further testing.

Downselection Studies

Mullite-3YSZ

The M3YSZ samples sintered at 1550°C were characterized by SEM. Microscopy of polished surfaces revealed very little porosity present but showed the present of large (10 to 50 μm) ball-like agglomerates (see Fig. 2a). EDS revealed that the balls were zirconia-rich. It was discovered that the Tosoh 3YSZ zirconia powders are spray-dried which creates ball-like agglomerates of up to 55 microns in size with an actual crystallite size of 30 to 100 nm.[14,15] Attrition milling mullite and the 3YSZ powder together was not efficiently breaking up the spray-dried agglomerates. To solve this problem the 3YSZ powder was attrition-milled by itself. Fig. 2b shows a polished M3YSZ sample fabricated using the attrition-milled zirconia powder. As compared to Fig. 2a the presence of the zirconia-rich agglomerates was drastically reduced.

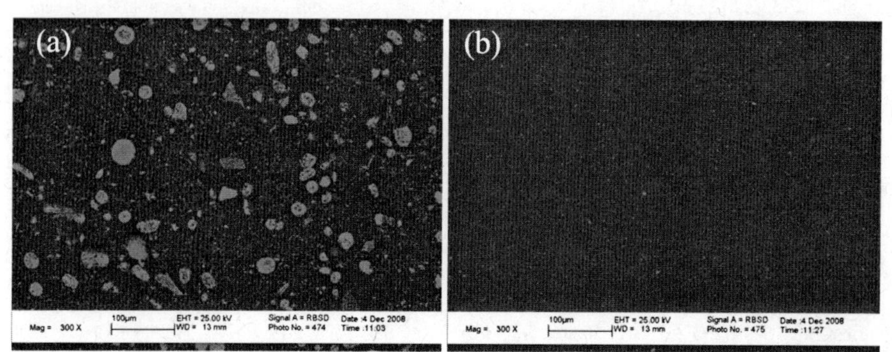

Figure 2. SEM micrographs of polished surfaces of M3YSZ samples showing the presence of zirconia agglomerates (Fig. a) when the 3YSZ powder was not attrition-milled by itself as it was in Fig. b.

Using this improved powder a billet of mullite-18 vol% 3YSZ was pressed and sintered for 6 h at 1550°C. Mechanical testing of bend bars cut from the billet showed that the mullite-zirconia composite had an average strength of 270 MPa and a SENB-measured fracture toughness of 3.1 MPa-m½. Material characterization was undertaken to understand why the measured fracture toughness was less than expected. Density measurements showed only a 96% T.D. Low cold-pressing pressures may have contributed to the reduction in porosity. SEM of the polished surfaces show an excellent dispersion of the zirconia in the mullite (see Fig. 3a) Thermal etching (see Fig. 3b) revealed that the zirconia particles were typically 1 to 1.5 microns in size and were found primarily between grains of mullite. Very little intergranular zirconia was observed and when it was it was typically to be observed as round particles in the interior of the mullite grains that were 0.2 to 0.5 microns in diameter. The mullite grains were showing signs of forming elongated grains with the typical cross section of the grain being one micron in size. Fractography of three different samples indicated all the fracture origins were associated with agglomerates 20 to 50 microns in size.

To improve the mechanical properties of the M3YSZ samples several different steps can be taken. First of all, the 3YSZ powder can be attrition-milled better perhaps using alcohol instead of water.[8] Higher cold-pressing pressures should lead to better densification and improved mechanical properties. Also optimizing the zirconia particle size may improve toughness. The as-received 3YSZ powder had a monoclinic zirconia content of about 10% (as determined by the Garvie and Nicholson method[16]). XRD of an as-fired surface of a bend bar showed that sintering did not change the amount of monoclinic zirconia present. Polishing the surface of the bend bar lead to only a small increase in the amount of monoclinic present (18%). This data is implying that the critical size of the zirconia for stress-induced transformation is larger than the average zirconia particle size in the M3YSZ composites. With the proper heat treatment the average zirconia particle size could be increased so that under stress the particles would have a greater propensity to transform to tetragonal zirconia which would enhance the toughness of the composites via stress-induced transformation toughening.

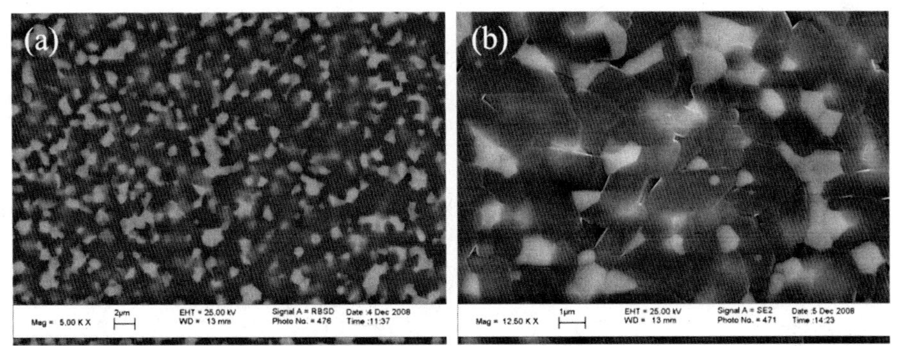

Figure 3. SEM micrographs of thermally-etched surfaces of M3YSZ showing (a) the good dispersion of zirconia particles (white phase) which are located (b) between grains of mullite (gray phase).

Mullite-CeSZ

The mullite-ceria-zirconia system was selected for further evaluation due to the excellent densification of the MCeSZ samples and their potential of high fracture toughness. However, the formation of an unknown Ce-Zr-Si liquid phase had to be prevented. To avoid the possibility of forming such a phase (Ce_2O_3-SiO_2-ZrO_2 ternaries are used for liquid-phase sintering of Si_3N_4)[17] it was decided to tie up the ceria with the zirconia by using a 10 mol% ceria-stabilized zirconia powder. Another benefit of using partially stabilized 10CeSZ powder is that these zirconia ceramics exhibit fracture toughness values as high as 18 MPa-m½.[18] Test samples made with 18 vol% 10CeSZ powder (M10CeSZ) were sintered at 1500°C for 4 h. No liquid phase was observed on the surface and XRD scans indicated only the presence of mullite and zirconia (see Table 1). SEM fractographs (see Fig. 4a) showed that the ceramic fractured transgranularly which is indicative of ceramics with glass-free grain boundaries. Samples sintered to 98% T.D. and had a fracture toughness of 4.8 MPa-m½.

Because of these excellent results a billet of M10CeSZ was fabricated using the identical sintering parameters. Mechanical testing of bend bars cut from the billet showed that the mullite-zirconia composite had an average strength of 340 MPa and a SENB-measured fracture toughness of 4.7 MPa-m½. Material characterization was undertaken to understand why the mechanical properties were better than the M3YSZ mullite-zirconia composites. Density measurements indicated that the billet sintered well achieving a relative theoretical density of 98% which was better than the M3YSZ sample. Fractography did not detect a consistent pattern for failure as most of the bars failed at large processing pores or due to the presence of machining cracks. Thermally-etched polished surfaces showed that the zirconia was well dispersed between mullite grains and had a typical grain size of 1-2 microns (see Fig. 4b). Typically, no particles of zirconia were found within the mullite grains. The cross section of the mullite grains were similar in size to those observed in the M3YSZ samples. However, the elongation of the mullite grains was more pronounced in the M10CeSZ samples. This could make the mullite matrix itself tougher which enhances the toughness benefits of the zirconia particles.[5] Also signs of microcracking were observed on polished surfaces of M10CeSZ samples which could contribute to the higher toughness of the material. The microcracking is due to the increased propensity of the zirconia particles to transform from tetragonal to monoclinic. This is evidenced by XRD data of the monoclinic to tetragonal ratio in the starting powder as compared to the as-fired sample as compared to the polished surface of the sample. The as-received 10CeSZ powder is 67% tetragonal. Upon firing the amount of tetragonal phase drops to 58% showing a conversion of some of the tetragonal phase to the monoclinic phase during processing. The amount of tetragonal phase drops further by 50% to 29% showing that there are a significant number of zirconia grains that are of critical size for a ceria-partially stabilized zirconia that allows them to transform from the monoclinic phase to the tetragonal phase under stress. However, the magnitude of the contribution from transformation toughening is unknown. Optimization of the mullite matrix microstructure and size of the dispersed zirconia grains via different sintering parameters could further increase the strength and the toughness of the 10CeSZ mullite-zirconia composites.

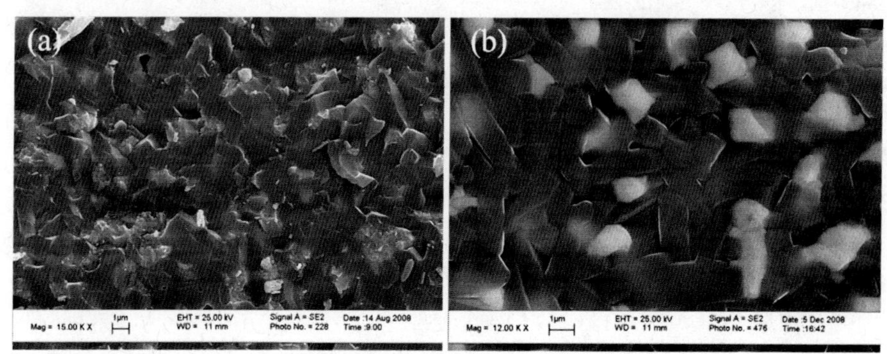

Figure 4. SEM micrographs of 10CeSZ of (a) the fracture surface showing mostly transgranular fracture and (b) the thermally-etched polished surface showing the mullite and zirconia grains.

CONCLUSIONS

Six different mullite-zirconia composite systems were tested to see if they had the potential to meet the stringent design requirements for a miniature 4 hp recuperated gas turbine engine. Design requirements called for the use of commercially available powders and pressureless sintering. Fabrication of the composites was done via conventional solid state reaction sintering of zirconia and mullite or *in-situ* reaction sintering of alumina and zircon. *In-situ* reaction sintering of just alumina and zircon or alumina and zircon doped with magnesia lead to incomplete reaction and porous compacts. Doping with dysprosia resulted in fully-reacted mullite-zirconia composites but with a lower than expected fracture toughness of 3.2 MPa-m$^{1/2}$. Doping with ceria resulted in a very dense fully reacted mullite-zirconia composite but lead to the formation of a liquid phase that formed on the top and bottom surfaces of the composite. However, due to the toughness potential this system was downselected for further evaluation. The other system downselected for further evaluation was the mullite-18 vol% 3YSZ system. Conventional solid state sintering lead to an almost fully-densified composite that contained no unwanted secondary phases with a toughness of 4.0 MPa-m$^{1/2}$.

Further mechanical testing of M3YSZ showed that the composite had a strength of 270 MPa and a toughness of 3.1 MPa-m$^{1/2}$. Fractography indicated that the fracture origins were associated with large agglomerates (20-50 microns in size) which may have contributed to the lower-than-expected measured fracture toughness. Further mechanical testing was carried out on the mullite-18vol% 10CeSZ system. The 10CeSZ samples had an excellent strength of 340 MPa and a fracture toughness of 4.7 MPa-m$^{1/2}$. Optimization of the zirconia particle size and the mullite grain morphology of both systems could enhance the contributions of crack deflection toughening and stress-induced transformation toughening leading to a composite with a toughness > 5 MPa-m$^{1/2}$. A higher toughness would improve the mechanical integrity and reliability of the designed high temperature miniature gas turbine engine.

REFERENCES

[1]S. Prochazka and F.J. Kug, Infrared-Transparent Mullite Ceramic, *J. Am. Ceram. Soc.*, **66**, 874-80 (1983).
[2]I.A. Askay, D.M. Dabbs, and M. Sarikaya, Mullite for Structural, Electronic, and Optical Applications, *J. Am. Ceram. Soc.*, **74**, 2343-58 (1991).
[3]P.A. Lessing, R.S. Gordon, and K.S. Mazdiyasni, Creep of Polycrystalline Mullite, *J. Am. Ceram. Soc.*, **58**, 149 (1975).
[4]G. Orange, G. Fantozzi, F. Cambier, C. Leblud, M.R. Anseau, and A. Leriche, High Temperature Mechanical Properties of Reaction-Sintered Mullite/Zirconia and Mullite/Alumina/Zirconia Composites, *J. Mat Sci.*, **20**, 2533-40 (1985).
[5]P.F. Becher and T.N. Tiegs, Toughening Behavior Involving Multiple Mechanisms: Whisker Reinforcement and Zirconia Toughening, *J. Am. Ceram. Soc.*, **70**, 651-54 (1987).
[6]M. Ishitsuka,T. Sato, T. Endo, and M. Shimada, Sintering and Mechanical Properties of Yttria-Doped Tetragonal ZrO_2 Polycrystal/Mullite Composites, *J. Am. Ceram. Soc.*, **70**, C-342-C-346 (1987).
[7] J.-M. Wu and C.-M. Lin, Effect of CeO_2 on Reaction-Sintered Mullite-ZrO_2 Ceramics, *J. Mater. Sci.*, **26**, 4631-36 (1991).
[8]S. Prochazka, J.S. Wallace, and N. Claussen, Microstructure of Sintered Mullite-Zirconia Composites, *J. Am. Ceram. Soc.*, **66**, C-125-C-127 (1983).
[9]G.R. Anstis, P. Chantikul, B.R. Lawn, and D.B. Marshall, A Critical Evaluation of Indentation Techniques for Measuring Fracture Toughness: I. Direct Crack Mechanisms, *J. Am. Ceram. Soc.*, **64**, 553-58 (1981).
[10]N. Claussen and J. Jahn, Mechanical Properties of Sintered, In Situ-Reacted Mullite-Zirconia Composites, *J. Am. Ceram. Soc.*, **63**, 228-29 (1980).
[11]P. Pena, J.S. Moya, S. Aza, E. Cardinal, F. Cambier, C. Leblud, and M.R. Anseau, Effect of Magnesia Additions on the Reaction Sintering of Zircon/Alumina Mixtures to Produce Zirconia Toughened Mullite, *J. Mat. Sci. Lett.*, **2**, 772-74 (1983).
[12]K. Das and G. Banerjee, Mechanical Properties and Microstructures of Reaction-Sintered Mullite-ZrO_2 Composites in the Presence of an Additive- Dysprosia, *J. Eur. Ceram. Soc.*, **20**, 153-157 (2000).
[13]J.S. Moya and M.I. Osendi, Effect of ZrO_2 (ss) in Mullite on the Sintering and Mechanical Properties of Mullite/ZrO_2 Composites, *J. Mat. Sci. Lett.*, **2**, 599-601 (1983).
[14]J.A. Brito-Chaparro, A. Aguilar-Elguezabal, J. Echeberria, M.H. Bocanegra-Bernal, Using High-Purity MgO Nanopowder as a Stabilizer in Two Different Particle Size Monoclinic ZrO_2: Its Influence on the Fracture Toughness, *Mater. Chem. Phys.*, in Press.
[15]E. Ozkol, J. Ebert, K. Uibel, A.M. Watjen, and R. Telle, Development of High Solid Content Aqueous 3Y-TZP Suspensions for Direct Inkjet Printing Using a Thermal Inkjet Printer, *J. Eur. Ceram. Soc.*, **29**, 403-09 (2009).
[16]R.C. Garvie and P. Nicholson, Phase Analysis in Zirconia Systems, *J. Am. Ceram. Soc.*, **55**, 303-05 (1972).
[17]S. Zec, S. Boskovic, M. Hrovat, and M. Kosec, Contribution to Phase Equilibria in the Ce_2O_3 Rich Part of the Ce_2O_3-SiO_2-ZrO_2 System, *J. Eur. Ceram. Soc.*, **27**, 523-26 (2007).
[18]S. Deville, H.E. Attaoui, and J. Chevalier, Atomic Force Microscopy of Transformation Toughening in Ceria-Stabilized Zirconia, *J. Eur. Ceram. Soc.*, **25**, 3089-96 (2005).

Comparison in Foreign Object Damage between SiC/SiC and Oxide/Oxide Ceramic Matrix Composites

Sung R. Choi[†], Donald J. Alexander, and David C. Faucett,
Naval Air Systems Command, Patuxent River, MD 20670

ABSTRACT

Foreign object damage (FOD) of two different ceramic matrix composites (CMCs), melt-infiltrated (MI) SiC/SiC and oxide/oxide, was assessed via impact testing at ambient temperature using impact velocities ranging from 100 to 400 m/s by 1.59mm-diameter steel ball projectiles. The extent of impact damage as well as post-impact strength degradation of both CMCs increased with increasing impact velocity. The degree of relative post-impact strength degradation of the MI SiC/SiC composite was similar to that of the oxide/oxide counterpart. Both of the CMCs were able to survive a very high impact velocity of 400 m/s, without complete structural failure, which is in a notable contrast with gas-turbine grade monolithic silicon nitrides. The damage of the oxide/oxide was characterized with its unique compaction/densification of the material just beneath the impact sites, attributed to the material's soft and open structure. Prediction of quasi-static impact force was made based on the static indentation data and was in reasonable agreement with the experimental data for the oxide/oxide composite.

INTRODUCTION

Monolithic ceramics or ceramic matrix composites, because of their brittle nature, are susceptible to localized surface damage and/or cracking when subjected to impact by foreign objects. It is also true that ceramic components may fail structurally by soft particles when the kinetic energy of impacting objects exceeds certain limits. The latter case has been often encountered in aeroengines in which combustion products, metallic particles or small foreign objects ingested cause severe damage to airfoil components, resulting in serious structural and functional problems.

In the previous studies [1-2], FOD behavior of two representative gas-turbine grade ceramic matrix composites (CMCs) – melt-infiltrated (MI) Sylramic™ SiC/SiC and N720™/aluminosilicate (AS) oxide/oxide – was determined using a rectangular plate target configuration. CMC targets were impacted at their centers by hardened chrome-steel ball projectiles with a diameter of 1.59 mm in a velocity range from 100 to 440 m/s. Unlike monolithic silicon nitrides observed from the previous studies [3-4], the SiC/SiC and the oxide/oxide CMCs exhibited monotonic strength degradation with increasing impact velocity without complete structural-type of failure even at a very high impact velocity of 400 m/s. The degree of FOD, when compared with respect to the types of target specimen support, was significantly greater in partial support than in full support [1-2].

The objective of the current work was to compare FOD behavior between the SiC/SiC and the oxide/oxide CMCs in terms of impact morphologies and strength degradation, based on the previous work [1-2]. In addition, the response to static indentation of the two composites with respect to deformation was also characterized using the same steel balls that were employed in the previous FOD testing. The static indentation results were then used in order to make an attempt of quasi-static prediction of impact force as a function of impact velocity involved in the FOD testing.

[†] Corresponding author; email address:sung.choi1@navy.mil

EXPERIMENTAL PROCEDURES

Target Materials and Test Specimens

The composites used in this study were described in detail elsewhere [1-2]. The MI SiC/SiC composite was fabricated by GE Power System Composites. Briefly, Sylramic™ fibers, produced in tow form by Dow Corning (Midland, MI) were woven into 2-D 5 harness-satin cloth and then converted to Sylramic™ iBN fibers. The Sylramic™ iBN cloth was cut into a pertinent size, which was 0/90°, 8 ply-stacked and chemically vapor infiltrated with a thin BN-based interface coating followed by SiC matrix over-coating. Remaining matrix porosity was filled with SiC particulates and then with molten silicon at 1400°C. The composite was composed of about 34 vol% SiC fibers. The nominal dimensions of each panel thus fabricated were about 230 mm by 150 mm with a thickness of about 2.2 mm.

Nextel™ 720 oxide fibers, produced in tow form by 3M Corp. (Minneapolis, MN), were woven into 2-D 8 harness-satin cloth. The cloth was cut into a proper size, slurry-infiltrated with aluminosilicate (AS), and 0/90°, 12 ply-stacked followed by consolidation and sintering. No interface fiber coating was employed. The fiber volume fraction of the oxide/oxide composite panels thus fabricated (by GE) was about 0.45. Porosity was about 20-25 %.

Rectangular plates of about 8-10 mm in width, about 45-50 mm in length, and about 2.2 mm (for SiC/SiC) or 3.0 mm (for oxide/oxide) in as-furnished thickness were cut from the composite panels for FOD as well as static indentation testing. Table 1 summarizes basic physical and mechanical properties of the two composites.

Table 1. Basic mechanical and physical properties of target and projectile materials at ambient temperature [1-4]

Material		Architecture	Fiber/matrix	Fiber volume fraction	Elastic modulus[1] E (GPa)	Flexure strength[2] (MPa)
Targets	MI SiC/SiC	2-D woven	Sylramic™ SiC/SiC	0.34	220	578±56
	Oxide/oxide	2-D woven	Nextel™ 720 /aluminosilicate	0.45	67	141±4
Projectile	Chrome steel (SAE52100)	-	-	-	200	>900[3]

1. By the impulse excitation technique, ASTM C 1259 [5]; 2. By four-point flexure testing with 20/40 mm spans; 3. Tensile data from the manufacturer.

Foreign Object Damage Testing

Foreign object damage (FOD) testing was carried out using a ballistic impact gun. Detailed descriptions of the impact apparatus have been described elsewhere [1-4]. Briefly, a hardened chrome-steel (SAE52100) ball projectile with a diameter of 1.59 mm was inserted into a 300 mm-long gun barrel with an inner diameter of 1.59 mm. A helium-gas cylinder and relief valves were used to pressurize the reservoir to a specific level, depending on prescribed impact velocity. Upon reaching a specific level of pressure, a solenoid valve was instantaneously opened accelerating the ball projectile

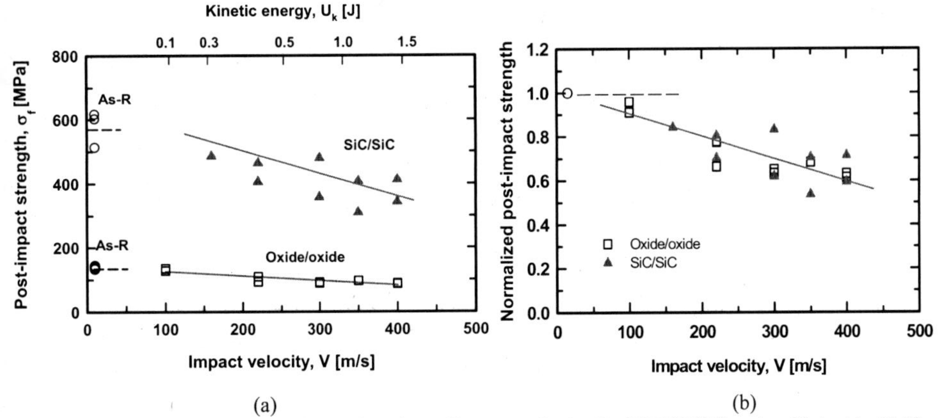

(a) (b)

Figure 1. Post-impact strength as a function of impact velocity for MI SiC/SiC and oxide/oxide CMCs impacted by 1.59mm-diameter steel ball projectiles: (a) Non-normalized; (b) Normalized with respect to as-received ('As-R') strength.

through the gun barrel to impact a target specimen. The target specimen was rigidly supported through a rigid steel block. Each target specimen was aligned such that the ball projectile impacted at the center of the target specimen with a normal incidence angle. Impact velocity of each projectile was determined using two pairs of laser transmitter and receiver. The range of impact velocity employed in this work was from 100 to 400 m/s. A total of 3 to 10 target specimens were used at each velocity for a given material. Some limited properties of the steel ball projectiles are also provided in Table 1.

Strength Testing

Strength testing for impacted target specimens was carried out at ambient temperature to determine the severity of impact damage, using a four-point flexure fixture with 20-mm inner and 40-mm outer spans. Each impacted specimen was loaded in the flexure fixture such that its impact site was subjected to tension within the inner span. An electromechanical test frame (Model 8562, Instron, Canton, MA) was used in displacement control with an actuator speed of 0.5 mm/min.

Static Indentation Testing

The indent sides of composite test specimens were polished with 600 SiC paper to provide a smooth surface for indentation. Indentation testing was performed with an electromechanical test frame using 1.59 mm-diameter, hardened (HRC≥60) chrome-steel ball indenters, the same ball projectiles that were utilized in FOD testing for the composites [1-2]. Indentation loads ranging from 0 to 2940 N were used, with typically a total of five indents for a given indentation load. Indentation load was applied onto the polished sides of the CMC specimens for about 20 s. In-situ indentation depth was determined using an LVDT to obtain load-versus-displacement curves.

RESULTS AND DISCUSSION

Post-Impact Strength

The results of strength testing for impacted target specimens are shown in Fig. 1, where post-impact flexure strength was plotted as a function of impact velocity for both of the CMCs. Included in the figure was their respective as-received ('As-R') flexure strengths. All the specimens impacted did fracture from impact sites. Post-impact strength decreased with increasing impact velocity, regardless of the type of materials, attributed to increased impact damage. Since the as-received strength (=578±56 MPa) of the MI SiC/SiC composite was much greater than that (=141±4 MPa) of the oxide/oxide composite, it was necessary for direct comparison to normalize the post-impact strengths with respect to their as-received strengths. Figure 1(b) compares normalized strength as a function of impact velocity for the two composites. For a given type of specimen support, the degree of strength degradation was almost identical in either the SiC/SiC or the oxide/oxide composite. The strength degradation at 400 m/s, for example, was about 40 % of their as-received strengths of the composites.

It has been observed that monolithic silicon nitride (AS800 and SN282) target specimens, with a flexure bar configuration of 3mm (height) x 4mm (impact side) x 25-50 mm (length), fractured upon impact catastrophically into two pieces when impact velocity was equal to or above their critical velocities of 300 m/s for SN282 and 400 m/s for AS800 [3]. By contrast, the two composites did not exhibit any of such catastrophic failure. Therefore, in this regard, the composites exhibited better FOD resistance over their monolithic counterparts.

Impact Morphology

1) Projectiles

The steel ball projectiles were flattened or severely deformed or fragmented, depending on impact velocity, upon impacting the SiC/SiC composite. By contrast, the projectiles that impacted the oxide/oxide composite were neither flattened nor noticeably deformed even at the highest impact velocity of 400 m/s, as shown in Fig. 2. This was due to the oxide/oxide composite's *soft* and *open* structure, compared to the hard-and-dense SiC/SiC counterpart. Often at higher impact velocities ≥ 300 m/s, steel ball projectiles were embedded into the oxide/oxide composite, similar to the case that 'sharp' brittle particles impacted ductile metal targets.

2) Impact Sites of Target Specimens

The front impact damages generated in target specimens were in the form of indents, craters, or spallation with their size being dependent on impact velocity and material. Figure 3 shows the front impact damage size as function of impact velocity. The SiC/SiC composite showed increased sensitivity of damage size to impact velocity while the oxide/oxide presented a monotonic increase in damage size with increasing impact velocity. The latter was again indicative of the oxide/oxide's softness with which impact size was developed early on at low impact velocities ≤ 200 m/s. However, it should be noted that the front impact-damage size alone does not contribute to post-impact strength since overall impact damage was associated with several different features such as front damage, internal cracking, and/or interlaminar delamination, etc.

The cross-sectional views of impact sites are shown in Fig. 4. It is noted from Fig. 4 that regardless of material, cone cracking was a common feature of impact damages of the two composites, which was also seen various silicon nitrides [1-4,6-8]. Because of its soft (open) structure due to significant porosity, the oxide/oxide material exhibited significant densification/compaction of the material just beneath the impact site. Some analysis on impact damage by optics and SEM has been performed previously [1-2]; however, more detailed analysis on impact damage of the two composites is needed using pertinent methodologies such as computed tomography (CT) and/or Pulsed Thermography techniques [9].

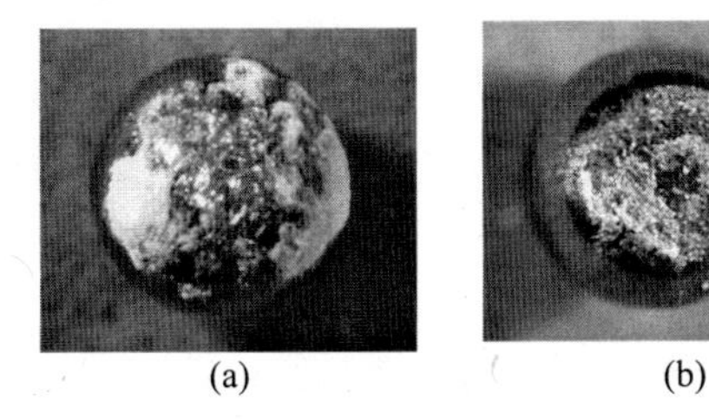

Figure 2. Typical examples of 1.59mm-diameter steel-ball projectiles impacted at a high velocity of 400 m/s on: (a) Oxide/oxide and (b) SiC/SiC composites. Note that some target materials were transferred to the projectiles.

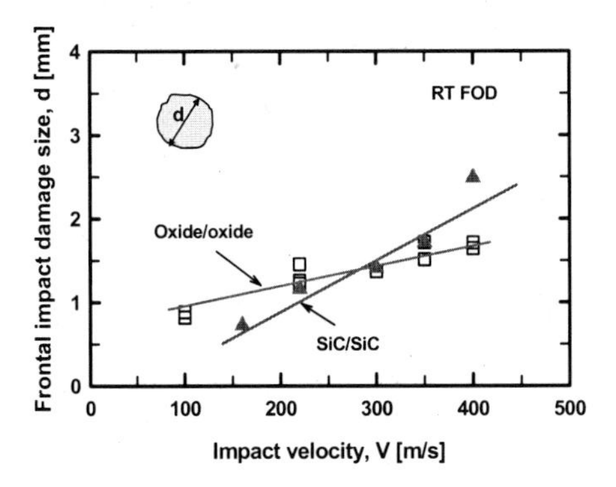

Figure 3. Front impact-damage size as a function of impact velocity for MI SiC/SiC and oxide/oxide composites impacted by 1.59mm-diameter steel ball projectiles.

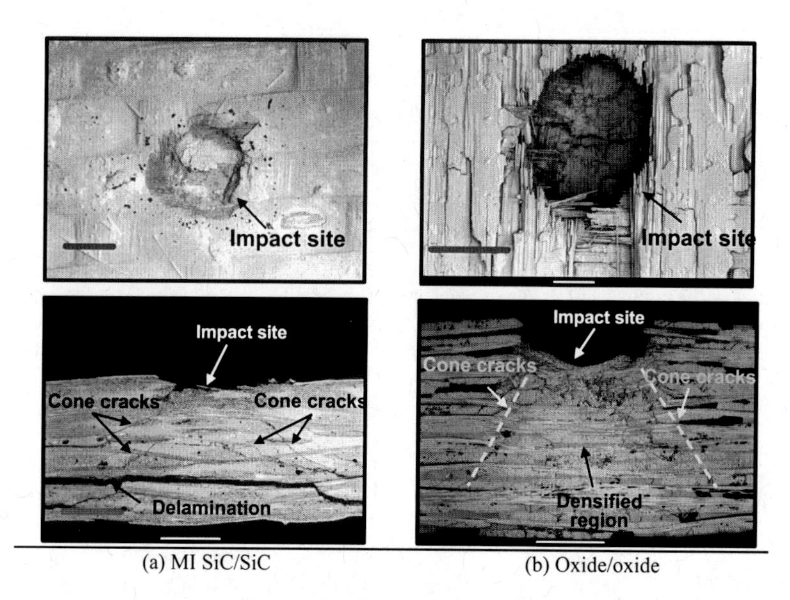

(a) MI SiC/SiC (b) Oxide/oxide

Figure 4. Impact damages showing the impact sites (top) and the cross-sectional views for (a) MI SiC/SiC and (b) oxide/oxide composites impact at 400 m/s by 1.59mm-diameter steel ball projectiles. Cone cracks for both composites and densified (compacted) region beneath the impact site for the oxide/oxide composite are clearly seen. Bar: 1000 μm.

Static Indentation Responses

Indentation response of the oxide/oxide composite was that permanent indentation deformation occurred only in test specimens but not in steel ball indenters, similar to the impact responses as aforementioned, due to significant softness of the oxide/oxide material. Static indentation typically generated permanent plastic deformation on the surfaces of test specimens, conformed to the ball configuration. By contrast, the SiC/SiC composite did not reveal any visible damage on its surface but the ball indenters were subjected to plastic deformation in a form of flattening in their contact areas.

Figure 5 shows typical examples of load-displacement curves during indentation obtained from the two composites. The curves were characterized by permanent deformation (and hysteresis) of the specimens (for the oxide/oxide) and the ball indenters (for the SiC/SiC). The unloading curve for each material was stiffer than the loading curve, due to the densification of the oxide/oxide composite beneath the ball indenter and to the geometrical change (flattening) of the ball indenter in the SiC/SiC composite.

A summary of the contact impression size (d) as a function of indentation load is shown in Fig. 6. Figure 7 shows a relationship between contact impression area ($A=\pi d^2/4$, calculated from the data in Fig. 6) and indentation load (P) determined for each composite. A linearity between P and A was well established for the ball indenters on the SiC/SiC and for the oxide/oxide by the ball indenters, with a coefficient of correlation (r_{coef}) of a linear fit all greater than 0.950. The linearity also implies that macroscopically the average 'contact yield pressure' ($=P/A$) almost remained constant regardless of the magnitude of indentation load. The contact yield pressure, p_y, is simply the inverse of the slope of each curve, so that:

$$p_y = \frac{dP}{dA} = \frac{\Delta P}{\Delta A} \tag{1}$$

The values of p_y were estimated to be:

$p_y = 3800$ MPa (For ball indenters on SiC/SiC)
$p_y = 1270$ MPa (For oxide/oxide by ball indenters)

Prediction of Impact Force

1) Impact Force Estimation

Ballistic impact is a complex dynamic phenomenon. Consequently its analysis is complex and therefore beyond of the scope of this paper. However, a first-order, quasi-static prediction of impact force will be described briefly and made in this paper based on a previous work [10]. Right before and after impact, the overall energy balance may hold [10]:

$$U_k = U_{el} + U_{pl} + U_L + U_{k,bc} \tag{2}$$

where U_k is the kinetic energy of the impacting projectile; U_{el} is the elastic strain energy of projectile and target; U_{pl} is the plastic strain energy of a projectile or a target; and U_L is the energy associated with damage/cracks generated on the target; $U_{k,bc}$ is the kinetic energy of a bouncing-back projectile. For a conservative estimation, assumptions were made such that $U_L \ll U_{pl}$ and $U_{el} \ll U_{pl}$. The resulting equation can be obtained from Eq. (2) as follows [10]:

$$m(1-e^2)V^2/2 = \pi\, p_y (Dz^2/2 - z^3/3) \tag{3}$$

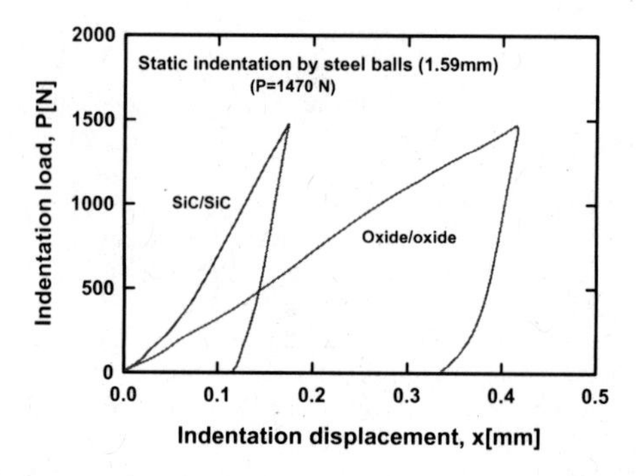

Figure 5. Typical curves of indentation load versus indentation displacement for MI SiC/SiC and oxide/oxide composites indented at 1470 N by 1.59mm-diameter steel balls.

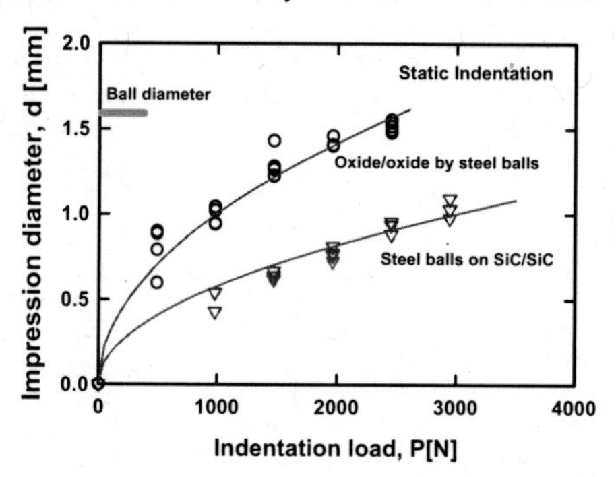

Figure 6. Impression diameter as a function of indentation load for MI SiC/SiC and oxide/oxide composites indented by 1.59mm-diamete steel balls. The lines represent the best fit.

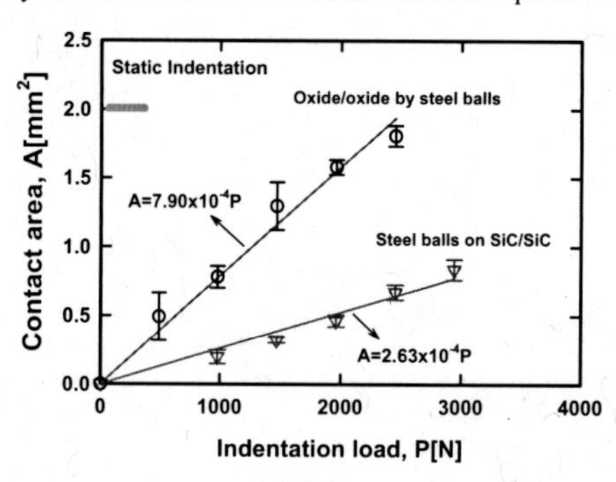

Figure 7. Contact area as a function of indentation load for MI SiC/SiC and oxide/oxide indented by 1.59mm-diameter steel balls. The lines represent the bet fit.

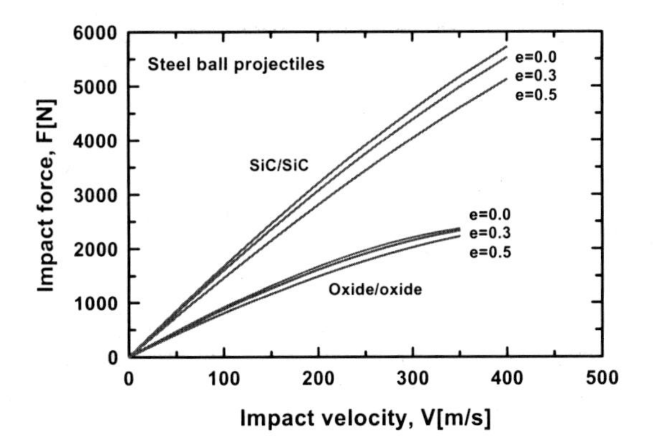

Figure 8. Predicted impact force as a function of impact velocity for different values of the coefficient of restitution (e) for MI SiC/SiC and oxide/oxide composites impacted by 1.59mm-diameter steel ball projectiles.

where m is the mass of the projectile, e is the coefficient of restitution, defined as $e = -V_{bc}/V$ with V_{bc} being the bouncing-back velocity of the projectiles, D is the projectile diameter, and z is the magnitude of plastic deformation of a projectile or a target. The value of z in Eq. (3) can be solved as a function of impact velocity V for a given p_y (Eq. (1)), m, and e. Once z is solved, the impact force F can be calculated using the following equation [10]

$$F = [\pi(Dz - z^2)]p_y \quad (4)$$

The resulting prediction of impact force within a geometrical limit ($D/z>2.0$) is shown in Fig. 8, where three different values of e=0.0, 0.3, and 0.5 were employed for each composite. As seen in the figure, F increases almost linearly with increasing V. For a given V, F depends on the value of e. However, the difference in F among the values of e chosen was insignificant (i.e., <10% between e=0 and 0.5). It should be noted that the impact force determined from the conventional *elastic-elastic* Hertzian contact/impact model [1,2,5,9] may not be appropriate in this case since significant plastic deformation either of projectiles or targets was involved in impact.

2) Verification

Verification of the impact-force analysis thus presented was not feasible since no measurements of impact force have been made. Instead, the size of surface impact damage on either projectiles or targets will be utilized to validate the analysis. Figure 9 shows the predicted impact-damage size (d) as a function of V, determined from Eq. (3) for the two composites. Also included was the FOD data (shown in Fig. 3) on the oxide/oxide composite. Despite its simplicity in analysis, the plastic model agrees very well with the experimental data on the oxide/oxide composite, except some slight deviation at higher velocities of $V\geq400$ m/s. Sensitivity of e to the solution was not significant. The reasonably good agreement is also indicative that the energy loss (U_L) appeared to be trivial, as

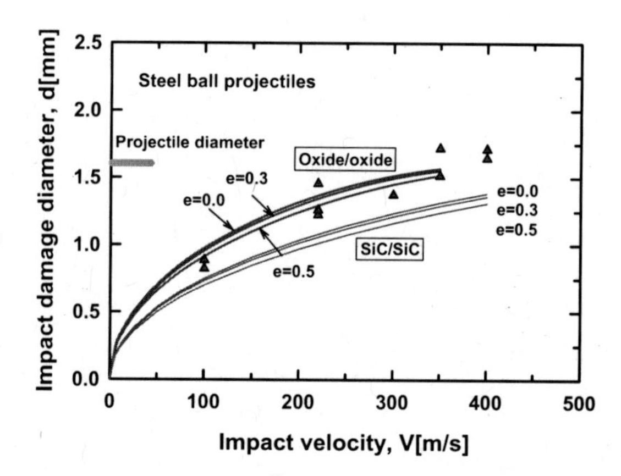

Figure 9. Predicted impact damage size as a function of impact velocity for different values of the coefficient of restitution (e) for MI SiC/SiC and oxide/oxide composites impact by 1.59mm-diameter steel ball projectiles. The experimental data (in triangle symbols) on the oxide/oxide composite from Fig. 3 were included for comparison.

assumed. Unfortunately, verification for the SiC/SiC impact was not feasible due to the unavailability of experimental data on projectiles or targets.

The simplicity of this plastic model is that all the parameter that is required is p_y, which can be determined through routine static indentation testing, as also demonstrated for various ductile projectiles (steels and brass) impacted on silicon nitride targets [10]. However, pertinent experimental techniques should be sought to determine more accurately the related dynamic parameters such as impact force, stresses, deformation, duration of impact, coefficient of restitution, and stress wave propagations, etc. Frictional constraint by property mismatch between projectiles and targets of dissimilar materials needs to be taken into account in some cases.

CONCLUSIONS

1) The overall impact damage of the MI SiC/SiC and the oxide/oxide composites was increased monotonically with increasing impact velocity without a catastrophic type of fracture up to 400 m/s. The degree of relative post-impact strength degradation of the MI SiC/SiC composite was similar to that of the oxide/oxide composite.
2) The impact damage of the steel ball projectiles was negligible in the oxide/oxide due to the composite's soft and open structure, as compared to that in the hard-and-dense SiC/SiC composite.
3) The front contact stresses by steel-ball projectile impact played a major role in generating impact damages such as craters, cone cracking, compaction/densification (for the oxide/oxide), and spallation, etc.

4) The prediction of quasi-static impact force based on the static indentation data yielded at least a first-order approximation, as verified for the oxide/oxide composite. More rigorous estimation of impact force awaits experimental data and advanced measurement techniques as well.

Acknowledgements

This work was supported by the Aircraft Propulsion Materials Project, the Office of Naval Research (Dr. Dave Shifler). The FOD work on the MI SiC/SiC was performed at the NASA Glenn (Cleveland, OH).

REFERENCES

1. (a) S. R. Choi, R. T. Bhatt, J. M. Perrira, and J. P. Gyekenyesi, Foreign Object Damage Behavior of a SiC/SiC Composite at Ambient and Elevated Temperatures, ASME Paper No. GT2004-53910 (2004); (b) S. R. Choi, Foreign Object Damage Phenomenon by Steel Ball Projectiles in a SiC/SiC Ceramic Matrix Composite at Ambient and Elevated Temperatures, *J. Am. Ceram. Soc.*, **91**[9] 2963-2968 (2008).
2. S. R. Choi, D. J. Alexander, and R. W. Kowalik, Foreign Object Damage in an Oxide/Oxide Composite at Ambient Temperature, ASME Paper No. GT2008-50505 (2008); also in print, *J. Eng. Gas Turbines & Power,* **130** (2009).
3. (a) S. R. Choi, J. M. Pereira, L. A. Janosik, and R. T. Bhatt, Foreign Object Damage of Two Gas-Turbine Grade Silicon Nitrides at Ambient Temperature, *Ceram. Eng. Sci. Proc.*, **23**[3] 193-202 (2002); (b) S. R. Choi et al., Foreign Object Damage in Flexure Bars of Two Gas-Turbine Grade Silicon Nitrides, *Mater. Sci. Eng.*, **A 379**, 411-419 (2004).
4. (a) S. R. Choi, J. M. Pereira, L. A. Janosik, and R. T. Bhatt, Foreign Object Damage of Two Gas-Turbine Grade Silicon Nitrides in a Thin Disk Configuration, ASME Paper No. GT2003-38544 (2003); (b) S. R. Choi et al., Foreign Object Damage in Disks of Gas-Turbine-Grade Silicon Nitrides by Steel Ball Projectiles at Ambient Temperature, *J. Mater. Sci.*, **39**, 6173-6182 (2004).
5. ASTM C 1259, "Test Method for Dynamic Young's Modulus, Shear Modulus, and Poisson's Ratio for Advanced Ceramics by Impulse Excitation of Vibration," Annual Book of ASTM Standards, Vol. 15.01, ASTM, West Conshohocken, PA (2008).
6. Y. Akimune, Y. Katano, and K. Matoba, "Spherical-Impact Damage and Strength Degradation in Silicon Nitrides for Automobile Turbocharger Rotors," *J. Am. Ceram. Soc.*, **72**[8] 1422-1428 (1989).
7. A. G. Evans, and T. R. Wilshaw, "Dynamic Solid Particle Damage in Brittle Materials: An Appraisal,"*J. Mater. Sci.*, **12**, 97-116 (1977).
8. A. D. Peralta and H. Yoshida, Ceramic Gas Turbine Component Development and Characterization, van Roode, M, Ferber, M. K., and Richerson, D. W., eds., Vol. 2, pp. 665-692, ASME, New York, NY (2003).
9. L. M. Cosgriff, R. Bhatt, S. R. Choi, and D. S. Fox, "Thermographic Characterization of Impact Damage in SiC/SiC Composite Materials," Proc. SPIE, Vol. 5767, pp. 363-372 in Nondestructive Evaluation & Health Monitoring of Aerospace Materials, Composites, and Civil Structure IV, 2005.
10. S. R. Choi, "Foreign Object Damage Behavior in a Silicon Nitride Ceramic by Spherical Projectiles of Steels and Brass," *Mat. Sci. Eng.* **A497**, 160-167 (2008).

$Ti_3(Si,Al)C_2$ FOR NUCLEAR APPLICATION: INVESTIGATION OF IRRADIATION EFFECTS INDUCED BY CHARGED PARTICLES

Marion LE FLEM, Xingmin LIU, Sylvie DORIOT, Théodore COZZIKA, Fabien ONIMUS, Jean-Luc BECHADE
CEA Saclay, DEN/DMN/SRMA, Gif Sur Yvette, FRANCE

Isabelle MONNET
Centre de recherche sur les Ions, les Matériaux et la Photonique, CEA/IRAMIS/CIMAP, Caen, FRANCE

and Yanchun ZHOU
Shenyang National Laboratory for Materials Science, Institute of Metal Research
Chinese Academy of Sciences, Shenyang, CHINA

ABSTRACT

In the frame of material development for the future nuclear reactors, MAX phases, especially Ti_3SiC_2-based materials have been recently considered as candidate for the core components in Gas Fast Reactors because of good thermal conductivity, neutron transparency and tolerance to damage. Nevertheless, their behaviour under irradiation has never been investigated until now (including the change in the lamellar microstructure and then a possible loss of the damage tolerance properties). In parallel to progressing neutron irradiations, CEA launched in 2006 irradiations with charged particles of high kinetic energy to precise the evolution of several Ti_3SiC_2 grades. In this work, $Ti_3(Si,Al)C_2$ fabricated by hot-pressing method were irradiated with Kr and Xe ions at 20°C and 500°C. This allowed to explore both the influence of electronic (near surface) and nuclear (implantation zone) interaction in the same experiment. The effect of dose and temperature on the microstructure and hardness was investigated. No change in hardness or microstructure was detected up to 10^{13}ions/cm^2. Higher doses induced hardness increase, but damage tolerance remained. This hardening should be correlated to atomic disorder and cell change highlighted by TEM and XRD. Annealing of the irradiation defects with temperature was highlighted and would start at 300°C. The formation of β-$Ti_3(Si,Al)C_2$ under irradiation is suggested.

INTRODUCTION

The most promising structural materials being able to reach the ambitious goals of dose (> 100 dpa) and temperature (> 1000°C) assigned to the Gas Fast Reactor (GFR) core components have been identified as carbides ceramics such as ZrC, SiC and related composites [1,2]. They can combine high melting temperature, good thermal conductivity, and neutronic compatibility with fast neutron spectrum. Conversely, refractory metals and their alloys obviously offer good structural properties at high temperatures but exhibit too low compatibility with fast neutron spectra. The main drawback of ceramics is their very poor damage tolerance and catastrophic brittle failure. Layered $M_{n+1}AX_n$, ternary compounds (where n=1, 2, or 3, M is an early transition metal, A is an A-group element, and X is either C or N [3]) recently attracted interest because they exhibit intrinsic damage tolerant properties, which are linked to specific nanolayered structure of atomic planes allowing deformation via delamination and kink bands formation [4,5]. Among them, Ti_3SiC_2 [6], and corrosion optimized grade $Ti_3(Si,Al)C_2$ [7,8], exhibit good thermal stability [9,10] that make them potential candidates for GFR in-core applications.

In parallel to efforts lead to enhance the creep behavior of layered ternary compounds at high temperature, it is also necessary to investigate the impact of irradiation on their specific microstructure and then physical and mechanical properties. To fulfill these non-existing data, CEA launched in 2006 both experimental neutron irradiations in French reactors and irradiation program with charged particles on Ti_3SiC_2 [11] and $Ti_3(Si,Al)C_2$ [12].

The present work introduces new results obtained after ion irradiation of $Ti_3(Si,Al)C_2$. 92 MeV Xe and 74 MeV Kr particles were used to produce damage between room temperature and 500°C. Nano-indentation measurements were used to estimate the change in mechanical properties, and TEM and XRD analysis were performed to follow the evolution of the microstructure. These characterizations allowed to precise the behavior of $Ti_3(Si,Al)C_2$ under irradiation and open new field of investigation in terms of fine structural changes.

EXPERIMENTAL PROCEDURE

The $Ti_3(Si,Al)C_2$ materials were fabricated at Institute of Metal Research, Chinese Academic of Sciences. To eliminate TiC impurity, 5, 7 and 10 at% of Al were added to substituted Si and real compositions are $Ti_3(Si_{0.95}Al_{0.05})C_2$, $Ti_3(Si_{0.93}Al_{0.07})C_2$ and $Ti_3(Si_{0.90}Al_{0.10})C_2$. The details for fabrication can be found elsewhere [7,8]. Briefly, the bulk ceramics were prepared by in-situ hot pressing solid–liquid reaction of elemental powders. Ti, Si, Al, and graphite elemental powders (stoichiometric proportions) were mixed in a polyurethane mill for 20 h and cold pressed in a graphite mold coated with a BN layer on the inner surface. The solid–liquid synthesis reaction and simultaneous densification were performed in a furnace using graphite as heating element in a flowing argon atmosphere. The compacted mixture was hot pressed at about 1520°C for 1 h with a uniaxial pressure of 30 MPa. This resulted in TiC free materials with very few amount of SiC and titanium silicides. A typical micrograph is shown in Figure 1.

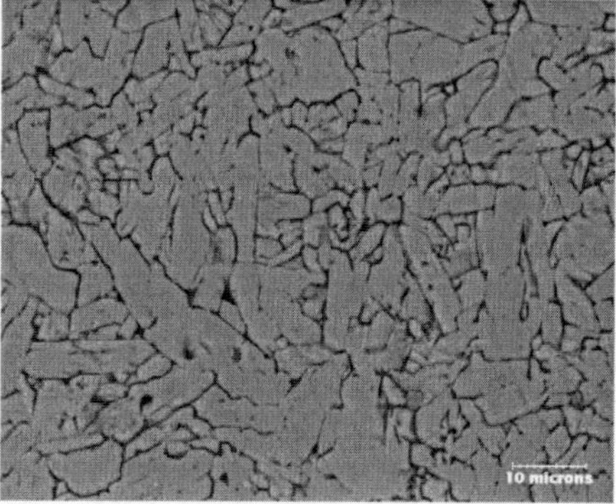

Figure 1. Typical SEM micrograph of $Ti_3(Si,Al)C_2$ (here $Ti_3(Si_{90\%}Al_{10\%})C_2$ with back-scattered electron imaging).

The irradiations were performed at GANIL, Caen, France (Grand Accélérateur National d'Ions Lourds) using IRRSUD line. The ions used were 74 MeV ^{86}Kr and 92 MeV ^{129}Xe. Bulk samples and perforated thin foils (ion milled) were irradiated at room temperature and at 500°C using a sample holder heated by a resistance method. The fluence ranged from 2×10^{11} to 2×10^{15} ions/cm^2. In order to limit the heating by the beam, the maximum flux used were 2×10^9 and 10^{10} ions.cm^{-2}s^{-1} for room temperature and 500°C, respectively. The electronic stopping power of ions, the damage induced by nuclear interaction and the penetration depth of both ions was estimated by SRIM2003 [13]. An example is shown in Figure 2 for a dose of 1×10^{15} ions/cm^2: the maximum damage was achieved between 6.5 and 7.5 µm to the surface (ballistic collisions and ion implantation) while the first few microns are less affected (region of maximum electronic loss which decreases along the penetration depth).

After irradiation, WDS and XPS chemical analysis showed no significant surface contamination by oxygen (oxide layer of less than 10 nm in thickness). For fluence of 1×10^{15} ions/cm^2 (and *a fortiori* higher), with Xe only, some hills were locally formed at the surface of the sample [14]: these point-modifications, attributed to electronic interactions [15], should remain at the very surface of the matter and should not impact the following nanoindentation and X-ray diffraction analysis.

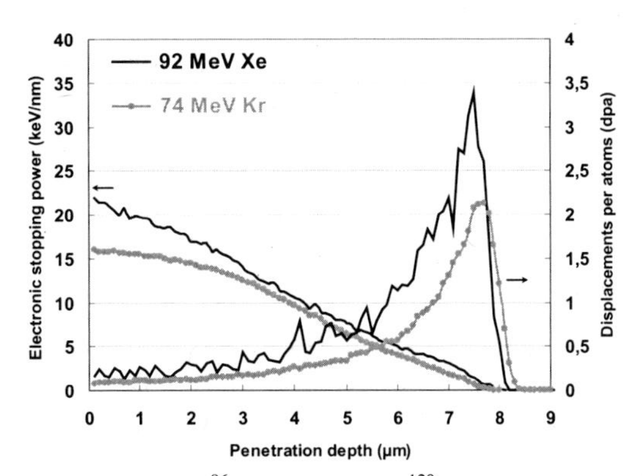

Figure 2. Damage induced by 74 MeV ^{86}Kr and 92 MeV ^{129}Xe at dose of 1×10^{15} ions/cm^2 (SRIM calculation).

Nanoindentation experiments were performed with a Berkovitch diamond indenter directly on the irradiated surface of the samples. A constant tip penetration of 1100 nm was used to sample a reproducible volume of matter, relevant of the damaged layer (indeed, the Berkovich diamond indenter tip sampled the hardness in the region of the indent and extending down seven times the indenter's contact depth [16]), without contribution of non-irradiated substrate. The hardness was extracted from the unloading curves. Basically 21 prints (step of 30 µm) were acquired in the middle of irradiated surface and bad points were deleted by observation of load-displacement curves: at least 17 prints were used to get the average hardness value. It must be noticed that nanoindentation could not provide a measurement of the true hardness because it hardly depends on loading [17]. It must then be considered as a tool to estimate the relative hardening effect as a function of irradiation dose. Field emission gun-scanning electron microscopy (FEG-SEM) was finally used to observe the indentation prints.

Transmission electron microscopy exams (TEM) were performed (with a FEG STEM Jeol 2010F microscope operating at 200 kV) on irradiated pre-perforated thin foils and on cross section samples machined from the bulk specimens (mechanical grinding and ion milling) irradiated at 20°C with Xe.

X-ray diffraction analysis (XRD) were performed on a classical θ/θ type goniometer (D8 Bruker axs) using Cu-Kα radiation on Kr irradiated samples. The analysed thickness was estimated to be 8 μm, i.e. in the range of ion penetration maximum depth: there might be a small contribution from the non-irradiated substrate to the XRD signal. A preliminary Rietveld analysis was done to investigate the change in the unit cell, i.e. lattice parameters and microdistorsions of the lattice (these last were suggested by a Hall and Williamson [18] analysis lead on virgin and 1×10^{15} ions/cm^2 irradiated samples). The Fullprof software was used with considering a two-phased material: affected material from the irradiated layer, and virgin material from the non irradiated substrate (Pseudo Voigt peaks). The space group for $Ti_3(Si,Al)C_2$ was $P6_3/mmc$.

RESULTS

Nanoindentation analysis

Before irradiation, no significant effect of Al content on $Ti_3(Si,Al)C_2$ hardness was noticed: an average hardness of 8 GPa could be considered. This is consistent with previous work [19] showing no change in mechanical properties with Al content at room temperature. For both $Ti_3(Si_{0.93}Al_{0.07})C_2$ and $Ti_3(Si_{0.90}Al_{0.10})C_2$ grades, the change in hardness after irradiation at 20°C with Kr is shown in Figure 3 (measurements on samples irradiated with Xe showed the same trends). Again, $Ti_3(Si_{0.93}Al_{0.07})C_2$ and $Ti_3(Si_{0.90}Al_{0.10})C_2$ exhibited the same trends suggesting that there is no obvious effect of Al content on behaviour under irradiation. Considering experimental errors, the hardness remains low up to 1×10^{13} ions/cm^2. Then it dramatically increases for 1×10^{14} ions/cm^2 and reaches 18-21 GPa at 1×10^{15} ions/cm^2 that is more than twice the hardness of the samples before irradiation. Nevertheless, no cracks formed from the corner of indentation prints and deformation by basal plane slip is clearly visible on the micrograph of Figure 3: this suggests that despite the high hardening, the material exhibited some damage tolerance even after irradiation up to 1×10^{15} ions/cm^2.

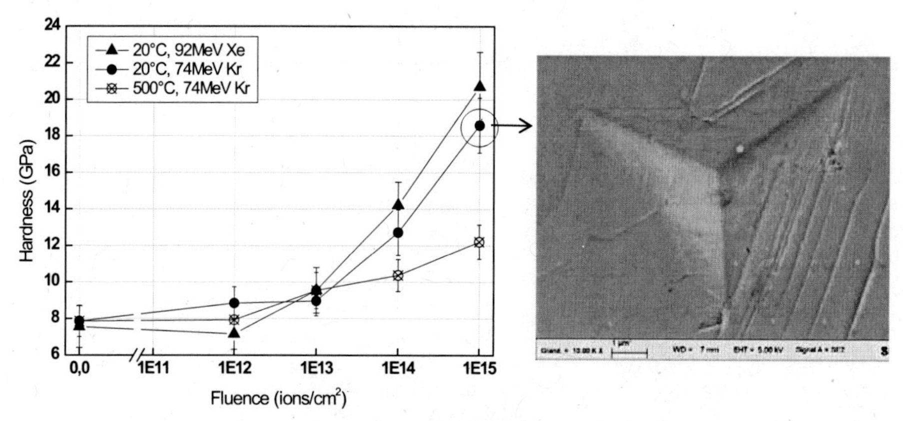

Figure 3. Change in hardness of $Ti_3(Si_{0.95}Al_{0.05})C_2$ with irradiation dose (a) and indentation print at high dose (b).

After irradiation at 500°C (open plot in Figure 3), the hardness at high fluence is considerably decreased compared to the results at 20°C: at 1×10^{15} ions/cm^2, the hardness is only

11 GPa, strongly demonstrating annealing of irradiation defects with temperature. Nevertheless, this annealing should not be complete since the hardness is still significantly higher than the starting value. Post irradiation annealing of 20°C-irradiated samples allowed to detect the beginning of defect annealing (i.e. drop in hardness) as low as 300°C while a complete recovery is suggested at 1000°C. In order to separate the increase in hardness due to modification induced by electronic interaction and nuclear collision, nanoindentation analysis along the penetration depth are in progress.

TEM examination

As shown in our previous work [12,14], no dramatic change in microstructure was detected in $Ti_3(Si,Al)C_2$ phase below 1×10^{13} ions/cm^2. Only some ions impacts could be observed in rare secondary silicide phases such as Ti_5Si_3 and $TiSi_2$. Previous exams of $Ti_3(Si_{93\%}Al_{7\%})C_2$ thin foils irradiated at 20°C up to 5.7×10^{14} ions/cm^2 allowed to detect formation of irradiation defects as shown on the dark field image of Figure 4. These so called black dots could not be identified as dislocation loops. Considering the thickness of the observed irradiated thin foil (< 200 nm), the observed defects should have been produced mainly by electronic interactions or by a low level of displacement per atom (< 0.1 dpa, see Figure 1).

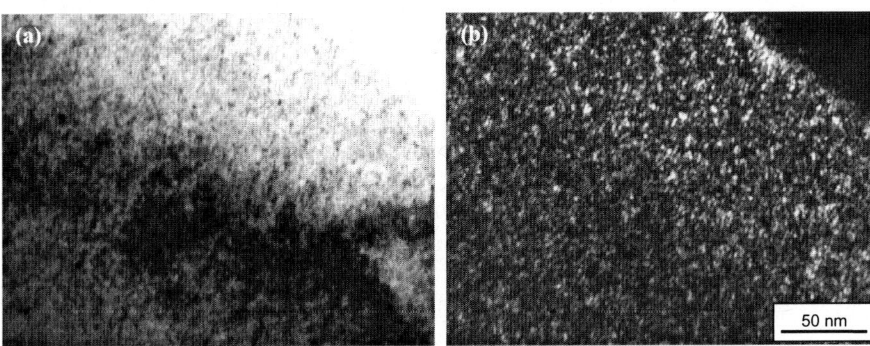

Figure 4. (a) Bright field and (b) weak beam dark field images of $Ti_3(Si_{0.93}Al_{0.07})C_2$ (20°C, 5.7×10^{14} ions/cm^2, 92 MeV Xe).

To visualise the high damage level microstructure induced by ballistic collision, a cross section specimen was fabricated from the corresponding bulk sample. The obtained micrograph is illustrated in Figure 5. The arrowed region of implanted Xe particles could be detected at about 8 μm to the surface, which is consistent with the prediction by SRIM. In this area, a $Ti_3(Si,Al)C_2$ grain is divided in two parts: A irradiated zone and B virgin area non affected by irradiation. The change in the contrast within the grain strongly suggests a high level of microstructure modification induced by irradiation: the A region exhibits a uniformly gray aspect with no modification by tilting which suggest an amorphous condition. From the corresponding diffraction patterns presented at the right of the Figure 5 (same grain, i.e. same orientation), an increase in diffuse scattering and decrease in intensity of some spots is obviously induced by irradiation, in agreement with amorphisation of Ti_3SiC_2 already suggested by XRD on irradiated samples [11,12]. The lattice dimensions deduced from the patterns clearly highlighted an increase in 'c' parameter with irradiation (~+5% in the region of Xe implantation) while 'a' is almost unchanged. The observed trends were confirmed by complementary examination of $Ti_3(Si_{90\%}Al_{10\%})C_2$ irradiated at 1×10^{15} and 2×10^{15} ions/cm^2. From Figure 5, it is obvious that the disorder is higher in the peak damage zone (higher dpa) than closer to the surface where the electronic stopping power is prepondernant and the dpa level is low. This modification in the implanted zone can be unambiguously related to ballistic collision (representative of neutron irradiation). For the zone where both electronic and nuclear interactions are present, it is more difficult to conclude (defect can be induced by ballistic collisions, by both nuclear and electronic interaction or by a synergetic or antagonist combination between the two of them [20]). Exams of samples irradiated at 500°C did not show any evidence of such an evolution and no obvious change in microstructure was seen.

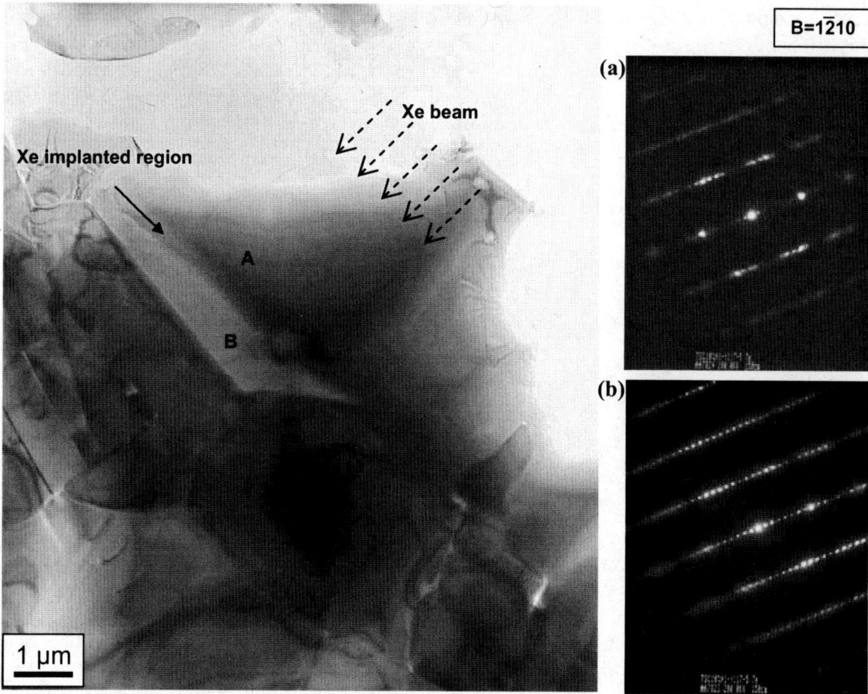

Figure 5. TEM micrograph of $Ti_3(Si_{0.93}Al_{0.07})C_2$ in the area of implanted ions (20°C, 5.7×10^{14} ions/cm^2, 92 MeV Xe) and corresponding diffraction patterns. (a) irradiated area and (b) virgin substrate.

XRD Analysis

The XRD diagrams obtained on virgin and Kr irradiated surface of the $Ti_3(Si_{95\%}Al_{5\%})C_2$ samples are shown in Figure 6. For irradiations at 20°C, the diagrams do not exhibit any evolution up to 1×10^{13} ions/cm^2 but changes are obvious for higher doses 1×10^{14} and 1×10^{15} ions/cm^2. First, accompanied with a decrease in intensity, a strong broadening of the peaks is observed: this could correspond to appearance of disorder which would be in agreement with TEM conclusions. A Hall and Williamson analysis [18] highlighted a dramatic increase in the microstrains in $Ti_3(Si_{0.95}Al_{0.05})C_2$ after irradiation to 1×10^{15} ions/cm^2. Second, a shift to low angles is observed for (008), (006) and (109) reflection peaks that should correspond to an increase in lattice parameter already suggested via grazing XRD [11]: reflection (008) is arrowed in Figure 6. A small contribution from the non irradiated bulk could be detected. After irradiation at 500°C to 1×10^{15} ions/cm^2, these broadening and shift are not so obvious and the starting microstructure seems recovered to a certain extend.

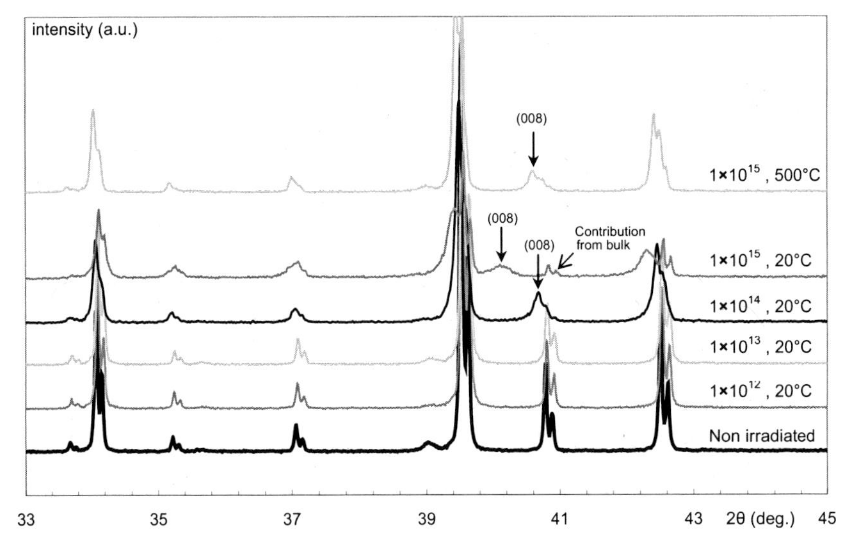

Figure 6. XRD patterns of $Ti_3(Si_{0.95}Al_{0.05})C_2$ before and after irradiation with 74 MeV Kr at 20°C and 500°C.

These trends were confirmed by preliminary Rietveld analysis of the diagrams with the assumption of microstrains as single origin of peak broadening in the irradiated layer of $Ti_3(Si_{95\%}Al_{5\%})C_2$ (the contribution from the unchanged underneath virgin $Ti_3(Si,Al)C_2$ was taken into account but is not discussed here). The results of the refinement are presented in Figure 7: at dose of 1×10^{15} ions/cm^2, the c parameter increase (average of the whole irradiated layer) is 1.7%. The recovery of the microstructure after irradiation at 500°C is obviously not complete. Effort is presently done to optimize the refinement quality: in particular, the assumption of a multiphase material is being demonstrated (see conclusion).

CONCLUSIONS

Grades of $Ti_3(Si,Al)C_2$ sample with various content of Al were irradiated at room temperature and 500°C by Kr and Xe up to 1×10^{15} ions/cm^2. The evolution of mechanical properties and microstructure was followed by nanohardness measurements, TEM exams and

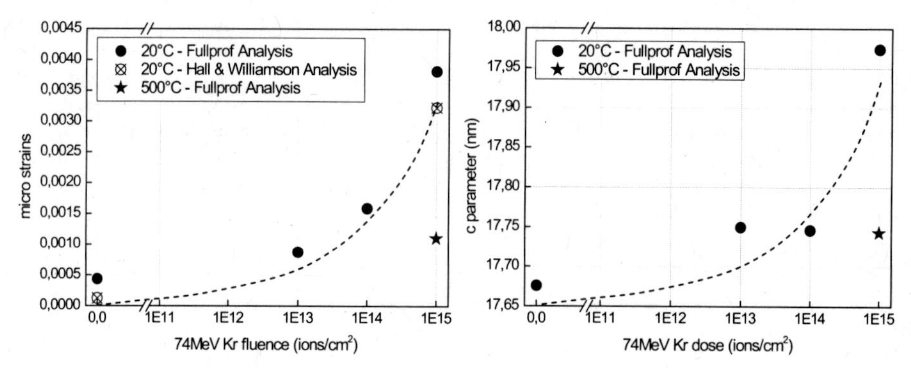

Figure 7. Effect of 74 MeV Kr irradiation on (a) microstrains and (b) 'c' lattice parameter of $Ti_3(Si_{0.95}Al_{0.05})C_2$ deduced from XRD patterns.

XRD analysis. These tools highlighted very consistent features allowing to precise the effect of charged particle irradiation on the stability of these ternary compounds. The main conclusions are the following:

- The amount of Al (5at% to 10at%) seemed to have no impact on the material evolution in the investigated range of dose and temperature.
- No significant evolution of the hardness or microstructure was observed up to 1×10^{13} ions/cm^2.
- For irradiation from 1×10^{14} ions/cm^2 up to 1×10^{15} ions/cm^2, change in microstructure was observed, i.e, appearance of disorder (black dots, partial amorphisation, microstrains) and increase in 'c' lattice parameter (more important in the highly damaged region of implanted ions).
- This change in microstructure induced a strong increase in hardness: the hardness doubled at 1×10^{15} ions/cm^2 but damage tolerance is remained.
- Irradiations at high temperature or post-irradiation annealing resulted in recovery of the microstructure and hardness. This recovery was incomplete after 500°C annealing.

Further Rietveld analysis on XRD patterns and TEM investigations are in progress on samples irradiated with Kr and Xe to doses ranging from 1×10^{14} to 2×10^{15} ions/cm^2. The earlier results suggest that the microstructure modification could be related to formation of β-Ti_3SiC_2 clusters induced by irradiation. The structure of this uncommon mestastable polymorph[21,22,23] consists in shifted Si planes regarding common α-Ti_3SiC_2. Complementary high resolution TEM, XPS and Raman spectroscopy analysis on irradiated $Ti_3(Si,Al)C_2$ grades are planned to precise its formation.

ACKNOWLEDGMENTS

The authors are very grateful to Mr. Jacques Pelé for his help in XRD analysis and Mr. Patrick Bonnaillie who performed SEM-FEG observations.

REFERENCES

[1] U.S. DOE Nuclear Energy Research Advisory Committee and the Generation IV International Forum, A Technology Roadmap for Generation IV Nuclear Energy Systems (2002).

[2] P. Yvon, F. Carré, Structural Materials Challenges for Advanced Reactor Systems, *Journal of Nuclear Materials* (2008), doi: 10.1016/j.jnucmat.2008.11.026.
[3] M. W. Barsoum, The $M_{N+1}AX_N$ Phases: A new class of solids; Thermodynamically stable nanolaminates, *Prog. Solid St. Chem.*, 28, 201-281 (2000).
[4] M. W. Barsoum, L. Farber and T. El-Raghy, Dislocations, kink bands, and room-temperature plasticity of Ti_3SiC_2, *Met. Mat. Trans. A*, 30, 1727-1738 (1999).
[5] Y. W. Bao, C. F, Hu and Y. C. Zhou, Damage tolerance of nanolayer-grained ceramic: a quantitative estimation, *Mater. Sci. & Tech.*, 22, 227-230 (2006).
[6] M. W. Barsoum, T. El-Raghy and M. Radovic, Ti_3SiC_2: A Layered Machinable Ductile Ceramic, Interceram 49, 226-233 (2000)
[7] Y. C. Zhou, H. B. Zhang, M. Y. Liu. J. Y. Wang and Y. W. Bao, Preparation of TiC free Ti_3SiC_2 with improved oxidation resistance by substitution of Si with Al, *Mater. Res. Innovat.*, 8, 97-102 (2004).
[8] H. B. Zhang, Y. C. Zhou, Y. W. Bao and M. S. Li, Mechanism for the enhanced oxidation resistance of Ti_3SiC_2 by forming a $Ti_3Si_{0.9}Al_{0.1}C_2$ solid solution, *Acta Mater.*, 52, 3631-3637 (2004).
[9] C. Racault, F. Langlais, and R. Naslain, Solid-state synthesis and characterization of the ternary phase Ti_3SiC_2, *J. Mater. Sci.*, 29, 3384-3392 (1994).
[10] R. Radakrishnan, J. J. Williams, and M. Akinc, Synthesis and high-temperature stability of Ti_3SiC_2, *J. Alloys Compd.*, 285, 85-88 (1999).
[11] J.C. Nappé, Ph. Grosseau, F. Audubert, B. Guilhot, M. Benabdesselam, M. Beauvy "Study of the irradiation damages in Ti_3SiC_2" E-MRS 2008 Spring Meeting , May 26-30 Strasbourg, France.
[12] X. Liu, M. Le Flem, J.L. Béchade, Y. Zhou, Th. Cozzika, S. Doriot, P. Forget, L. Gosmain, I. Monnet, TEM observations and nanoindentation measurements of Ti_3SiC_2 irradiated by charged particles, Proc. of the *2nd International Congress on Ceramics* June 29 – July 4, Verona, Italy.
[13] F. Ziegler, J.P. Biersack, U. Littmark, The stopping and range of ions in solids, New York, 1985.
[14] I. Monnet, X. Liu, J.C. Nappé, M. Le Flem, Ph. Grosseau, F. Audubert, Y. Zhou, Surface modification of Ti_3SiC_2 under heavy ion irradiation, *7th International Symposium Swift Heavy Ions in Matter*, June 2-5, 2008, Lyon, France.
[15] J.C. Nappé, Ph. Grosseau, B. Guilhot, F. Audubert, M. Beauvy, M. Benabdesselam, Heavy ions induced damages in Ti_3SiC_2: effect of irradiation temperature, these Proceedings.
[16] L. E. Samuels and T. O. Mulhearn, An experimental investigation of the deformed zone associated with indentation hardness impressions, J. *Mech. Phys. Solids,* 5, 125-134 (1957).
[17] P. Pampuch, J. Lis, L. Stobierski and M. Tymkiewiez, Solid combustion synthesis of Ti_3SiC_2, *J. Eur. Ceram. Soc.*, 5, 283-287 (1989).
[18] R. Guinebretière, Diffraction des rayons X sur échantillons polycristallins, ed. Hermès Science, 2002.
[19] H. B. Zhang, Ph. D Thesis. Inst. Met. Res. CAS June, 2006.
[20] A. Audren, I. Monnet, Y. Leconte, X. Portier, L. Thome, M. Levalois, N. Herlin-Boime, C. Reynaud, Structural evolution of SiC nanostructured and conventional ceramics under irradiation, *Nuclear Instruments and Methods in Physics Research B* 266, 2806–2809 (2008).
[21] L. Farber, I. Levin, M. W. Barsoum, T. El-Raghy, T. Tzenov, High-resolution transmission electron microscopy of some $Ti_{n+1}AX_n$ compounds: n=1, 2; A=Al or Si; X=C or N, *Journal of Applied Physics* 86, 5, 2540-2543 (1999).

[22] R. Yu, X.F. Zhang, L.L. He, H.Q. Ye, Topology of charge density and elastic properties of Ti_3SiC_2 polymorphs, (June 24, 2004). *Lawrence Berkeley National Laboratory.* Paper LBNL-55853.
[23] J.Y. Wang, Y. Zhou, Polymorphism of Ti_3SiC_2 ceramic: First-principles investigations, *Physical Review B. Condensed Matter and Materials Physics* 69, 14, 144108.1-144108.13 (2004).

HEAVY IONS INDUCED DAMAGES IN Ti_3SiC_2: EFFECT OF IRRADIATION TEMPERATURE

J.C. Nappé, Ph. Grosseau, B. Guilhot
École Nationale Supérieure des Mines, SPIN/PMMC, LPMG UMR CNRS 5148
Saint-Étienne, France

F. Audubert, M. Beauvy
CEA, DEN, DEC/SPUA/LTEC
St Paul lez Durance, France

M. Benabdesselam
Université de Nice - Sophia Antipolis, LPMC UMR CNRS 6622
Nice, France

ABSTRACT

Thanks to their refractoriness, carbides are sensed as fuel coating for the IVth generation of reactors. Among those studied, the Ti_3SiC_2 ternary compound can be distinguished for its noteworthy mechanical properties: the nanolamellar structure imparts to this material some softness as well as better toughness than other classical carbides such as SiC or TiC. However, under irradiation, its behaviour is still unknown. In order to understand this behaviour, specimens were irradiated with heavy ions of different energies, then characterised. The choice of energies used allowed separation of the effects of nuclear interactions from those of electronic ones. Thus, AFM, SEM and XRD techniques allowed to note an important spoiling due to nuclear collision whereas electronic interactions would induce the formation of hills and the expansion of the unit cell. Irradiations at higher temperatures allowed to study the effect of temperature on these results.

Key words: Ti_3SiC_2, irradiation, heavy ions, AFM, SEM, XRD

INTRODUCTION

As part of Generation IV International Forum (GIF), new systems are studied, from the point of view of the reactor as well as the fuel cycle. These systems are characterised by an increased security level, better economic competitiveness, and an ability to recycle all the fuel in order to upgrade to a fissionable material and minimize long-lived waste production by transmutation. Among the six systems considered by the GIF, the Gas Fast Reactor (GFR) is studied in France; it is designed to work under helium-pressure and high-temperature (1100-1300 K).

These working conditions led to the selection of non-oxide refractory ceramics as fuel coating. Thus, carbides turn out to be great candidates thanks to their remarkable mechanical and thermal properties. However, their behaviour under irradiation has to be studied in more details.

Among the studied carbides, the ternary Ti_3SiC_2 presents some interesting prospects. Indeed, in 1972 Nickl *et al.*[1] remarked that this material, which was synthesized for the first time by a Viennese group in 1967[2], is abnormally soft for a carbide. Actually, Pampuch *et al.*[3,4] and Lis *et al.*[5-7] demonstrated that this material is both stiff (Young's modulus of 325 GPa) and soft (Vickers hardness of 6 GPa). Moreover, like Goto *et al.*[8] some years before, they noted that the hardness of this carbide decreases as the applied load increases; this property led them to qualify Ti_3SiC_2 as a "ductile ceramic".

These interesting mechanical properties led to think about Ti_3SiC_2 as fuel coating material; actually, its ability to be damage tolerant against mechanical solicitations was expected to be true against irradiations. Nevertheless, besides one paper to be published[9], no information is available about its behaviour under irradiation.

Therefore, the aim of this study is a better knowledge of the behaviour under irradiation of

Ti_3SiC_2, and more particularly the effect of the irradiation temperature.

EXPERIMENTAL

The studied material is a polycrystalline commercial compound provided by 3-ONE-2 (Voorhees, NJ, USA). Pure bulk of Ti_3SiC_2 is quite difficult to synthesize: the as-received material consists of about 80 % Ti_3SiC_2, 15 % TiC, and 5 % $TiSi_2$ (estimation by X-Ray Diffraction). Figure 1 shows the surface state of the as-polished specimens through Atomic Force Microscopy (AFM); the lower hardness of Ti_3SiC_2 compared with the TiC and $TiSi_2$ phases can be noted.

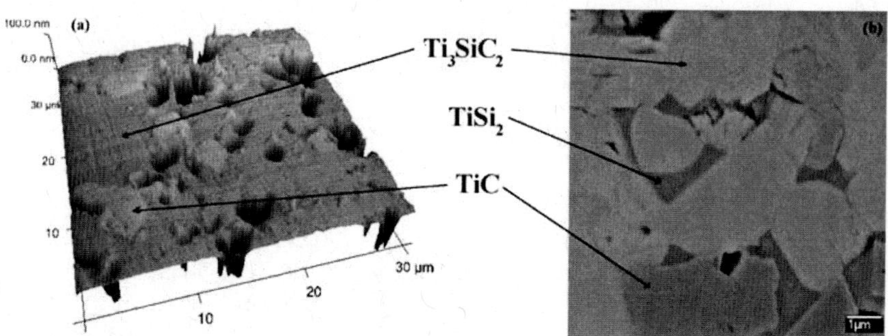

Figure 1. Surface of as-polished Ti_3SiC_2 by (a) AFM and (b) back-scattered electrons FEG-SEM.

The as-received specimens were polished with diamond suspensions down to 1 µm. They were subsequently irradiated with heavy ions. This kind of irradiation aims to simulate the interactions present in reactors (impacts of fission products, recoil atoms of alpha-decays, and neutrons) without the drawback of activating the specimens.

When heavy ions penetrate through a solid, they get decelerated *via* two different processes, namely, elastic collisions with target nuclei (with nuclear stopping power S_n) and inelastic collisions with target electrons (with electronic stopping power S_e). In order to understand the effects of both nuclear and electronic interactions on Ti_3SiC_2, two kinds of irradiation were performed.

The first one, named "high energy irradiation", was made with ions delivered by the IRRSUD accelerator at the GANIL (Caen, France). The energy of these ions reaches several dozen of MeV; for this range of energy, ion path consists in a long zone of electronic interactions (some microns) at the end of which nuclear collisions occur. Thus, studying only the first microns, effects of electronic interactions can be understood. In order to study the effect of electronic stopping power, two different ions were used at the GANIL: 74 MeV Kr and 92 MeV Xe ions, which respectively lead to a maximum electronic stopping power of 15.6 and 21.3 keV nm^{-1}. Irradiations with Kr were carried out at room temperature (RT) and 773 K whereas the ones with Xe were at RT and 573 K.

The second one, called "low energy irradiation", was achieved with the 4 MeV Au ions of the ARAMIS accelerator at the CSNSM (Orsay, France). This kind of irradiation essentially leads to nuclear collisions. This irradiation was carried out at RT, 773 K and 1223 K.

Each kind of irradiation was carried out for 4 ion fluences: 10^{12}, 10^{13}, 10^{14}, and 10^{15} cm^{-2}; the fluence is the number of ion by surface unit. The Table I gives some irradiation parameters determined by the TRIM code[10].

Table I. Irradiation parameters

Ion	Au	Kr	Xe
Mass (amu)	197	86	129
Energy (MeV)	4	74	92
Projected range (μm)	0.52	7.81	7.60
Nuclear stopping power (keV/nm)	3.29	0.07	0.16
Electronic stopping power (keV/nm)	4.0	15.6	21.3

To characterise surface modifications to the samples, two microscopic methods were used: Field Emission Gun Scanning Electron Microscopy (FEG-SEM) and AFM. The structure of the irradiated zone was analysed by Low Incidence X-Ray Diffraction (LI-XRD) technique.

RESULTS

Low Energy Irradiations

The irradiations performed with 4 MeV Au ions essentially lead to nuclear shocks, which generally cause an important disorder in the materials. As noted elsewhere[9], this disorder is visible on diffractogrammes from a fluence of 10^{14} cm^{-2} for irradiations performed at RT, whatever the phase.

The samples irradiated at high temperatures (773 K and 1223 K) have not yet been characterized by LI-XRD. Nevertheless, we hope to observe a decrease of disorder due to a greater mobility of atoms making up the samples and to an annealing of the defects by the temperature. This decrease would be a good result if we consider the nominal operating temperatures of the GFR.

In terms of surface conditions, no changes were found before a fluence of 10^{15} cm^{-2}. At room temperature, we noticed a microstructure that we attributed to a sputtering depending on the orientation of cristallites[9]. It is interesting to note that this microstructure is less pronounced during irradiation carried out at 773 K, regaining a microstructure similar to the as-polished specimen for irradiation at 1223 K (cf. Figure 2).

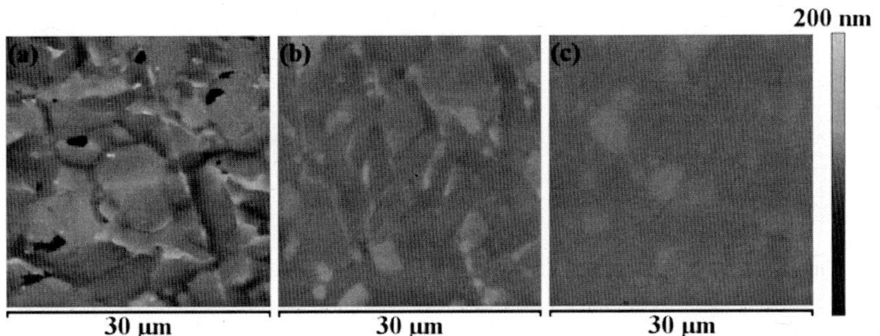

Figure 2. Surface of Ti_3SiC_2 after a 10^{15} cm^{-2} 4 MeV Au irradiation at (a) RT, (b) 773 K, and (c) 1223 K.

High Energy Irradiations

Concerning the high-energy irradiations, we noted by LI-XRD some disorder from 101^4 cm^{-2} for 92 MeV Xe irradiation at RT, but only for phase Ti_3SiC_2; this disorder is noticeable through a rise of the baseline. Moreover, we noticed an expansion of the hexagonal unit cell of Ti_3SiC_2 along c axis without change in the unit volume[9]. This result is also true for 74 MeV Kr irradiation at RT.

The temperature has two effects on XRD results. First, a drop-off of the rise of baseline shows a decrease of the disorder as a function of temperature. Second, the unit cell is less expanded for irradiation with 92 MeV Xe at 573 K to be almost unchanged compared with virgin sample for the irradiation with 74 MeV Kr at 773 K. Note that whatever the irradiation, the change of the unit cell parameters occurs at constant unit volume (cf. Figure 3).

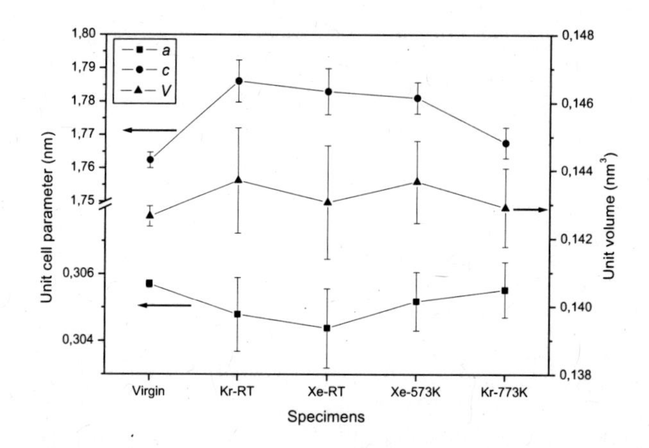

Figure 3. Evolution of Ti3SiC2 unit cell parameters as a function of irradiation parameters for high energy irradiation; lines are only to guide the eye.

With regard to the change in the surface, the first notable point that, unlike irradiation with 92 MeV Xe at RT, the 74 MeV Kr irradiation does not cause changes whatever the fluence or temperature. Concerning the 92 MeV Xe irradiation, we underlined the formation of hills on Ti_3SiC_2 phase and a revelation of the microstructure due to a different behaviour under irradiation of the secondary phases against Ti_3SiC_2[9]. The difference that is noted for these two high-energy irradiation suggests the existence of an electronic stopping power threshold beyond which there is modification of the surface of the samples.

At 573 K, the surface state of our samples after 92 MeV Xe irradiation is quite similar to irradiations at RT. However, if the revelation of microstructure is always obvious, we can notice that the shape of the hills slightly varies; actually, as shown in Figure 4, they seem larger than for irradiation at RT. This coalescence phenomenon could be compared with the one noted during sintering, suggesting a transfer of matter by temperature.

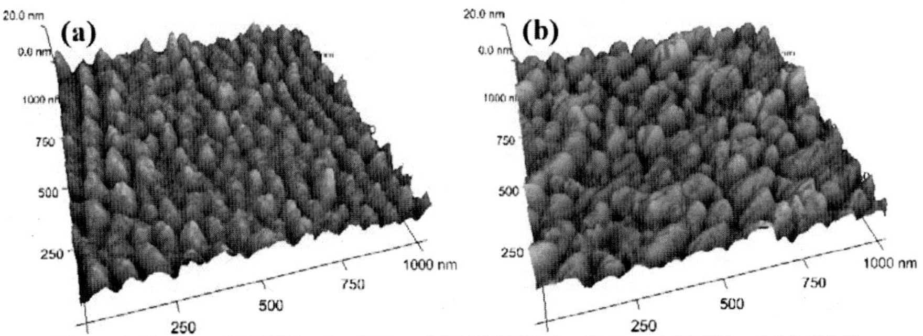

Figure 4. Surface of Ti_3SiC_2 after 10^{15} cm^{-2} 92 MeV Xe irradiation at (a) RT and (b) 573 K.

CONCLUSION

In this study, we demonstrated a different behaviour of Ti_3SiC_2 under low energy and high energy irradiation. Thus, if the nuclear shocks tend to create an important disorder as well as a sputtering of the surface, the electronic interactions essentially cause an expansion of the unit cell along c axis and the formation of hills beyond an electronic stopping power threshold between 15.6 and 21.3 keV nm^{-1}.

Temperature seems to have two different effects. The first one, as it can be noticed in other carbides like SiC[11-14], is an annealing effect of the irradiation defects: decrease of disorder whatever the irradiation, of unit cell expansion due to high energy irradiations, and of the surface changes for the 4 MeV Au irradiation. The second one is the enlarging of the hills. We understand this effect as a transfer of matter due to both the temperature and the electronic interactions.

ACKNOWLEDGEMENT

The authors would like to gratefully thank Isabelle Monnet (GANIL) and Lionel Thomé (CSNSM) for their great help during the irradiation experiments. This work was partly supported by the French research group MATINEX.

REFERENCES

[1]J.J. Nickl, K.K. Schweitzer, P. Luxenberg, Gasphasenabscheidung im Systeme Ti-C-Si, *J. Less-Common Metals*, **26,** 335-353 (1972).

[2]W. Jeitschko, H. Nowotny, Die Kristallstruktur von Ti_3SiC_2 - ein neuer Komplexcarbid-Typ, *Monatsh. Chem.*, **98**, 329-337 (1967).

[3]R. Pampuch, J. Lis, L. Stobierski, M. Tymkiewicz, Solid Combustion Synthesis of Ti_3SiC_2, *J. Eur. Ceram. Soc.*, **5**, 283 (1989).

[4]R. Pampuch, J. Lis, J. Piekarczyk, L. Stobierski, Ti_3SiC_2-based materials produced by self-propagating high temperature synthesis and ceramic processing, *J. Mater. Synth. Process.*, **1**, 93 (1993).

[5]J. Lis, R. Pampuch, L. Stobierski, Reactions during SHS in a Ti-Si-C system, *Int. J. Self-Propag. High-Temp. Synth.*, **1**, 401 (1992).

[6]J. Lis, R. Pampuch, J. Piekarczyk, L. Stobierski, New ceramics based on Ti_3SiC_2, *Ceram. Int.*, **19**, 219 (1993).

[7]J. Lis, Y. Miyamoto, R. Pampuch, K. Tanihata, Ti_3SiC_2-based materials prepared by HIP-SHS techniques, *Mater. Lett.*, **22**, 163-168 (1995).

[8]T. Goto, T. Hirai, Chemically vapor-deposited Ti_3SiC_2, *Mater. Res. Bull.*, **22**, 1195-1201 (1987).

[9]J.C. Nappé, Ph. Grosseau, F. Audubert, B. Guilhot, M. Beauvy, M. Benabdesselam, I. Monnet,

Damages induced by heavy ions in titanium silicon carbide: effects of nuclear and electronic interactions at room temperature, *J. Nucl. Mater.*, in press.
[10]J.F. Ziegler, http://www.srim.org/.
[11]S. Zinkle, L. Snead, Influence of irradiation spectrum and implanted ions on the amorphization of ceramics, *Nucl. Instrum. Methods Phys. Res. Sect. B-Beam Interact. Mater. Atoms*, **116**,92-101 (1996).
[12]W. Weber, N. Yu, L.M. Wang, N.J. Hess, Temperature and dose dependence of ion-beam-induced amorphization in α-SiC, *J. Nucl. Mater.*, **244**,. 258-265 (1997).
[13]W. Weber, W. Jiang, S. Thevuthasan, Accumulation, dynamic annealing and thermal recovery of ion-beam-induced disorder in silicon carbide, *Nucl. Instrum. Methods Phys. Res. Sect. B-Beam Interact. Mater. Atoms*, **175-177**, 26-30 (2001).
[14]W. Weber, L.M. Wang, N. Yu, N.J. Hess, Structure and properties of ion-beam-modified (6H) silicon carbide, *Mater. Sci. Eng., A*, **253**, 62-70 (1998).

TITANIUM CARBIDE AND SILICON CARBIDE THERMAL CONDUCTIVITY UNDER HEAVY IONS IRRADIATION

J. Cabrero[1,2], F. Audubert[2], P. Weisbecker[1], A. Kusiak[3], R. Pailler[1]

1: LCTS, 3 allée de la Boétie, 33600 Pessac, France
2: CEA, DEN 13108 Saint Paul lez Durance, France
3: TREFLE, Esplanade des Arts et Métiers 33405 Talence Cedex, France

ABSTRACT

SiC_f/SiC ceramic matrix composites (CMC) are considered as structural materials in next generation fission nuclear reactors. However, thermal conductivity of SiC is reduced on the one hand at the highest temperatures, but also under irradiation. Titanium carbide, because of its peculiar thermal properties is an attractive material to be used as a matrix in a CMC to enhance the thermal conductivity of CMC under irradiation and at high temperature.

In this study, we performed irradiation experiments on TiC, TiC_xSiC_{1-x} and SiC samples, with heavy ions at room temperature (74 MeV Kr, fluence from 10^{13} to 10^{15} ions/cm^2). This energy results in an irradiated layer of about 7 µm for TiC. Thermal conductivity of the irradiated layer is measured using IR radiometry as a function of fluence and composition. The structural evolution of the irradiated samples was investigated by Raman micro spectroscopy and transmission electron microscopy.

INTRODUCTION

Silicon carbide possesses many desirable benefits for high temperature applications in a neutron radiation environment[1]. Monolithic SiC is a brittle ceramic, so the structural applications will likely rather involve SiC fiber-reinforced ceramics[2]. Continuous fiber-reinforced composite can provide good thermomechanical properties to materials. Nevertheless, their behaviour under irradiation is a key point. Indeed, the thermal conductivity of the SiC_f/SiC composites currently being studied is reduced at the highest temperatures, but also under irradiation [1,3-5]. Thermal conductivity of SiC_f/SiC has to be improved to be used as cladding materials in next generation fission reactor. One proposes to act on the intrinsic conductivity of the matrix by introducing more or less large fraction of transition metal carbides in the matrix. Transition metal carbides, because of their metallic bond, have a real potential for nuclear environment. One thinks of TiC, whose conductivity is partially electronic and increases with temperature[6-7]. The purpose of the study is thus to develop a SiC_f/TiC_x-SiC_{1-x} composite, for which the irradiation is less detrimental to the thermal conductivity of material than for SiC_f/SiC. However, prior to elaborate such fiber-composite, understanding the fundamental mechanisms for radiation damage in monolithic TiC is necessary, especially the TiC thermal conductivity dependence with irradiation and temperature.

The main irradiation damage in a nuclear core of inert materials such as silicon carbide and titanium carbide is primarily due to atomic displacement by fast neutrons. Such damages are roughly simulated by the irradiation with heavy ions. Simulation of neutron damage using ions has been considered as a prominent alternative due to the limitations and difficulties generally encountered with neutron irradiation. Extensive studies were carried out on microstructure[8-10], amorphisation[11-14], annealing process[15-16], recrystallization[7-21] and thermal conductivity[22] of ion irradiated SiC. To our knowledge, there are no such studies on titanium carbide. We propose to implant 74 MeV Kr ions into titanium carbide and silicon carbide to

simulate neutron irradiation. However, as heavy ions are implanted only near the surface of the irradiated sample (depending on the ion, its energy and the sample to be implanted), the thermal conductivity analysis of the irradiated layer has to be suitable for thin layers.
The present work is focussed on the thermal conductivity measurement of thin TiC, TiC-SiC and SiC irradiated layers. The effects of fluence and composition are presented.

EXPERIMENTAL

Samples

TiC-SiC samples (with different SiC volume contents, from 0% to 50%) have been obtained by Spark Plasma Sintering (SPS, 1800 °C, 75 MPa, holding time 5 min in a graphite die) using commercially available nano-powders of TiC (d = 30-80 nm, 97.5%) and β-SiC (d = 45-55 nm, 97.5%). The powders with different TiC contents were mixed by a mortar and pestle with medium of alcohol. After stir-drying at 100°C for 2 h, the mixture was reground in mortar (Retsch MM200, 30 min, 18Hz). The relative density of the samples is between 97% and 95% versus the composition. Commercial α-SiC (d = 96%) is also studied to compare to TiC behaviour under irradiation.

Ions implantation

The pellets were irradiated at room temperature with 74 MeV Kr ions and fluences from 10^{13} to 10^{15} ions/cm^2 on the IRRSUD line of the GANIL facility. The samples were fixed on a water-cooled plate in order to minimize the heating of the sample under the ion beam. To a first approximation, the initial damage has been estimated with SRIM[23] program, with threshold displacement energies of, respectively, 25, 25 and 35 for Si[24], C[24] and Ti[25]. The damage curves are reported in fig. 1.

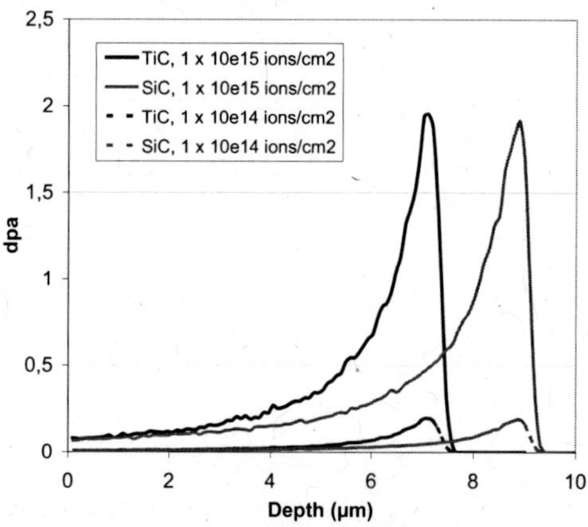

Figure 1. SRIM estimation of damage in TiC and SiC implanted with 10^{15} ions/cm^2 (full line), and 10^{14} ions/cm^2 (broken line) 74 MeV Kr

The IRRSUD line of the GANIL facility allows the observation of the combined electronic and nuclear interactions with matter, with typical ion energies in the 0.3–1 MeV/u range. The number of dpa varie according to the depth and is higher in the nuclear cascade region, which corresponds to the region where incidents ions mainly interact with the atoms of the solid via nuclear collision and finally stop in the lattice. For 74 MeV Kr ions, number of dpa is 2 at 7

μm for a fluence of 10^{15} ions/cm^2. The number of dpa is smaller in the track region, between the surface and the nuclear cascade region; the track region corresponds to the zone where ions slow down mainly by electronic energy loss process. Several studies about SiC can be found in the literature[26-27]. These studies reveal that SiC is not disordered by electronic interactions. It is even found that ionizing radiation can promote the recovery of displacement damage[28].
The number of dpa is increasing with increasing fluence. For a depth of 7 μm, the number of dpa increases from 0.02 at 10^{13} ions/cm^2 to 0.2 dpa at 10^{14} ions/cm^2 and about 2 dpa at 10^{15} ions/cm^2. Under neutron or ion irradiations, the disorder increases with the irradiation fluence to reach amorphous state as soon as the fluence exceeds an amorphization threshold. For monocrystalline 6H-SiC[6], this amorphization threshold was determined to be in the 0.2– 0.5 dpa range at room temperature.

Thermal conductivity measurement

The range of 74 MeV Kr ions is equal to 7.5 μm for TiC and 9.5 for SiC. As shown on fig. 2, the irradiated samples are considered as coating (the implanted layer of few μm), on a substrate (the unimplanted material).

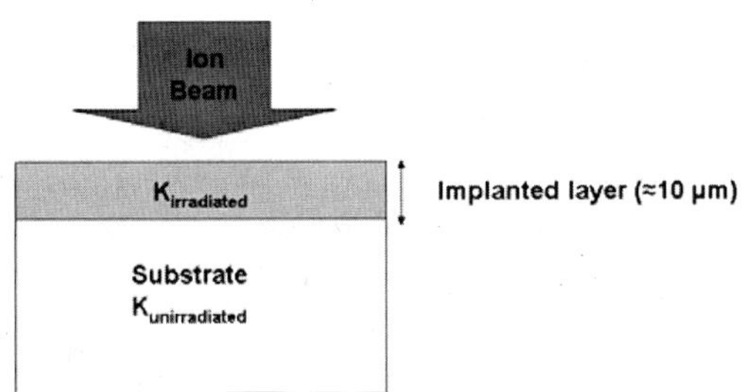

Figure 2. Schematic representation of the irradiated materials

Evolution of the thermal properties of irradiated materials is estimated using modulated photothermal IR radiometry experiment. It is based on gain measurement from an infrared photothermal measurement[29]. The thermal excitation is generated on the irradiated front face from a single line laser (488 nm, 2.5 W). The laser is modulated by an acousto-optic modulator from 10 Hz to 3 kHz. The thermal response is measured by a HgCdTe based photovoltaic infrared photodetector, cooled down by liquid nitrogen. A measure of the gain between the thermal perturbation and the response of the samples was used as experimental data. According to the knowledge of the substrate thermophysical properties (thermal conductivity, density, thickness), and to the knowledge of the irradiated layer (thickness, density), it is possible to estimate the thermal conductivity of the irradiated layer. A significant contrast between the thermal conductivity of the irradiated layer and the substrate is necessary to identify thermal conductivity of the irradiated layer.

Thermal diffusivity of unirradiated SPS samples (α) was measured at room temperature using a laser flash apparatus. Thermal conductivity values were deduced from equation (1).

$$K = \alpha \cdot \rho \cdot C_p \quad (1)$$

According to Lee[30], the effect of irradiation on the specific heat of SiC was negligibly small. There is no data on TiC. The specific heat of SiC and TiC is therefore assumed to be unchanged by irradiation. The density remains unchanged in the area between the surface and the nuclear cascade region (the zone where ions slow down mainly by electronic energy loss process), but decreases in the nuclear cascade area. The variation is found to be about 1%[1] (depending on the dose) if SiC remains crystallized, and about 10% if SiC becomes amorphous [31]. In our case, and for the highest dose (10^{15} ions/cm^2), the amorphous area is not expected to be more than 1 μm (for an irradiated layer of 9 μm). In such conditions, the density variation is lower than 2% and thus remains negligible in a first approximation. Density variation of TiC is also estimated unchanged. The specific heat capacity (*Cp*) of SiC and TiC was evaluated from Snead[1] and Lengauer[6] studies respectively. The specific heat capacity of TiC-SiC composites was calculated using the law of mixture (Table 1). Thicknesses of the implanted layers were calculated by SRIM and are in good agreement with the measure performed by TEM and SEM. Physical properties and thermal conductivity are reported on the table 1.

Table I. Properties of materials

	α-SiC	TiC-50vol.% SiC	TiC-25vol.%SiC	TiC-10vol.%SiC	TiC
Elaboration	-	SPS	SPS	SPS	SPS
Porosity (%)	3.5%	5%	3%	3%	3%
ρ (kg/m^3)	3200	3847	4340	4590	4656
Cp (J.Kg^{-1}.K^{-1})	680	623	595	577	566
K (W/mK at RT)	160 ± 20	27 ± 1	22 ± 1	20 ± 1	17± 1
Implanted layer thickness (SRIM)	9.3	8.3	7.9	7.8	7.5
Implanted layer thickness (TEM, SEM)	9.6	8.4	8.1	7.9	7.5

Transmission electron microscopy

Following irradiation, cross section transmission electron microscopy (TEM) specimens were prepared by gluing the irradiated disks to polished unirradiated disks, sectioning, grinding down to a thickness of 100 μm, and finally ion beam thinning (Ion Slicer JEOL) with 6 kV Ar ions until perforation occurred near the interface. The specimens were cleaned using 2 kV Ar ions and then examined using a conventional bright field and dark field imaging technique with a Philips CM30ST microscope.

Raman

The raman analyses were performed with a Labram HR (Jobin Yvon) microspectrometer (λ = 632.8 nm, spectral resolution≈ 2 cm^{-1}, lateral resolution ≈ 1μm). A motorized micro-displacement stage (± 0.1 μm) was used for the linescan measurement throw the cross section of the irradiated specimens.

RESULTS AND DISCUSSION

Fig. 3 shows the variation of the thermal conductivity as a function of fluence for α-SiC. For a fluence of 10^{13} ions/cm^2, SiC thermal conductivity is estimated to 90 W/m.K that is about half of the value measured on a non-irradiated material (160 W/m.K). Thermal conductivity has already significantly decrease under irradiation at 10^{14} ions/cm^2. Beyond

5.10^{14} ions/cm², thermal conductivity remains nearly constant at about 22 W/m.K. This corresponds to a drop of 86% as compared to the non irradiated material.

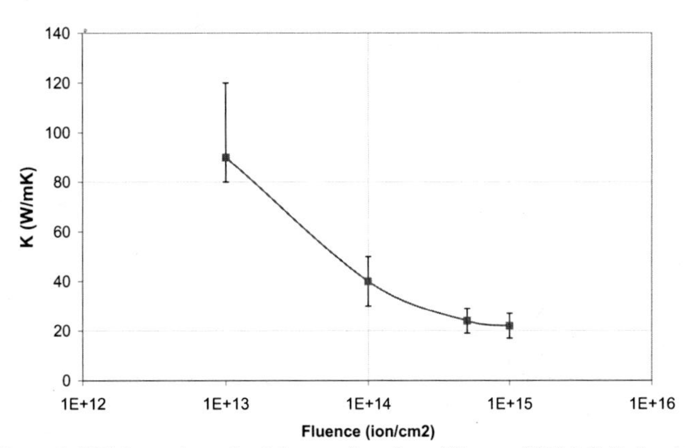

Figure 3: SiC thermal conductivity as a function of fluence (74 MeV Kr ions)

Raman spectra of the unirradiated and room temperature Kr-irradiated α-SiC specimens, at various fluencies are shown in figure 4. The spectra are recorded in the peak damage region. The spectrum of the original material shows two main sharp and intense peaks ar 790 cm^{-1} and 865 cm^{-1}, corresponding to the TO and LO modes characteristic of crystalline SiC. The intensity of the SiC peaks decreases, and their width increases with increasing dose. The evolution of the spectra indicates a local structure modification induced by irradiation. New peaks at 580 cm^{-1} and 850 cm^{-1} are detected and are associated to a distorted / disordered SiC structure corresponding to an intermediate state between amorphous and crystallized SiC, involving short range defects[32-33]. Dose increase induces a decrease of both the crystalline peaks (700 – 1000 cm^{-1}) and disorder peaks intensities. These peaks vanish beyond 5.10^{14} ions/cm² (1 dpa), and two broad bands appear around 520 and 1420 cm^{-1}. They are characteristic of Si-Si (520 cm^{-1}) and C-C homonuclear bonds (1420 cm^{-1}). As described by Sorieul[32], those peaks indicate that SiC is amorphous in this region.

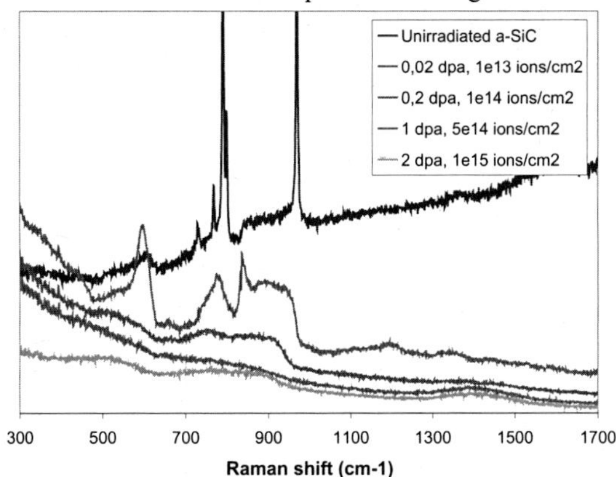

Figure 4. Raman spectra of Kr-ion irradiated SiC

In the nuclear collision region, amorphisation of SiC occurs for fluences larger than 5.10^{14} ions/cm^2. The Raman spectroscopy results are in good agreement with thermal conductivity measurements. It seems that both microstructure and thermal conductivity reach a saturated state beyond 5.10^{14} ions/cm^2, that corresponds to SiC amorphisation.

Thermal conductivity of amorphous SiC is expected to range from 3.6[31] to 5 W/m.K[34], that is significantly lower than our measured values. This probably results from the thermal conductivity of the whole implanted layer. Fig 5a shows the irradiated layer for a fluence of 10^{15} ions/cm^2. The amorphous area is clearly observed whereas a well ordered structure is observed in the electronic loss area: no damage is detected by TEM observation in this area. As the dpa number increases (as a function of depth, fig. 1), the lattice fringes are discontinuous and twisted because of the accumulation of irradiation-damages. In the peak damage region, which corresponds to the fully amorphous state under such irradiation, only a mottled contrast is observed (fig 5b).

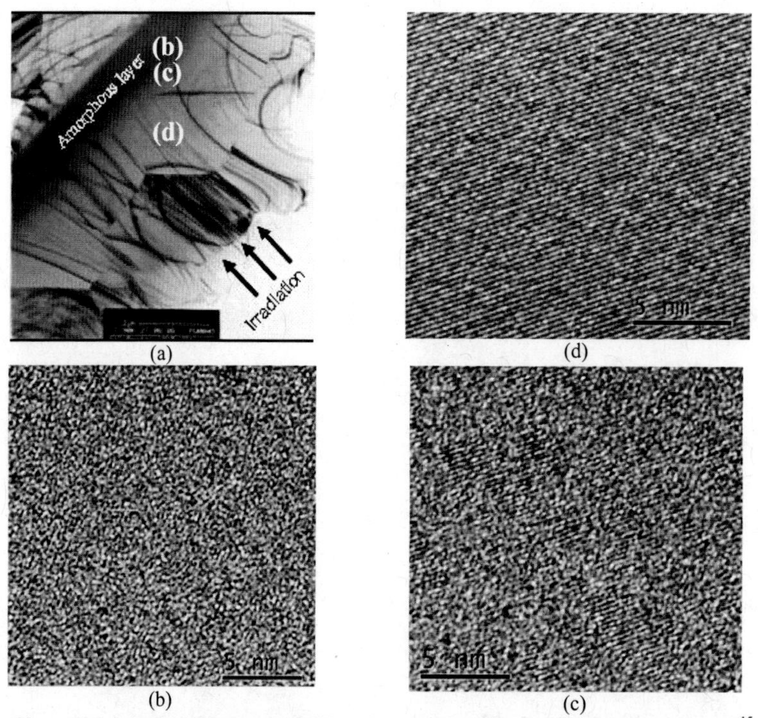

Figure 5. Bright field TEM and HRTEM observations of Kr 74 MeV irradiated SiC, 10^{15} ions/cm^2. (a) Implanted layer (b) Amorphous layer, (c) Limit electronic loss region / nuclear cascade region and (d) Electronic loss region

Raman spectra of Kr-irradiated (10^{15} ions/cm^2) α-SiC at room temperature are reported as a function of the depth (figure 6). In the electronic loss region (from the surface to about 7μm) the spectra reveal a moderate damage induced by irradiation. In the peak damage region (at 8.6 μm), two broad bands appear around 520 and 1420 cm^{-1}, indicating an amorphisation of SiC.

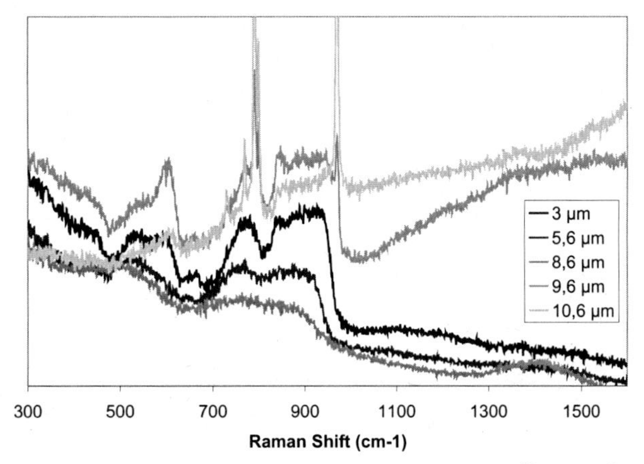

Figure 6. Irradiated SiC Raman spectra as a function of the depth (10^{15} ions/cm^2 74 MeV Kr)

It is obvious that such irradiations are responsible for a multi damaged layer. SiC is weakly damaged in the electronic loss region whereas SiC is totally amorphous in the nuclear cascade region. This explains the rather high thermal conductivity values measured in the whole damaged area (higher than amorphous SiC).

Fig. 7 shows the variation of thermal conductivity as a function of fluence for TiC. The variation is similar to that observed for silicon carbide. For a fluence of 10^{13} ions/cm^2, thermal conductivity is estimated at 14 W/m.K: the contrast between the substrate (non-irradiated layer, 17 W/m.K) and the irradiated layer is very weak. Beyond 5.10^{14} ions/cm^2, thermal conductivity of TiC is estimated to be 7 W/m.K, and remains nearly unchanged at 10^{15} ions/cm^2: the decrease is about 58% that is lower than the decrase observed for SiC.

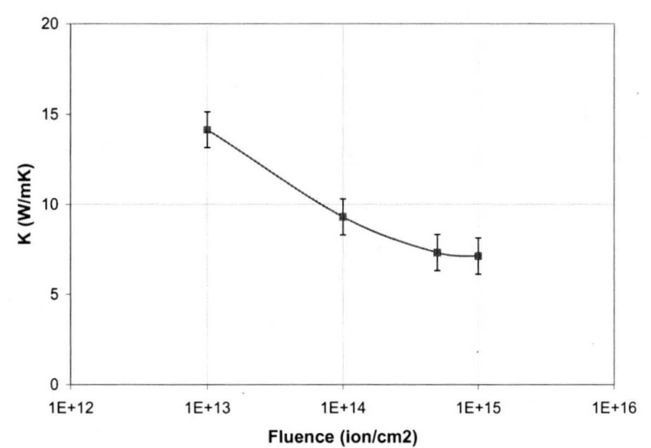

Figure 7. TiC thermal conductivity as a function of fluence (74 MeV Kr ions)

TEM observations have been carried out on materials irradiated at 10^{15} ions/cm^2. A high density of small defects with strong contrasts is observed (black dots). No such defects are detected in the electronic loss area (fig. 8). This suggests that, as observed in SiC, TEM observations do not reveal any damage in this area. However, this doesn't imply that there is no damage in the electronic loss region.

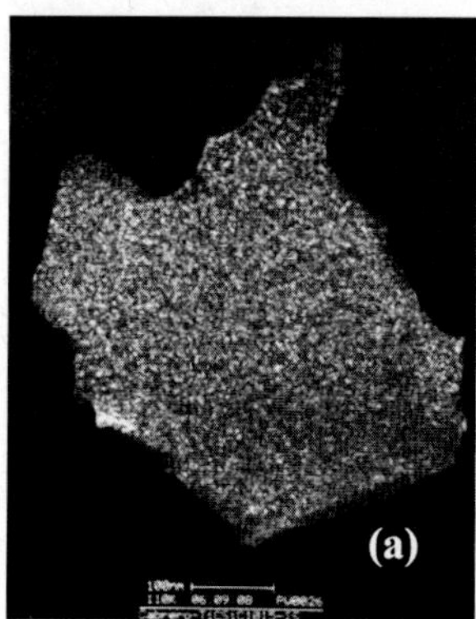

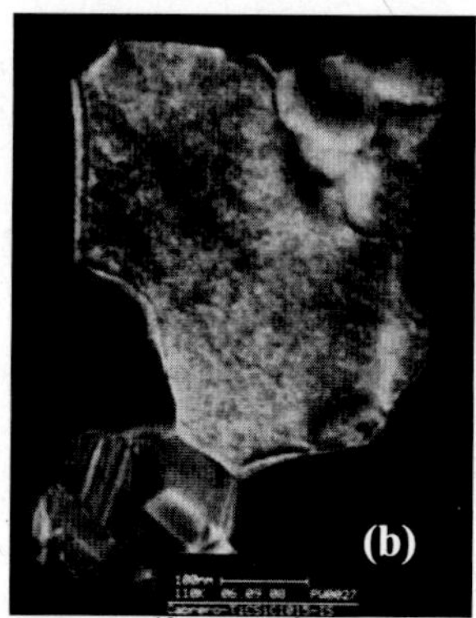

Fig 8. Dark field TEM observations of Kr 74 MeV irradiated TiC (10^{15} ions/cm^2). (a) Peak damage region, (b) Electronic loss region

High resolution micrographs (figure 9) show some very small disordered domains in the peak damage region, whereas under same conditions, SiC is completely amorphous. The size of the disordered region is of the order magnitude of the size of the cascade (few nm), and dislocations are observed in these area.

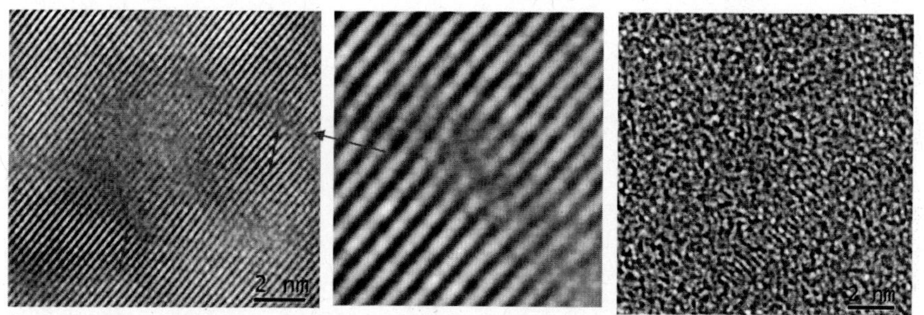

Fig 9. HRTEM observations of Kr 74 MeV irradiated TiC (left) and SiC (right), 10^{15} ions/cm^2 in the nuclear cascade region. Dislocation in TiC (middle)

Even if the thermal conductivity of TiC remains lower than thermal conductivity at room temperature of SiC after irradiation, the decrease is quite larger for SiC. Microstructural observations indicate that under same irradiation conditions, TiC seems to be less sensitive to irradiation and does not become fully amorphous.

Thermal conductivity of TiC-SiC samples with different SiC volume contents have been irradiated to study the influence of the composition. Fig. 10 shows the modification of the thermal conductivity induced by a 10^{15} 74 MeV-Kr irradiation. Thermal conductivity of all specimens is in the range 22-7 W/m.K. As shown in figure 10, the thermal conductivity decrease is larger for pure SiC. This can be explained by two effects. First, and as reported by Vaben[35] it is because in a rather defect free matrix (with a high thermal conductivity), the production of points defect (vacancies and interstitials) will have a more pronounced effect than in a matrix with a high density defect (which exhibits a low thermal conductivity). Secondly, and as shown previously, TiC is less damaged by Kr irradiation than SiC.

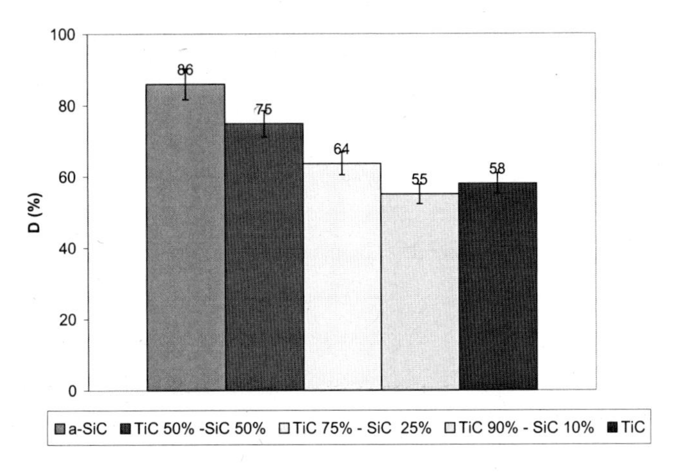

Figure 10. Drop of thermal conductivity after a 1 x 10^{15} 74 MeV Kr irradiation

By comparison with non-irradiated TiC-SiC composites, which present similar thermal conductivities, we observe that the higher the SiC content, the higher the decrease. Taking into account the measurement error, the decrease in thermal conductivity of TiC and $TiC_{90\%}$-$SiC_{10\%}$ samples are assumed to be equivalent. These observations are in good agreement with the previous results on TiC and SiC: SiC is more sensitive to 74 MeV-Kr irradiation than TiC.

CONCLUSION

α-SiC and TiC-SiC composites were irradiated with 74 MeV Kr ions with fluences ranging from 10^{13} to 10^{15} ions/cm^2. Thermal conductivity of silicon carbide decreases as fluence increases then remains constant above 5.10^{14} ions/cm^2. Raman and MET analyses reveal that for these fluences, SiC is amorphous in the peak damage region. At 10^{15} ions/cm^2 a thermal conductivity of 22 W/m.K is estimated, which is higher than that of amorphous SiC. This is because measures take into account two different areas, the nuclear cascade region where SiC is amorphous, and track region where SiC is only weakly damaged.

Titanium carbide exhibits the same behaviour than silicon carbide. Thermal conductivity decreases down to 7 W/m.K reached for a fluence of 10^{15} ions/cm^2. However, in contrast to SiC, TiC is not totally amorphous for a such fluence. Only small damaged domains are observed. Even if thermal conductivity of irradiated silicon carbide remains higher than titanium carbide, thermal conductivity measurement and microstructural analyses reveal that titanium carbide is less sensitive to irradiation than silicon carbide.

The value of k measured in this work takes into account the two different damaged layers induced by irradiation by energetic heavy ions. As the thermal conductivity of SiC is expected to be very high in the electronic loss region (about 50 W/m.K, and about 3 W/m.K in the peak damaged region), it is not surprising that the average value of k is higher for SiC than for TiC. Thus, in order to obtain a good estimation of the thermal conductivity of TiC and SiC in a nuclear core, it is necessary to measure thermal conductivity in the peak damage region for both SiC and TiC. This is the reason why irradiation with low energy heavy ions has been carried out at room temperature and at 900°C. Measurements of the thermal conductivity from room temperature to 1000°C of the above-mentioned samples are currently under progress.

ACKNOWLEDGEMENTS

The authors would like to thank I. Monnet, for her assistance during implantation experiments, G. Chollon, LCTS for his precious advices about Raman spectroscopy and J.L. Battaglia for helping during IR Radiometry experiments. Finally, MATINEX is greatly acknowledged for its financial support.

REFERENCES

[1]L.L. Snead, T. Nozawa, Y. Katoh, T.S. Byun, S. Kondo, D. A. Petti, Handbook of SiC properties for fuel performance modelling, *J. Nucl. Mat.*, 371, 329-377, (2007)

[2] R. Naslain and F. Christin, SiC-matrix composite materials for advanced jet engines, *MRS Bulletin*, 654–658, (2003)

[3] T. Maruyama, M. Harayama, Relationship between dimensional changes and the thermal conductivity of neutron irradiated SiC, *J. Nucl. Mat.*, 329-333, 1022-1028 (2004)

[4] M. Akiyoshi, I. Takagi, T. Yano, N. Akasaka, Y. Tachi, Thermal conductivity of ceramics during irradiation, *Fusion engineering and design*, 81, 321-325, (2006)

[5] Defect structure and evolution in silicon carbide irradiated to 1 dpa-SiC at 1100 °C *Journal of Nuclear Materials* 317, 145–159 (2003)

[6] W. Lengauer, S. Binder, K. Aigner, Solid state properties of group IVb carbonitrides, *Journal of Alloys and Compounds* 217, pp 137-147, (1995)

[7] E.R. Smith, D. Johnson, C. Brockway, J.K. Thompson, J.F. Lynch, *Engineering property data selected ceramics Vol. 2, Carbides, MCIC report,* (1979)

[8] J.A. Edmond, S.P. Withrow, H.S. Kong and R.F David, 1986, *Mater. Res. Soc. Porc.* 51 395-402

[9] V. Heera, J. Stoemenos, R. Kogler and W. Skorupa, Amorphization and recristallisation of 6H-SiC by ion beam irradiation, *J. Appl. Phys.* 77 2299-3009, (1995)

[10] M F Beaufort, F Pailloux, A Dclémy and J F Barbot, Transmission electron microscopy investigations of damage induced by high energy helium implantation in 4H–SiC, *J. Appl. Phys.* 94 7116-7120, (2003)

[11]L L Snead, and S J Zinkle, Mater. Res. Soc. Porc. 373 377-382 (1995)

[12] S J Zinkle and L L Snead, Influence of irradiation spectrum and implanted ions on the amorphization of ceramics, *Nucl. Instrum. Methods Phys. Res. B* 116 92-101, (1996)

[13] W Weber, L M Wang, N Yu and N J Hess, Structure and properties of ion-beam-modified (6H) silicon carbide, *Mater. Sci. Eng. A* 253 62-70, (1998)

[14] E Wendler, A Heft, and W Wesch, Ion-beam induced damage and annealing behaviour in SiC, *Nucl. Instrum. Methods Phys. Res. B* 141 105-117, (1998)

[15] J Grisola, B de Mauduit, J Gimbert, Th Billon, G Ben Assayag, C Bourgerette and A Claverie, TEM studies of the defects introduced by ion implantation in SiC, *Nucl. Instrum. Methods Phys. Res. B* 147 62-67, (1999)

[16] H Heera, J Stoemenos, R Kogler, M Voelskow and W Skorupa, Crystallization and surface erosion of SiC by ion irradiation at elevated temperatures, *J. Appl. Phys.* 85 1378-1386 (1999)

[17] V Heera, R Kögler and W Skorupa, Complete recrystallization of amorphous silicon carbide layers by ion irradiation, *Appl. Phys. Lett.* 67 1999–2001 (1995)

[18] A Heft, E Wendler, J Heindl, T Bachmann, E Glaser, H P Strunk and W Wesch, Damage production and annealing of ion implanted silicon carbide, *Nucl. Instrum. Methods Phys. Res. B* 113 239–43 (1996)

[19]Y Pacaud, J Stoemenos, G Brauer, R A Yankov, V Heera, M Voelskow, R Kogler and W Skorupa, Radiation damage and annealing behaviour of Ge^{+}-implanted SiC, *Nucl. Instrum. Methods Phys. Res. B* 120 177–80 (1996)

[20]A Höfgen, V Heera, F Eicchorn and W Skorupa, Annealing and recrystallization of amorphous silicon carbide produced by ion implantation, *J. Appl. Phys.* 84 4769–74 (1998)

[21]T Bus, A van Veen, A Shiryaev, A V Fedorov, H Schut, F D Tichelaar and J Sietsma, Thermal recovery of amorphous zones in 6H-SiC and 3C-SiC induced by low fluence 420 keV Xe irradiation, *Mater. Sci. Eng. B* 102 269–76, (2003)

[22] L. David, Développement de la microscopie thermique à sonde locale pour la détermination de la conductivité thermique de films minces, application aux céramiques pour le nucléaire, PhD, Institut national des Sciences Appliquées de Lyon, (2006)

[23] The Stopping and Range of Ions in Matter, http://www.srim.org/

[24] R. Devanathan a, W.J. Weber, Displacement energy surface in 3C and 6H SiC, *Journal of Nuclear Materials* 278 258-265 (2000)

[25] D. Gosset, M. Dollé, D. Simeone, G. Baldinozzi, L. Thomé Structural evolution of zirconium carbide under ion irradiation, *Journal of nuclear materials*, 373 pp123-129 (2008)

[26] Benyagoub A, Irradiation effects induced in silicon carbide by low and high energy ions , *Nucl. Instrum. Methods Phys.* Res. B 266 2008 2766-2771

[27] A. Audren, A. Benyagoub, L. Thomé and F. Garrido, Ion implantation of iodine into silicon carbide: Influence of temperature on the produced damage and on the diffusion behaviour, *Nuclear Instruments and Methods in Physics Research Section B: Beam Interactions with Materials and Atoms* Volume 266, Issues 12-13, 2810-2813 (2008)

[28]S.J. Zinkle, V.A. Skuratov, D.T. Hoelzer, On the conflicting roles of ionizing radiation in ceramics, *Nucl. Instr. And Meth.* B 191 (2002) 758-766.

[29] J.L. Battaglia, A. Kusiak, M. Bamford, J.C. Batsale, Photothermal radiometric characterization of a thin deposit using a linear swept-frequency heat flux waveform, *International journal of thermal sciences* 45 1035-1044 (2006)

[30] Lee C W, Pineau F J, Corelli J C, Thermal properties of neutron-irradiated SiC; effects of boron doping, *J. Nucl. Mater.* 108&109, 678 (1982)

[31] L.L. Snead, S.J. Zinkle, Structural relaxation in amorphous silicon carbide, *Nucl. Instr. And Meth.* B 191 497-503 (2002)

[32] S Sorieul, J-M.Costantini, L. Gosmain, L. Thomé and J-J. Grob, Raman spectroscopy study of heavy-ion-irradiated α-SiC, *J. Phys.: Condens. Matter* 18 5235–5251, (2006)

[33] G.Chollon, J.M. Vallerot, D. Helary, S. Jouannigot, Structural and textural changes of CVD-SiC to indentation, high temperature creep and irradiation, *J. Eur. Ceram. Soc.*, 27, 1503-11 (2007)

[34] J.P. Crocombette, G. Dumazer and N.Q. Hoang, Molecular dynamics medeling of the thermal conductivity of irradiated SiC as a function of cascade overlap, *Journal of applied physics* 101 023527 (2007)

[35] R. Vaben, D. Stöver, Processing and properties of nanophase non oxide ceramics, *Materials Science And Engineering* A301 59-68 (2001)

CORROSION RESISTANCE OF CERAMICS IN VAPOROUS AND BOILING SULFURIC ACID

C.A. Lewinsohn, H. Anderson, and M. Wilson
Ceramatec Inc.
Salt Lake City, UT, 84119

M. Sunderberg, J. Brangefalt
Kanthal AB,
Hallstahammar, Sweden

Ceramic microreactors are suitable for containing a variety of high temperature reactions for applications such as hydrogen production, fuel processing and reforming, synthetic chemistry, and point-of-use synthesis of hazardous material. In many of these applications, not only are the high-temperature structural properties of ceramics beneficial, but also their resistance to various forms of chemical corrosion. Although a number of materials are candidates for use in these demanding applications, the behavior of many of these materials under conditions anticipated during operation are unknown. Therefore, high temperature exposure tests were used to characterize the effects on mechanical strength and the formation of corrosion products on silicon-based ceramic materials and on certain intermetallic materials. Corrosion was investigated in vapour environments, and in boiling, liquid-sulfuric acid. Based on the results that will be discussed, non-oxide ceramics that form protective silica layers in oxidizing environments show promise for use in corrosive environments, especially at high temperatures.

INTRODUCTION

In 1979, Fernanda Coen-Porisini of the Commission of the European Communities JRC Ispra Establishment performed corrosion studies in sulfuric acid at 800°C and found that whereas all the metals tested displayed considerable to severe corrosion, alumina, mullite, and zirconia (ZrO_2), which were the few ceramics tested, were unchanged by exposure or had only coatings on the surface[1]. Irwin and Ammon of the Westinghouse Electric Corporation, in 1981, found that silicon and materials containing significant amounts of silicon, such as silicon carbide (SiC) and silicon nitride (Si_3N_4), have the greatest resistance to attack by boiling sulfuric acid.[2] Also in 1981, Tiegs identified silicon carbide as having the greatest corrosion resistance at 1000° and 1225°C in the simulated sulfuric acid decomposition environment[3], as compared to Sialon, MgO, ZrO_2(MgO) and ZrO_2(Y_2O_3). Tiegs recommended further testing for the SiC materials at conditions more representative of an actual sulfuric acid decomposition environment, that is, at temperatures of 800 to 900°C and pressures up to 3 MPa. Ishiyama et al. reported that SiC was the most corrosion resistant followed by Si-SiC and then by Si_3N_4 in high-pressure boiling sulfuric acid, as measured by the weight change and corrosion rate of samples exposed for 100 hours. Ishiyama also rated materials after 1000 hours of exposure and the three materials mentioned above were listed as all being the least affected by the long exposure.

In earlier studies[4,5], Lewinsohn et al. investigated the weight change and affect on strength of silicon carbide and silicon nitride after exposure to: (1) vaporous sulfuric acid at temperatures ranging from 850-950°C, approximately 86 kPa; (2) liquid sulfuric acid at 375-400°C, 14 bar; and (3) submersion in liquid, the interface between liquid and vapour, and the vapour at 250°C and approximately 86 kPa. These earlier studies showed that SiC and Si_3N_4 behave similarly. At elevated temperatures, greater than 850°C in the studies performed, protective silica (SiO_2) layers form on the outside of the materials and protect them from additional reaction at elevated temperatures. Based on measurements of the strength of Si_3N_4 in vaporous sulfuric acid at 250°C, at intermediate temperatures the growth rate of the SiO_2 film may not be sufficient to protect the materials from corrosion. Herrmann has also shown that there are conditions at which passivating layers cannot form during the corrosion of grain-boundary phases from the triple-points of silicon nitride[6]. Elevated pressure (1.4 MPa) increased the corrosion rate of both SiC and Si_3N_4, but both were still an order of magnitude lower than superalloy materials. In this work, the effect of exposure to vaporous and boiling sulfuric acid on two intermetallic materials, Ti_3SiC_2 (Maxthal 312) and Ti_2AlC (Maxthal 211) was measured. Results from these tests will be presented in this paper and, in some cases, compared with results from silicon carbide and silicon nitride obtained previously[4,5]. In addition, since the materials under investigation are candidates for microchannel components that must be bonded together for assembly, the effects of exposure to vaporous and boiling sulfuric acid on a polymer-derived joining material were examined.

METHODS

Samples were exposed to sulfuric acid at two conditions of temperature and pressure:

1. 850-950°C, atmospheric pressure (approx 86 kPa)
2. 250°C, atmospheric pressure (approx 86 kPa).

The corrosive environments for these exposure tests were selected to mimic the decomposition and boiling environments of sulfuric acid, respectively. The intermetallic materials were pre-oxidized, at 900°C in air for 24 h, since preliminary experiments indicated significant improvements in corrosion resistance relative to un-oxidized material. Moreover, the operating environment under consideration would be oxidizing and pre-oxidation could easily be performed prior to operation. The thickness of the resulting oxide layers were measured using micrometers. The thickness of the oxide layer for the Ti_3SiC_2 material was approximately 4 μm-thick; Ti_2AlC, 11 μm-thick. The effect of pre-oxidation on the resistance to corrosion from boiling sulfuric acid environments was unclear, and so both un-oxidized and pre-oxidized specimens were tested.

ASTM Standard C 1161-02C bend bars, type B (45 mm^L x 4.0 mm^W x 3.0 mm ^{H}) were scribed, weighed and randomly positioned within three small sample cups for sulfuric acid exposure. The simulated vapour conditions were 60% H_2SO_4, 30% H_2O and 10% air at 900°C. These sample cups were then loaded into a vertical quartz tube, contained by a clamshell furnace and heated to 900°C with flowing argon gas. Once at temperature, the simulated sulfuric acid environment was attained by switching over to air from argon and by dripping in the acid solution. At predetermined intervals (100, 500 and 1000 hours), samples were removed, weighed and fractured according to ASTM Standard C 1161-02C procedures. Table I describes the materials that were tested.

Experiments in boiling sulfuric acid, at atmospheric pressure, were conducted in a glass, reflux unit. The unit consisted of a heated flask and a reflux condenser cooled by recirculating, chilled water. ASTM Standard C 1161-02C bend bars were scribed, weighed and randomly positioned in a quartz fixture with three levels. Samples on the lowest level were submerged in sulfuric acid; the upper, only in contact with vapour. On the middle level, the interface between the liquid sulfuric acid and the vapour was at approximately the mid-plane of the samples. After 100 hours of exposure the samples were removed, weighed again, fractured, and examined by microscopy.

Table I
Materials Investigated

Material	Fabrication Process	Vendor	Identifier
Silicon Nitride	Hot Pressed	Ceradyne	HPSN
Silicon Carbide	Pressureless Sintered	Ceramatec	SiC
Maxthal 312 (Ti_3SiC_2)	Pressureless Sintered	Kanthal	312
Maxthal 211 (Ti_2AlC)	Pressureless Sintered	Kanthal	211

To evaluate the effect of exposure to sulfuric acid environments on a candidate joining material, silicon carbide samples consisting of 12.7 mm outer diameter, 1.6 mm wall thickness, and approximately 25.4 mm high tubes were bonded to square tiles approximately 25.4 mm^L x 25.4 mm^W x 6.3 mm^H. A paste containing a preceramic polymer, allyl-hydrido polycarbosilane (Starfire Systems, Watervliet, NY), and SiC powder was applied to the base of the tubes. A single tube was positioned on the 25.4 mm x 25.4 mm surface of a tile and placed on a supporting setter. Three assemblies of tubes and tiles were positioned on the setter in such a way that when force was transmitted to them through an articulated plate they received equivalent loads. The three samples were loaded into a resistively heated furnace with an alumina load rod attached to an air cylinder that was used to generate compressive, axial load on the joints during heat treatment. Joining was performed at approximately 1200°C using approximately 1.38 MPa pressure. Before exposure to sulfuric acid containing environments, the leak rate of the samples was measured using a helium leak checker to test whether they were hermetic. Approximately 83% of the samples were hermetic (15/18) and, hence, subsequently used for exposure testing.

Torsional shear-strength testing was performed using a custom apparatus that applied a rotational force, through a lever arm, on the hexagonal cap of a bolt glued into the open end of the tube using epoxy (**Figure 1**). The force was applied and measured by an electromechanical test frame (Instron, Model 5566). The square tile was constrained from rotating by stiff, metal supports and the force on the lever was measured by an electronic load cell using a strain gauge. Three samples were tested at each condition.

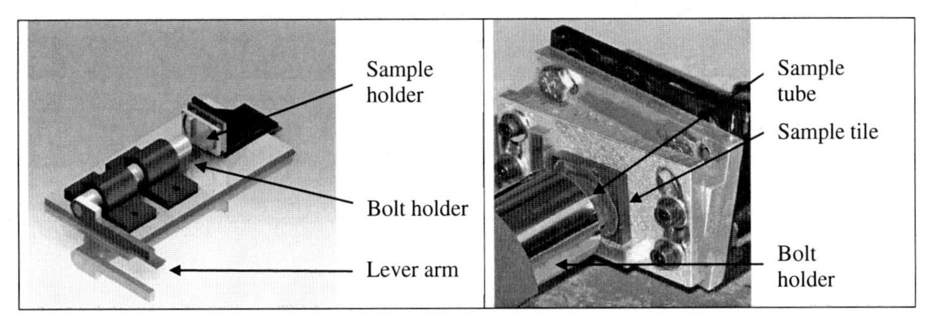

Figure 1 Torsional shear-strength test apparatus.

RESULTS AND DISCUSSION

Intermetallics: Vapour exposure

As reported earlier[5], SiC and Si_3N_4 experience an increase in weight as a result of exposure to a sulfuric acid decomposing environment at 900 °C. X-ray photoelectron spectroscopy (XPS) and Energy dispersive X-ray (EDX) analysis on a scanning electron microscope (SEM) indicated that SiO_2 (silica) was adhered to the surfaces of the samples[4]. Detailed results from these earlier studies are found elsewhere[4,5] and will not be reviewed in this section; nevertheless, some earlier results will be included for comparison purposes. The weight change, measured after various exposure times, for the pre-oxidised, intermetallic materials are shown in Figure 2. Both materials exhibited an increase in weight. The Ti_3SiC_2 (312) material gained significantly more weight than the Ti_2AlC (211) material, although the thickness of the oxide layers on both materials after 1000 h of exposure was similar. Since the weight gain of the Ti_3SiC_2 material was linear, and continuously increasing throughout the experiment, it appears that the oxide layer on the Ti_3SiC_2 material was not passivating.

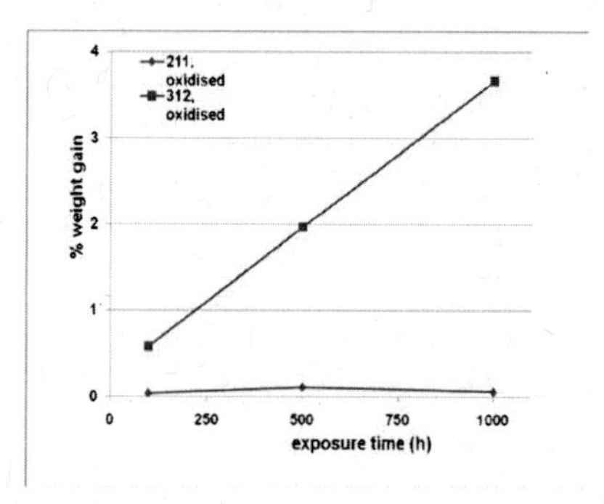

Figure 2 Weight change of Ti_3SiC_2 and Ti_2AlC after exposure at 900°C with H_2SO_4 vapour.

After exposure, both the silicon carbide and silicon nitride specimens exhibited a slight increase in strength[4,5], presumably due to the blunting of surface flaws by the formation of silica. For the intermetallic materials, the measured strength values after 100 h of exposure were not statistically different from the room-temperature strength value reported by the manufacturer.

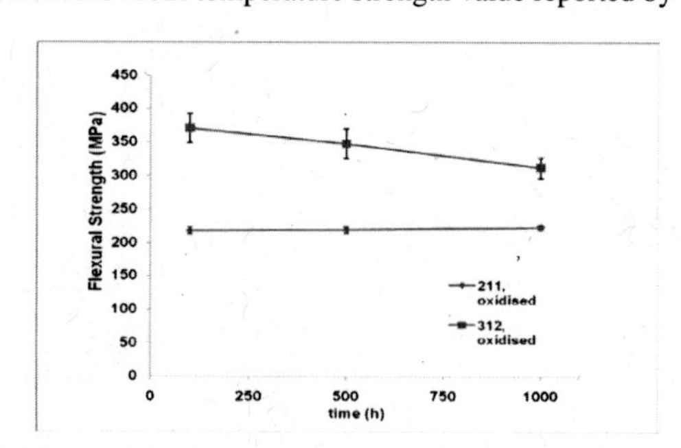

Figure 3 Strength of materials after exposure at 900°C with H_2SO_4 vapour.

The strength of Ti_2AlC was barely affected by the exposure, but the strength of Ti_3SiC_2 was lower after 1000 h of exposure, as shown in Figure 3. This supports the observation that the oxide layer on the Ti_3SiC_2 material was not completely passivating.

Intermetallics: Boiling, liquid exposure

Both materials exhibited significant weight loss after exposure to boiling sulfuric acid, Figure 4. Pre-oxidized Ti_3SiC_2, however, was extremely resistant to corrosion in these conditions. Exposure at the interface between the liquid and the vapour appeared to be slightly more aggressive than exposure either in the pure liquid or vapour environments. The strength of the materials after exposure showed similar trends to the weight change data: the strength of pre-oxidized Ti_3SiC_2 was not strongly affected by exposure and was only slightly less than that of silicon carbide, Figure 5.

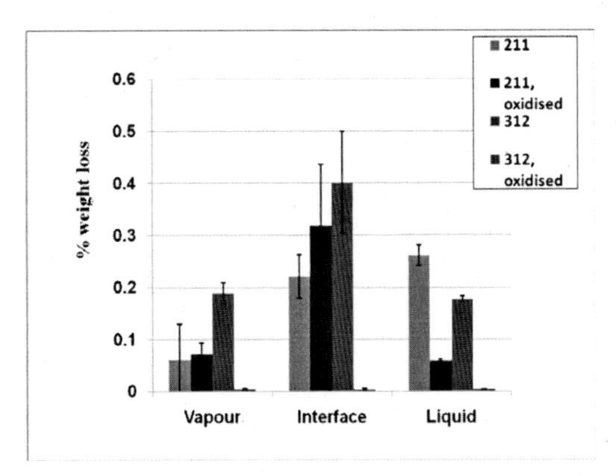

Figure 4 Weight change of intermetallic materials after exposure in boiling sulfuric acid environments.

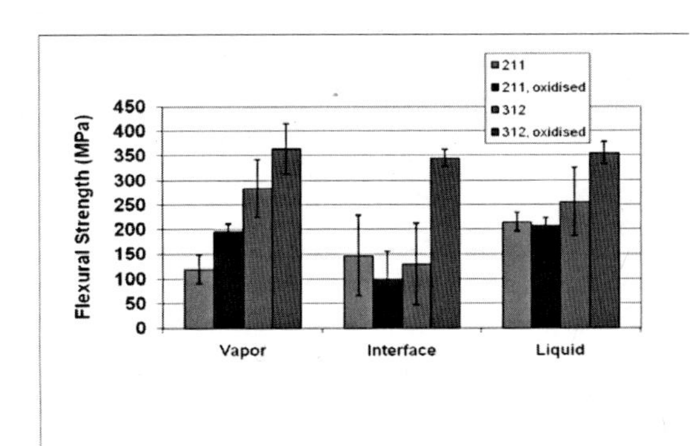

Figure 5 Strength of intermetallic materials after exposure in boiling sulfuric acid environments.

Ceramic-ceramic joining

Samples joined as described previously were exposed at 900°C in an environment containing 60 mol% H_2SO_4, 30 mol% H_2O, 9.5 mol% O_2, balance N_2 at 900°C for up to 1000 hours. After exposures of 100, 600, and 1000 h, specimens were retrieved and the weight, leak rates, and torsional strengths were measured for comparison to the unexposed condition. The weight change results, Figure 6, showed that the samples exhibited a small weight gain after exposure. Despite some apparently anomalous data, the weight gain due to exposure in the environment containing sulfuric acid was similar to that after exposure to ambient air, at 900°C. These results are similar to those observed for exposure of silicon carbide and silicon nitride materials reported earlier[4,5]. Despite some scatter in the data, the strength of the joined specimens did not appear to be affected by the exposures either, Figure 7. Most importantly, all the specimens remained hermetic after exposure also. Therefore, it appears that this bonding method is suitable for assembling ceramic, microchannel heat exchangers.

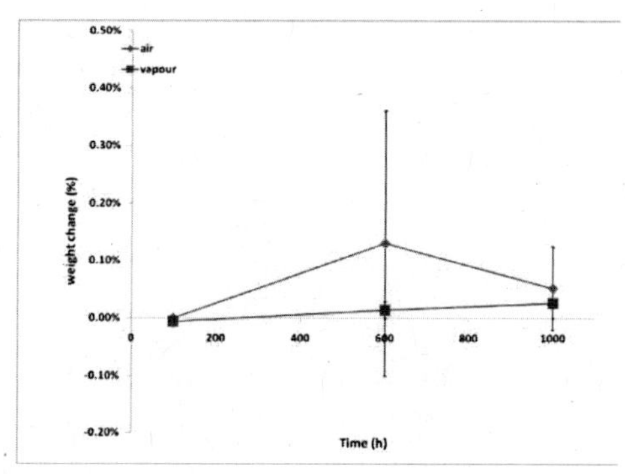

Figure 6 Weight change of joined specimens after exposure to sulfuric acid containing environments.

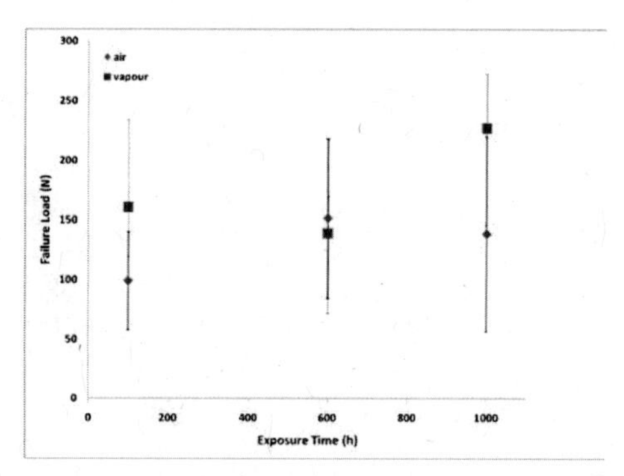

Figure 7 Torsional strength of joined specimens after exposure to sulfuric acid containing environments.

SUMMARY

Due to minimal changes in weight and strength after exposure to sulfuric acid, pre-oxidized Ti_3SiC_2 offers promise for use in boiling sulfuric acid environments. On the other hand, Ti_2AlC appears to be more suitable at higher temperatures in environments containing sulfuric acid vapour. The strength of Ti_3SiC_2 and Ti_2AlC are slightly lower than that of silicon carbide, but they possesses ductility at high temperatures and a lower sintering temperature. Therefore, in certain applications, these materials are candidates for use in ceramic microchannel reactors. Furthermore, the results indicated that the joining method based on conversion of polycarbosilane to a covalently-bonded solid is capable of producing leak tight joints that are also resistant to sulfuric acid. This joining method enables the fabrication of heat exchanger stacks from mass-produced, microchannel plates.

REFERENCES

[1] Coen-Porisini, Fernanda. *Corrosion Tests on Possible Containment Materials for H_2SO_4 Decomposition.* Jt. Res. Counc., ERATOM, Ispra, Italy. Advances in Hydrogen Energy (1979), 1(Hydrogen Energy Syst., Vol. 4), 2091-112.

[2] Irwin, H.A., Ammon, R. L. *Status of Materials Evaluation for Sulfuric Acid Vaporization and Decomposition Applications.* Adv. Energy Syst. Div., Westinghouse Electric Corp., Pittsburg, PA, USA. Advances in Hydrogen Energy (1981), 2(Hydrogen Energy Prog., Vol. 4), 1977-99.

[3] Tiegs, T. N., "*Materials Testing for Solar Thermal Chemical Process Heat.*" Metals and Ceramics Division, Oak Ridge National Laboratory. Oak Ridge, Tennessee. ONRL/TM-7833, pp. 1-59, 1981.

[4] C.A. Lewinsohn, H. Anderson, M. Wilson, T. Lillo, and A. Johnson, "Corrosion Resistance of Ceramics in Vaporous and Porous Sulfuric Acid," Cer. Engng. & Sci. Proc., vol. 29, pp.xxx, 2008.

[5] C.A. Lewinsohn, H. Anderson, and M. Wilson, "Corrosion Resistance of Ceramics in Sulfuric Acid Environments at High Temperature," Cer. Engng. & Sci. Proc., vol. 28, pp.289-295, 2007.

[6] Herrmann, M., Schilm, J., "Shape dependence of corrosion kinetics of Si_3N_4 ceramics in acids," Ceramics International, v35, pp. 797-802 (2009).

UNLUBRICATED CLUTCH SYSTEM BASED ON THE FUNCTION RELEVANT FRICTION PAIRING ADVANCED NON-OXIDE CERAMIC VS. STEEL

A. Albers, S. Ott, and M. Mitariu

IPEK – Institute of Product Development

University of Karlsruhe (TH)

Karlsruhe, Germany

ABSTRACT

Within the frame work of the Centre of Excellence in Research CER 483 "High performance sliding and friction systems based on advanced ceramics" one approach is to apply advanced ceramics as friction material for e.g. a dry running motor vehicle clutch using the ceramic specific benefits as wear and temperature resistance combined with lightweight design to fulfil today's demands as e.g. functionality, life time and reliability.

CLUTCH SYSTEM WITH ADVANCED CERAMICS

Product Development Process

The "ceramic specific" product development process is based on designing and dimensioning guidelines and tools for ceramic specific designs and integration into the dry running clutch systems. In combination with SPALTEN, a problem solving methodology in product development [1], and the Contact & Channel Model [2], a methodology for analyzing and generating system functionality by working-surface-pairs and channel-support-structures, new dry running clutch system designs are developed and analyzed by virtual simulation methods as e.g. Finite-Element techniques.

In the following, the developed clutch system based on the "ceramic specific" product development process is presented concerning its bearing, spring and fixing concepts for an entire clutch system with advanced engineering ceramics. The developed clutch system (Fig. 1) is divided into sub-systems: the fly wheel, the clutch disk and the clutch cover.

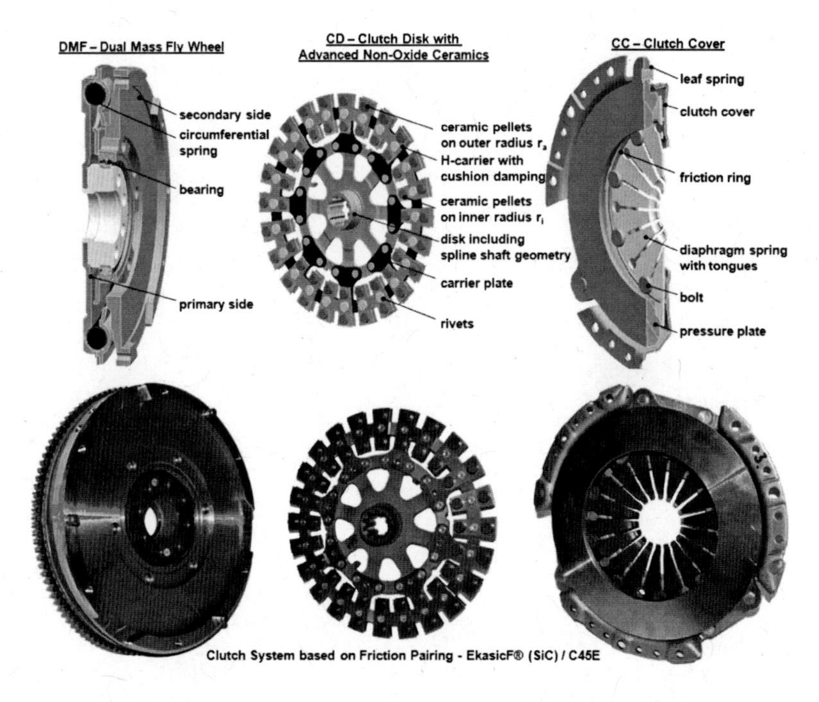

Fig. 1: Sub-systems of the clutch system with advanced ceramics

The recent design of the clutch disk, developed in the context of the "ceramic specific" product development process, is optimized with regard to increased transmissible torque, improved wear resistance and weight (fig. 1). The clutch disk is composed of a weight optimized disk including spline shaft geometry for the transmission input shaft of testing vehicle at IPEK laboratory, a carrier ring and the H-shaped cushion damping device with the advanced engineer ceramic. When engaging, the cushion damping generates the designated contact pressure adaptation up to the operation point - the modulation - and an equal and wear reduced carrying of the pellets. Additionally, the cushion damping interacts with the deterioration improved pellet shape and avoids high abrasive wear of the metallic friction partners. The dry running friction contact consists of the non-oxide ceramic SiC (2540±10 HV) [EkasicF®, ESK Inc.] and of the normalized steel C45E (206±6 HV) integrated within the developed clutch system solution.

Engineering-ceramic materials excel in their high temperature resistance, compressive strength, rigidity and hardness while having relatively low density. These advantages with regard to pressure and temperature are accompanied by disadvantages such as low bending stress, shear stress and tensile stress. Thus, the integration of the ceramic into a metallic environment is a big challenge for the design. Different thermal expansion coefficients for example can lead to temperature-dependent pressures or constructive clearances in the contact area of the working surface pairs. Distortion by temperature gradients and manufacturing deviations from parallelism would lead to bending stress in the ceramic ring when the clutch is closed. Furthermore, the supporting surface part during the sliding phase is small, due to low yielding and therefore "adaptability" of the ceramic. For these reasons the round-shaped facing which is common in most clutches cannot be substituted directly by an engineering-ceramic. This shows the effects of material choice on the constructive shape.

Experimental Analysis within Different Levels

Experimental analysis and examination are carried out in different levels of the "Tribological Testing Chain" [3] according to Czichos and Habig to gain different aspects and to deepen the comprehension of dry running friction contact.

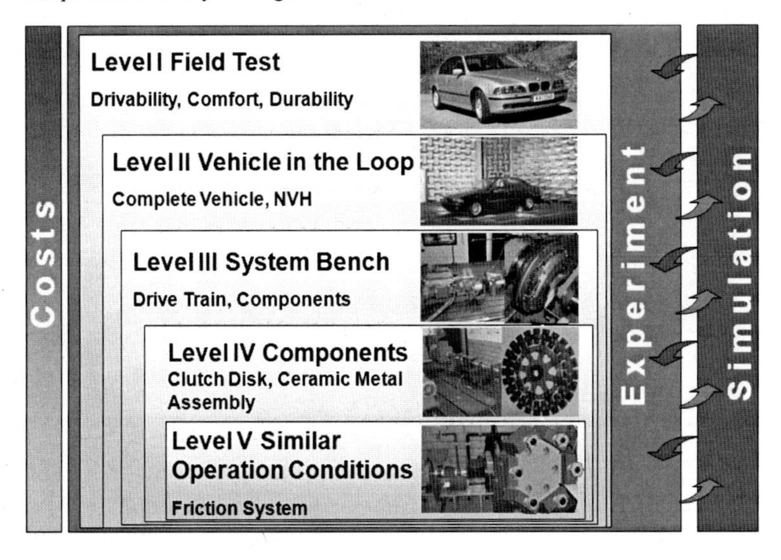

Fig. 2: IPEK interpretation of the "Tribological Testing Chain"

The IPEK interpretation of the "Tribological Testing Chain" combines experimental analysis with parallel simulation techniques within the different levels (Fig. 2). Within testing level V the focus of the examination is friction and wear behaviour of the specimen. The focus of testing level IV is the interaction of design and tribological performance. Friction and wear behaviour are still detectable, so good transferability between levels can be realized. Within testing level III the focus of analyses are design aspects (e.g. integration ceramic, steel, cushion damping) as well as comfort aspects (e.g. clutch induced vibrations, noise, drivability). The debris is measured and analyzed to examine the reliability of the clutch system as well as to ensure the transferability and the application of the proper wear mechanism within lower levels.

EXPERIMENTAL RESULTS CONCERNING COMPONENTS AND CLUTCH SYSTEM PERFORMANCE

The experimental results within level IV "Components" visualize the high potential of ceramic material for the dry running friction contact (Fig. 3). The focus is the analysis of the interaction of friction behaviour and clutch disk design. The operation parameters are a drive off speed, n_{An} = 1500 rpm, and the axial load, F = 6100 N. This synchronization considers a regular drive-off characteristic [4]. The demanded high coefficient of friction for the ceramic/steel pairing is realized by SSiC/C45E. In comparison to state of the art friction pairings it is visible that the ceramic/steel pairing provides a coefficient of friction higher than $\Delta\mu = 0.15$ at slow sliding velocities. Parallel the interaction of friction behaviour and system design shows a further possibility to improve dry running friction systems. The clutch disk of the 2nd generation enhances the friction behaviour due to new developed cushion damping.

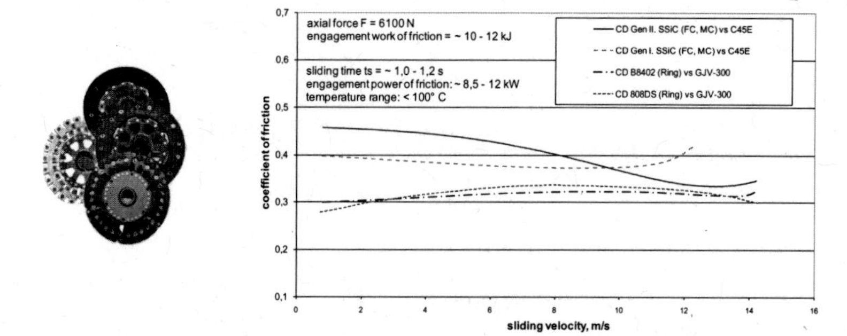

Fig. 3: Testing Level IV - Experimental results of the coefficient of friction over sliding velocity for different clutch systems with different friction pairings

Stepping up to level III "System Bench" the complete clutch system (fig. 1) is analyzed instead of single clutch disks. New design aspects as e.g. integration of the steel plates and the spring and damping devices inside the clutch can now be analyzed under real operation conditions. With regards to transferability the operation conditions are similar to level IV, drive off speed, n_{An} = 1500 rpm, and a axial load, F = 6100 N. The axial load is applied by the diaphragm spring of the clutch system. Concerning comfort first results can be gained regarding the effects of clutch induced vibrations and noise. Further, first estimations concerning drivability (start-behaviour) are realized. Considering comfort aspects the developed clutch system with advanced ceramics (friction pairing: EkasicF® (SiC) / C45E) induces high vibrations within the drive train measureable at the output torque of the system configuration. These torsional vibrations would induce longitudinal vibrations within a vehicle application and would influence strongly the comfort.

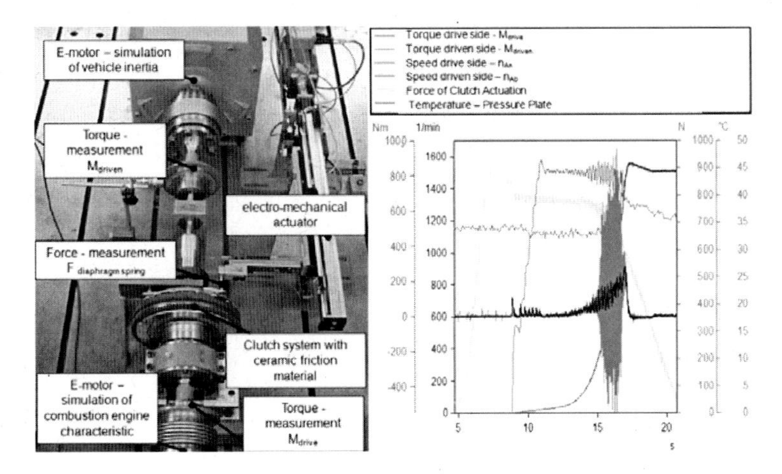

Fig. 4: Testing Level III - Experimental results of clutch system with advanced ceramics (Friction pairing: EkasicF® (SSiC) / C45E)

CONCLUSIONS

Within this paper the result of the "ceramic specific" product development process is presented concerning the realization of the second generation of a prototype of a clutch disk with advanced ceramics and the first prototype of a complete clutch system concerning bearing, spring and fixation designs. Due to comfort by the designated force modulation as well as wear resistance by single spring adaptation for each ceramic pellet a ceramic specific clutch disk design is developed to perform demanded functionality and to apply the ceramic specific properties as e.g. high wear and temperature resistance for an automotive application.

In terms of validation within the IPEK interpretation of the "Tribological Testing Chain" according to Czichos and Habig the experimental evaluation of different friction pairings concerning friction behaviour is documented. Further, the experimental evaluation of the prototypes of a clutch system is presented on level IV "Components" and level III "System Bench" under real operating conditions. The effect concerning the interaction of design and tribology is focused on the testing level IV visualizing the design influence and the potential of advanced ceramics in comparison to today´s state of the art friction facings in terms of coefficient of friction. The results of the comfort analyses on level III parallel indicate the further development needs in terms of vibration reduction. Further investigations will be realized avoiding vibrations by a further development of the friction partners (multilayered ceramic with friction modifiers) as well as the development of vibration counter steering control units within a ceramic specific clutch actuation system.

REFERENCES

[1] Albers, A.; Burkardt, N.; Mebold, M.: SPALTEN – problem solving methodology in the product development, Proceedings of Conference on Engineering Design, ICED 2005 Melbourne, 2005.

[2] Albers, A. ; Matthiesen, S.: Konstruktionsmethodisches Grundmodell zum Zusammenhang von Gestalt und Funktion technischer Systeme - Das Elementmodell "Wirkflächenpaare & Leitstützstrukturen" zur Analyse und Synthese technischer Systeme; Konstruktion, Zeitschrift für Produktentwicklung; Band 54; Heft 7/8 - 2002; Springer-VDI-Verlag GmbH & Co. KG; Düsseldorf 2002, Seite 55 – 60.

[3] Czichos, H.; Habig, K.-H.: Tribologie Handbuch, Reibung und Verschleiß, 2. Auflage, Wiesbaden, Vieweg, 2003.

[4] Albers, A., Ott, S., Mitariu, M.: Dry running clutch system with advanced ceramics for automotive applications. Proceedings of 13th Nordic Symposium on Tribology, Tampere, Finland, June, 10.-13., 2008.

NONDESTRUCTIVE INSPECTION OF CERAMIC BEARING BALLS USING PHASED ARRAY ULTRASONICS

J.G. Sun, E.R. Koehl, S. Steckenrider,
Argonne National Laboratory, Argonne, IL, USA

Charlotte Vieillard,
SKF Research & Development Centre, The Netherlands

Ton Bayer,
SKF Manufacturing Development Centre, Gothenburg, Sweden

Hybrid bearings, consisting of silicon-nitride ceramic rolling elements (e.g., balls) and a metal race, are a critical enabling technology for advanced gas turbines. A primary factor for reliable manufacturing of ceramic bearing balls is to establish an affordable nondestructive inspection method to detect flaws on the ball surface and subsurface. The current inspection method is based on fluorescent penetrant examination under ultra-violet lights, which is a slow process. A nondestructive evaluation (NDE) method, based on ultrasonic Rayleigh waves that are generated/detected by phased array probes, is being investigated for detection and characterization of these flaws, including Hertzian C-cracks. Both detection sensitivity and speed are important parameters to evaluate the performance of this NDE method. This paper describes the principle of this method and presents recent results on detection capability for various flaws in bearing ball specimens.

INTRODUCTION

Hybrid bearings, comprised of silicon-nitride ceramic rolling elements (e.g., balls) and a metal race, have significant advantages over traditional steel bearings, including increased bearing life, reduced friction and heat generation, increased bearing stiffness, improved corrosion resistance, larger operating temperature range, and increased rotational speed [1]. These advantages have led to large-volume commercial use of hybrid bearings as instrument and machine tool spindle bearings, and recently in more critical/demanding applications such as in racing cars and in wind and gas turbines [1-2].

Because of the important roles played by hybrid bearings, the quality of ceramic bearing components (balls) is critical to assure their safety and reliability in those demanding applications. A silicon-nitride ceramic bearing ball may contain various types of surface and subsurface flaws. One particular type of flaw is the surface-breaking Hertzian C crack [3-4] that is normally induced during the manufacturing process and in handling. These cracks and flaws can significantly reduce the reliability and lifetime of the bearing components [5-7]. A critical factor for reliable manufacturing of ceramic bearing balls is to establish an affordable nondestructive inspection method to detect all flaws on the ball surface and subsurface. The current inspection method is based on fluorescent penetrant examination under ultra-violet lights, which is a slow and labor-intensive process. A nondestructive evaluation (NDE) method, based on ultrasonic Rayleigh waves generated/detected using phased array probes, is being investigated for detection and characterization of these flaws, including Hertzian C-cracks. For this method, both detection sensitivity and speed are important parameters to evaluating its performance. This paper describes the principle of this NDE method and presents recent results on its detection capability for various flaws in bearing ball specimens.

ULTRASONIC PHASED-ARRAY EXPERIMENTAL SYSTEM

Ultrasonic Rayleigh waves, also called surface acoustic waves (SAW), are defined as a guided wave that propagates along the surface of a medium. Like other guided waves (such as Lamb waves), SAWs travel in the longitudinal direction, and are most often used for NDE in a single-sided pulse-echo technique. This makes SAWs an attractive means of detecting defects that are perpendicular to a material's surface, such as surface breaking cracks. SAWs are suitable for detecting near surface features because the energy of the waveform is concentrated within a shallow surface layer. The SAW travels along the surface, but its amplitude decays within the material. The decay depth is on the order of the wavelength of the propagating waveform. The asymptotic decay of the waveform amplitude means that the majority of the wave's energy is contained within a single wavelength from the material's surface.

A SAW can be generated by a longitudinal ultrasonic transducer by positioning the transducer at a critical angle $\theta_C = \sin^{-1}\left(V_{water} / V_{Rayleigh}\right)$ with respect to the local surface normal (see Fig. 1), determined by the acoustic velocities of the material and the coupling medium (water) [8]. If a surface breaking crack is present, a portion of the waveform's energy is reflected and can be detected by the transducer in the pulse-echo mode. As the wave propagates along the surface, SAWs can be used on non-planar geometries, such as pipes and weld radii.

To further increase the sensitivity of the system, Argonne National Laboratory (ANL) developed a specialized phased-array probe for generation and detection of Rayleigh waves in bearing balls (see Fig. 2). The probe consists of transducer elements arrayed along a concave surface. Each transducer element is aligned normal to that concave surface and produces an ultrasonic wave that interacts with the surface of the ball at the critical angle to form a Rayleigh wave within the ball's surface that propagates along the probe axis direction. The firing of the various transducer elements is timed so that the Rayleigh waves from all elements are superimposed in-phase on the ball's surface, thereby dramatically boosting beam power, and therefore flaw detection sensitivity [9].

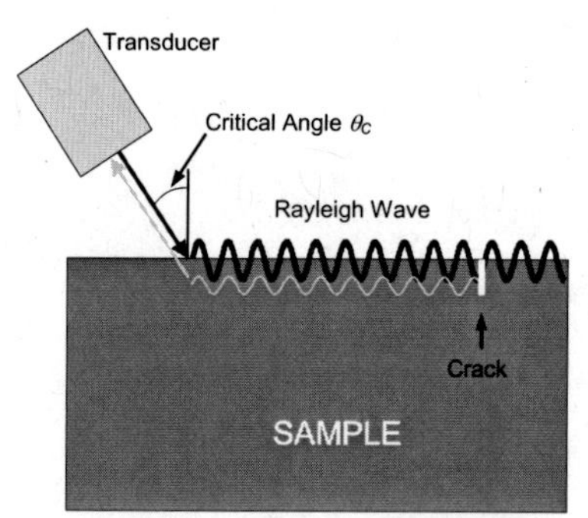

Fig. 1. Schematic diagram showing the creation and detection of SAW wave using an ultrasonic longitudinal transducer.

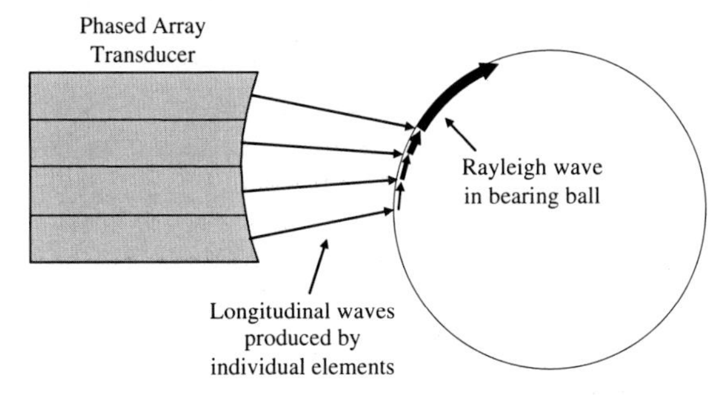

Fig. 2. Schematic diagram showing how the waves produced by phased-array elements will generate a stronger Rayleigh wave.

Fig. 3. Experimental setup for scanning a bearing ball.

Figure 3 is a photograph of the experimental setup for ultrasonic inspection of a ceramic bearing ball. Two 32-element ultrasonic phased-array probes with a nominal frequency of 16.5 MHz (made by Imasonic S.A., Besançon, France) were orientated in horizontal and vertical planes, respectively, relative the vertical rotation axis of the bearing ball. The probe oriented in the horizontal direction is used to perform a line scan. Once fired, this probe produces a SAW traveling along the horizontal circumference (e.g., equator line) of the bearing ball. Because the SAW propagation path is parallel to the motion direction of the surface during the ball rotation, only a single circumferential line is scanned by the ultrasonic SAW from the horizontally-oriented probe. This scan configuration is used to establish detection sensitivity of the probe to detect defects (positioned along the path of the SAW propagation) as a function of distance. On the other hand, the vertically-orientated probe produces a SAW that travels from the equator upward toward the top dead center (TDC) of the ball. As the ball is rotated about the vertical axis during the test, the upper half surface of the ball is scanned by the ultrasonic SAW. Therefore, this scan configuration is used to detect all flaws located in the upper half surface of a bearing ball. Obviously, if the vertically-oriented probe is positioned to produce a downward SAW, the lower half surface of the ball can be scanned. It should be noted that although the scan results from the two configuration setups appear to be similar, they have completely different interpretations as stated above.

The ANL phased-array system is composed from two commercially available systems. The data acquisition and data analysis was done with a RD Tech Focus LT system (Olympus NDT, Québec, Canada), which is capable of a data acquisition rate of 100 kHz. The automated motion system was based on a Sonix (Sonix, Inc., Springfield, VA) C-scanning system and included gantry, motors, rotation stage, and drivers. The two systems were coupled with a specially designed encoder system that enabled the two computer-based systems to communicate.

CERAMIC BEARING BALL SPECIMENS

Two 1-inch-diameter silicon-nitride bearing balls with various defect features were used to determine the detection sensitivity and performance of the NDE technique based on ultrasonic Rayleigh waves using the phased array probes. Ball #1 contains several surface/subsurface flaws. Of particular interest to this study is a large machined gouge shown in Fig. 4a. It was characterized to have a length of >1.5 mm, a width of 0.287 mm, and a shallow depth. The gouge is indicated by a marking that consists of four Vicker's indents (each ~70 μm in size; three in a line and the fourth offset) located at ~7 mm away from the gouge.

Bearing Ball #2 contains an artificial C-crack and another small surface feature. The artificial C-crack, shown in Fig. 4b, has a chord length of 1.32 mm with a radius of 0.745 mm, and a crack mouth opening at middle arc of <0.7 μm. The markings for the C-crack and the surface feature also consist of four Vicker's indents and located ~6.5 mm from the corresponding features.

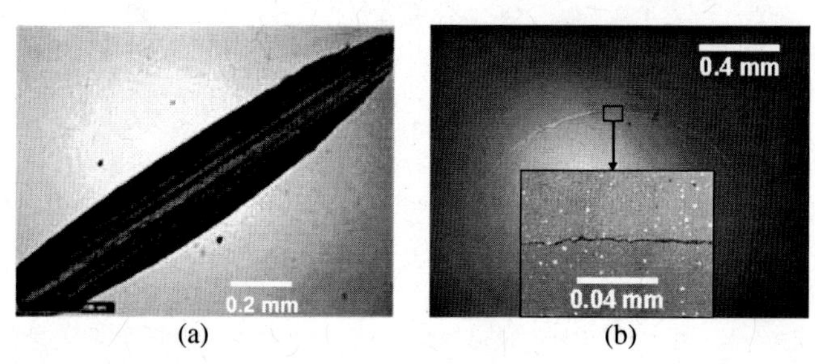

Fig. 4. Micrographs of (a) large gouge in bearing Ball #1 and (b) artificial C-crack in bearing Ball #2.

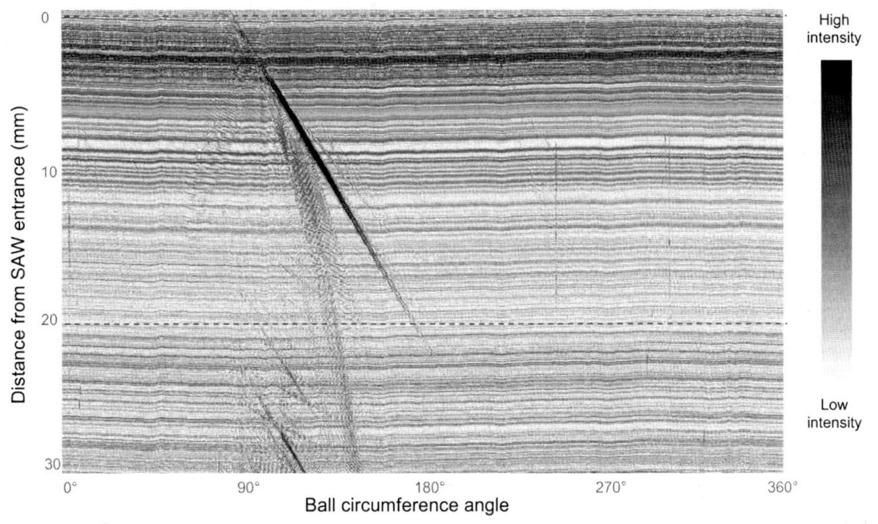

Fig. 5. Scan image of large gouge in bearing Ball #1 using horizontally-oriented probe.

EXPERIMENTAL RESULTS

The large gouge in bearing Ball #1 was used to establish the detection sensitivity of a flaw as a function of the distance between the SAW initiation position (i.e., SAW entrance position) and the flaw position using the horizontally-oriented probe. The gouge was initially placed on the equator and at 90° behind the point where the SAW was initiated. The major axis of the gouge was positioned perpendicular to the SAW propagation direction in order to maximize the reflection signal. During the scan, the ball was rotated in the direction of SAW propagation. The image from this scan is shown in Fig. 5. The vertical axis indicates the detection distance from the initiation point (at top of the image) along the equator of the ball. The horizontal axis shows the angular position as the ball was rotated. Correspondingly, the gouge was moved first toward (beginning at 90°) and then away from the initiation point as ball rotated. The detected SAW signal reflected from the gouge can be clearly seen as a diagonal line across the image. It is seen that the reflected signal from the gouge could be detected up to a distance of >22 mm away from the SAW initiation point. However, as would be expected, the detection sensitivity (or detected signal intensity) was dependent upon the flaw distance from the SAW entrance point: the detection sensitivity is highest within the distance range of 4-15 mm. In addition, artefacts were observed when the gouge was around 10 mm of the SAW entrance point. These artifacts can be seen in Fig. 5 as a series of dark lines immediately behind and at >20 mm below the detected gouge signal. It is not currently known what caused these artefacts.

The experimental configuration with horizontally-oriented probe was only used to establish the Rayleigh wave detection sensitivity as a function of the distance between SAW entrance point and flaw location. To detect flaws on the ball surface, the entire ball surface needs to be scanned. This is achieved by using the vertically-oriented probe configuration (see Fig. 3). In this configuration, the phased-array probe was oriented vertically to produce a SAW traveling from the equator upward towards the top dead center (TDC) of the ball (see Fig. 2). As the ball is rotated for a complete 360° revolution, the entire top-half surface of the ball is scanned (i.e., area scan). For this configuration, the Rayleigh wave detection sensitivity is required to cover at least the distance between the equator and the TDC, which is <20 mm for a 1-inch-diameter ball. As shown in Fig. 5, this required detection sensitivity is achievable using the 32-element curved phased array probe. Typical scan results using this area-scan configuration are presented in the following.

Figure 6 shows a scan image for a top-half surface of bearing Ball #1. In this image, the top horizontal line represents the equator line where the SAW enters the ball and propagates towards the TDC which, although showing in the image as a horizontal line, should be a single point. The bearing ball #1 has several flaw features, including the large gouge shown in Fig. 4a. In this scan, the large gouge was initially positioned at about -160° from the angular position of the probe and ~10 mm above the ball equator. In Fig. 6, it is interesting to note that only the two ends of the large gouge were detected, which was positioned at an angle with respect to the SAW propagation direction. Besides the large gouge, a marking for another flaw was detected in the image (at the upper left corner of the image). In addition, an electronic noise was shown in the image, which generally appeared randomly in scan images.

Figure 7 shows a typical scan image for a top-half surface of Ball #2. In this scan, the artificial C crack was initially positioned at about -80° from the angular position of the probe and ~10 mm above the ball equator. It is seen that the artificial C crack and the small surface feature were clearly detected. The markings (Vickers indents) for the crack and the surface feature were also detected. Again, an electronic noise appears in the image.

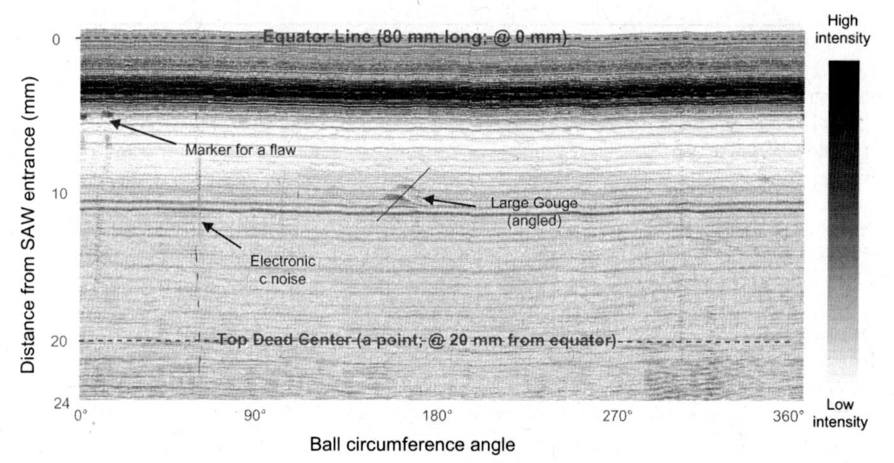

Fig. 6. Scan image of a top-half surface of bearing Ball #1.

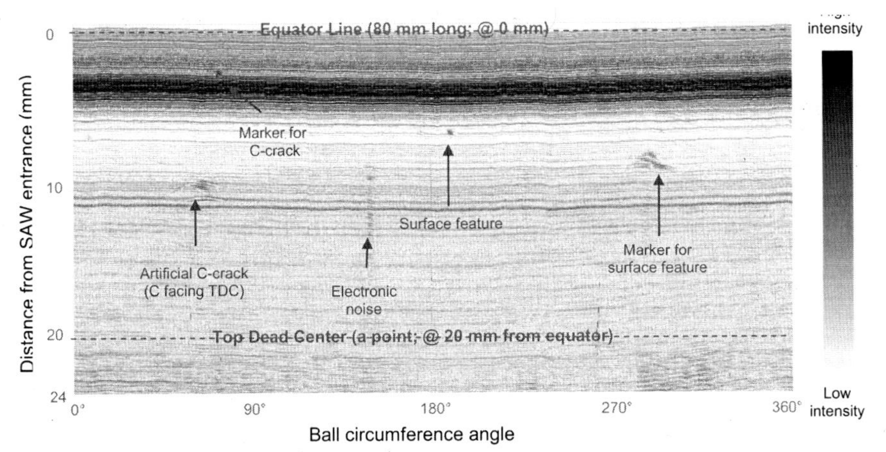

Fig. 7. Scan image of a top-half surface of bearing Ball #2 with an artificial C-crack.

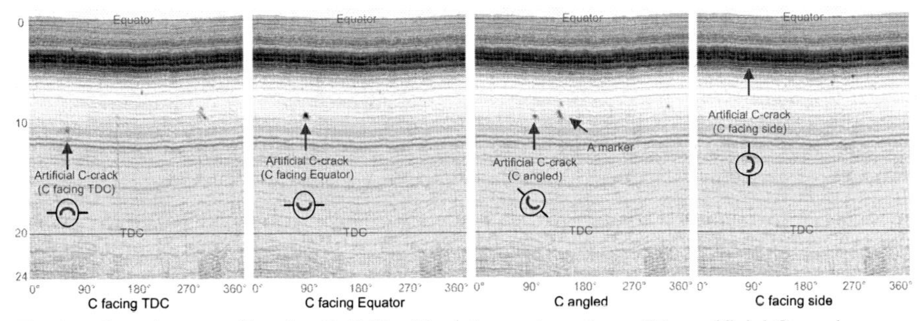

Fig. 8. Scan images of bearing Ball #2 with various orientations of the artificial C-crack.

Figure 8 shows four scan images for the top-half surface of bearing Ball #2. In these images the artificial C-crack has been placed in different orientations in respect to the propagation direction of the SAW, coming from the ball equator and propagating towards the top dead center of the ball. When the crack presents a concave arc to the SAW propagation direction, the detection of the crack is much stronger than when the crack presents a convex arc. This is possibly due to the fact that the reflected SAW signal from a convex arc is scattered in various directions. When the crack chord length is aligned to the SAW propagation direction, the detected signal is much weaker and can be faded under the background signal if the crack is located within the high background-signal band (see Fig. 8, right image).

CONCLUSION

An ultrasonic Rayleigh wave detection system was developed and tested for fast NDE inspection of ceramic bearing balls. The Rayleigh wave system uses a phased-array probe consisting of a single assembly of 32 elements at a frequency of 16.5 MHz. The detection sensitivity is affected by three parameters: (1) the distance between the SAW entrance point and the flaw position, (2) the angle between the SAW propagation direction and the major flaw axis, and (3) flaw size and subsurface depth. It was identified that the Rayleigh wave detection sensitivity is adequate in the distance range of 0 to 22 mm between the SAW entrance point and the flaw position. The detection sensitivity required for inspecting 1-inch-diameter bearing balls is <20 mm. The scan results showed that the ultrasonic Rayleigh wave technology can detect various flaws in the surface and subsurface of 1-inch-diameter ceramic bearing balls, including a Hertzian C crack with small crack opening on the surface.

ACKNOWLEDGMENT

This work was partially sponsored by the SKF Engineering and Research Center, The Netherlands. Research was supported by the Heavy Vehicle Propulsion Materials Program, DOE Office of FreedomCAR and Vehicle Technology Program, under contract DE-AC05-00OR22725 with UT-Battelle, LLC.

REFERENCES

1. http://www.cbrbearing.com.
2. http://productsearch.machinedesign.com/FeaturedProducts/Detail/Solutions_for_wind_power_generator_applications/70224/0.
3. H. Hertz, "Hertz's Miscellaneous Papers" (Macmillan, London, 1896) Ch. 5 and 6.
4. M. Bashkansky, M.D. Duncan, M. Kahn, D. Lewis III, and J. Reintjes, "Subsurface defect detection in ceramics by high-speed high-resolution optical coherent tomography," Optics Letters, Vol. 22, No. 1, pp. 61-63, 1997.
5. S.K. Lee, S. Wuttiphan, and B.R. Lawn, "Role of microstructure in Hertzian contact in silicon nitride: I, mechanical characterization," Journal of the American Ceramic Society, Vol. 80, No. 9, pp. 2367-2381, 1997.
6. P. Zhao, M. Hadfield, Y. Wang, and C. Vieillard, "Subsurface propagation of partial ring cracks under rolling contact Part I. Experimental studies," Wear, Vol. 261, pp. 382-389, 2006.
7. P. Zhao, M. Hadfield, Y. Wang, and C. Vieillard, "Subsurface propagation of partial ring cracks under rolling contact Part II. Fracture mechanics analysis," Wear, Vol. 261, pp. 390-397, 2006.
8. Yoseph Bar-Cohen, Ajit K. Mal. "Ultrasonic Inpsection." In *ASM Handbook, Volume 17, Nondestructive Evaluation and Quality Control*, by P. J. Blau and ASM Handbook Committee. American Society for Metals, 1989.
9. W.A. Ellingson, C. Deemer, and S. Steckenrider, "Method and Apparatus for Ultrasonic Phased Array Testing of Bearing Balls," Argonne National Laboratory invention report, US patent pending, 2006.

THERMAL EXPANSION COEFFICIENT OF SiO_2-ADDED LEUCITE CERAMICS

J. P. Wiff, Y. Kinemuchi, S. Naito, A. Uozumi and K. Watari
National Institute of Advanced Industrial Science and Technology (AIST)
463-8560 Nagoya, Aichi, Japan

ABSTRACT

In this work the influence of SiO_2 additions in leucite ceramics on the linear thermal expansion coefficient (TEC) during the phase transition has been studied. Thermal expansion and X-ray diffraction measurements at high temperatures were carried out to characterize the tetragonal–cubic phase transition. TEC of control and SiO_2-added leucite samples exhibited similar behavior as a function of temperature. Before and after the phase transition, the TEC values were similar to those observed in non-added samples, whereas during the phase transition, a maximum TEC value was observed and it tends to decrease as the SiO_2 addition increases. This behavior could be caused by the formation of an intermediate phase with an extremely high TEC ($70\times10^{-6}\,^{\circ}C^{-1}$) during the phase transformation. Furthermore, the results indicate that both the intermediate phase is partially suppressed and the cubic phase is partially stabilized at temperature as low as 200 or 300°C via SiO_2 addition.

INTRODUCTION

Leucite powders ($KAlSi_2O_6$) have been extensively used in dental applications to reduce the mismatch between the linear thermal expansion coefficient (TEC) of porcelain and metal reinforcement, thus enhancing the mechanical strength[1,2,3]. However, they have been rarely used as bulk material because they exhibit a phase transition with different TEC values[4,5].

Tetragonal leucite or low leucite (*$I/4_1a$*, N°88 a=13.090 and c=13.750, TEC ~$20\times10^{-6}\,^{\circ}C^{-1}$) is stable from room temperature up to approximately 600°C[4], whereas cubic leucite or high leucite (*Ia-3*, N°206 a=13.574, TEC ~$8\times10^{-6}\,^{\circ}C^{-1}$) is stable over 650°C[4].

Leucite structure consists of an aluminosilicate framework, which is linkage of four- and six-membered rings of the $(Si,Al)O_4$ tetrahedra. The framework contains cation sites of two different sizes: one has 12-coordination (W-sites) and the other has 6-coordination (S-sites)[5]. Normally, K^+ ions occupy the W–sites, but depending on the nature of the sample, synthetic or natural, sometimes K^+ ions can be slightly displaced from the W-sites, changing the thermal expansion coefficient in leucite ceramics.

The formation of an intermediate tetragonal phase (ITP) during the phase transition has been previously reported[5,6,7]; however, there are no reports about physical or mechanical properties of this phase. The identification of this ITP by XRD, is quite difficult because the main difference with the tetragonal $I4_1/a$ is the ordering of the Si–Al tetrahedra; Grögel et al. found the existence of this ITP by neutron diffraction of a natural leucite sample around the phase transition temperature[7]. They also obtained a continuous Si–Al ordering from room temperature until the phase transition temperature, whereas after this threshold, an abrupt collapse of ordering was observed, which is coincident with the end of transition to cubic phase.

The phase transformation in samples composed by solid solution between natural leucite including iron-leucite, $KFe^{3+}Si_2O_6$, and orthoclase feldspar, $KAlSi_3O_8$, has been previously studied[8]. It was found that the c/a lattice ratio decreases from 1.053 to 1.050 as the excess SiO_2 amount increases from 0.1 to 0.5 moles, which can be explained by the inclusion of extra Si in the leucite framework. In addition, it was found that the tetragonal and cubic phases can coexist within a range of around 100°C, and the onset temperature of the transition is slightly shifted to lower temperatures with increase in the excess SiO_2[8] and probably induces a change in TEC by SiO_2 addition.

Therefore, this work focuses on the influence of small lattice deformations caused by SiO_2 addition in synthetic leucite ceramics. In particular, it is expected that small changes in the lattice structure of leucite induced by SiO_2 additions can allow the control of the ITP formation and subsequently the regulation of TEC.

EXPERIMENTAL PROCEDURE

K_2CO_3 (Kanto Chemicals, #32323-00, 99.5%, Japan), SiO_2 (Quartz, Kanto Chemicals #37974-00, Japan), and γ-Al_2O_3 (AKP-G015, 99.995%, Sumimoto Chemical, Japan) powders were mixed in stoichiometric amounts and then milled in a polyethylene bottle containing ethanol and ZrO_2 balls for a night. Then, powders were dried and kept under vacuum (~ 1 Pa) at 70°C for several hours. Later, all powders were calcinated twice at 1000°C and 1100°C for 8 h, respectively, to promote their complete reaction. Afterwards, powders were milled in a planetary ball milling containing Al_2O_3 balls, at 300 rpm, until we obtained 1 μm average particle size. Finally, powders were cold pressed isostatically at 100 MPa for 1 min and sintered using a heating and cooling rate of 10°C/min and at different conditions in order to maximize their densification (Table I). All SiO_2-added samples exhibited densities up to 95% with respect to the nominal leucite density (2.46 g cm^{-3}). Nominal stoichiometric leucite or control samples (with no SiO_2 addition) were fabricated by using pulsed electric current sintering (PECS) at 1200°C for 5 min to enhance their density around 95%.

Table I. Samples description

Sample	Sintering temperature (°C)	Sintering time (h)
Control leucite*	1200	0.1
Leucite + $0.5SiO_2$	1550	8
Leucite + $1.0SiO_2$	1400	8
Leucite + $1.8SiO_2$	1200	8

* Sintered by using PECS

X-ray diffraction (XRD) was performed for identifying crystallographic phases and calculating lattice parameters using Rietveld refinement[9]. The XRD patterns were recorded in θ–2θ configuration, under air, using CuK_α radiation and from room temperature until 800°C in a diffractometer Rigaku RINT ULTIMA+.

Linear thermal expansion coefficient (TEC) was measured on pieces of 14 mm × 5 mm × 5 mm between room temperature and 800°C both in heating and cooling conditions. The TEC measurements were performed in a BRUKER MTC1000SA dilatometer, using an Al_2O_3 bar as standard for calibration.

RESULTS AND DISCUSSION

Figures 1a-i and 1a-ii correspond to XRD patterns at room temperature for control and Leucite+1.8SiO_2 samples after sintering, respectively. In both cases, a main phase of tetragonal leucite (PDFCD #38-1423) and a secondary one of kalsilite ($KAlSiO_4$, PDFCD #11-0579) were indexed[10]. Al_2O_3 was used as internal standard for high-temperature measurements. The kalsilite content was higher in control samples than in SiO_2-added ones and it tends to decrease as the SiO_2 content increases. The SiO_2 addition induces a liquid phase formation at high temperatures[11], and the reaction between liquid and kalsilite may be enhanced, hence reducing its content in the sample. Figures 1b-i and 1b-ii correspond to XRD patterns of control and Leucite+1.8SiO_2 samples at 500°C. In both cases, leucite was indexed as main phase; however, cubic phase (PDFCD #85-1421)[10] was found in SiO_2-added samples, and tetragonal phase was only observed in the control sample. In addition, no extra phases were detected and the reduction of kalsilite content was also observed in SiO_2-added samples.

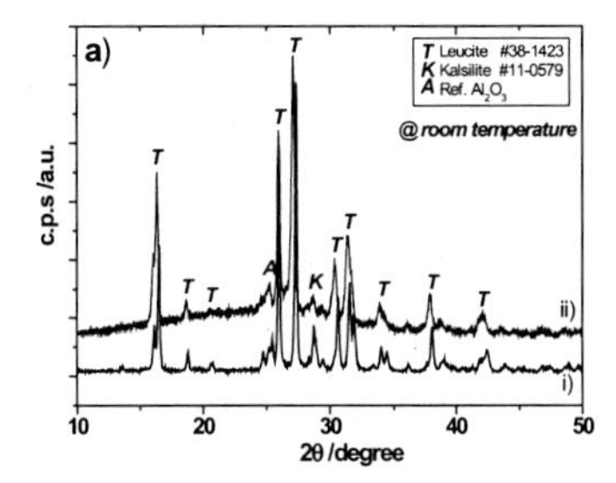

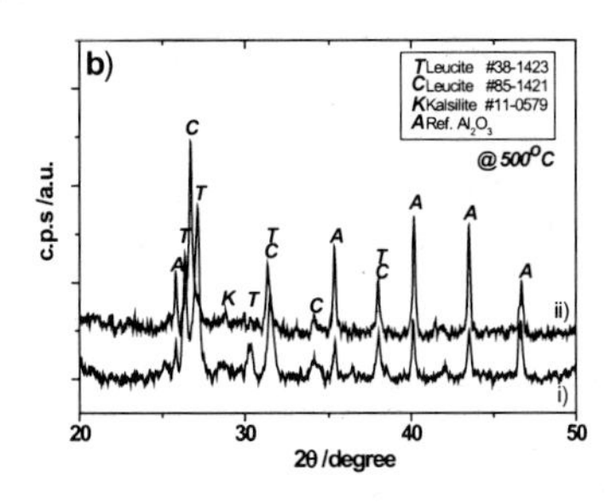

Figures 1a and 1b correspond to XRD patterns at room temperature and 500°C for control (i) and Leucite+1.8SiO_2 (ii) samples, respectively

Figure 2 shows the leucite lattice parameters as a function of SiO_2 addition. Lattice parameters for the control sample are matched with reported values[4], whereas in SiO_2-added samples, the c parameter decreases and a parameter increase compared with those in the control sample. This indicates that SiO_2 addition induces a lattice compression along the c-axis. However, regardless of SiO_2 addition the c/a lattice ratio is always close to 1.046; this value is lower with respect to the previously reported ones (1.050)[8]. At higher temperature (not shown), the c/a lattice ratio systematically decreases, indicating a compression along the *c*-axis caused to the transformation into cubic phase, as it has been described elsewhere[5].

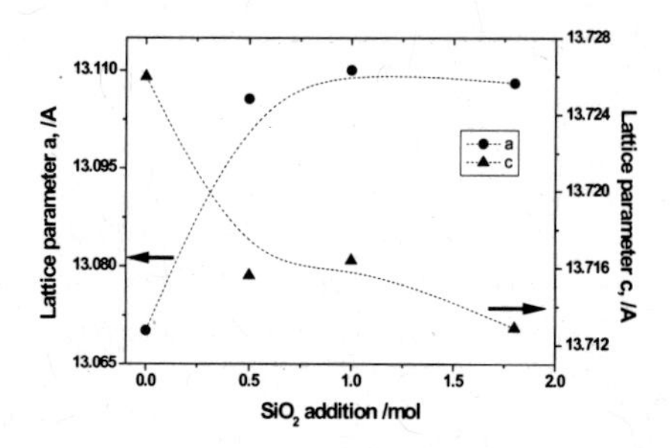

Figure 2 Lattice parameters as a function of SiO_2 addition. Lines are used as a reference

Figures 3a and 3b correspond to XRD patterns between 25° to 29° before (i), during (ii), and after (iii) phase transformation from control and Leucite+1.8SiO_2 samples, respectively. Al_2O_3 peak was used as reference for high-temperature measurements. In the control sample under 600°C, no cubic phase was detected, whereas in Leucite+1.8SiO_2 sample, the formation of a cubic phase was detected early to be around 300°C.

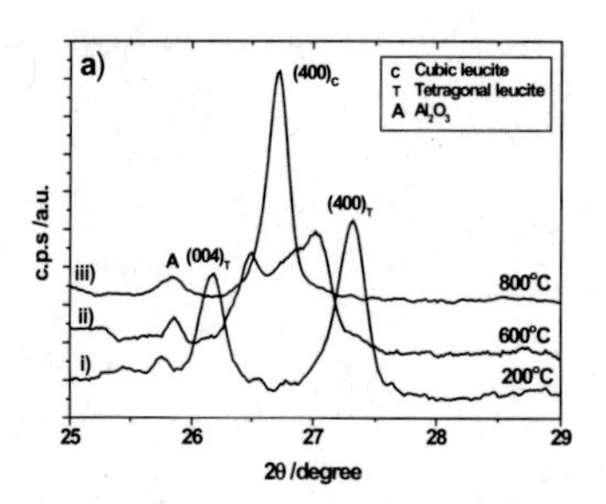

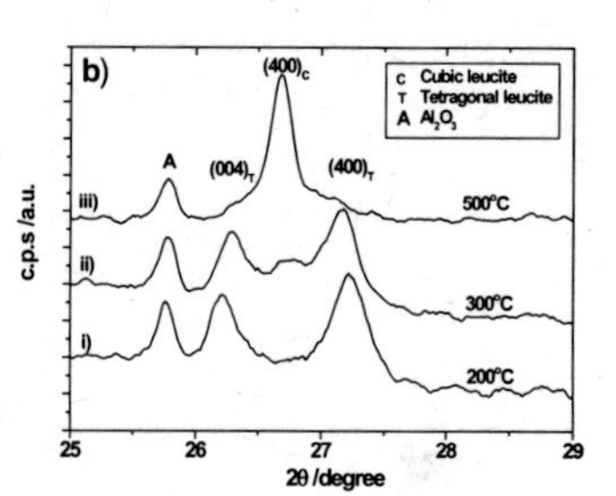

Figures 3a and 3b correspond to XRD patterns at different temperatures for control and Leucite+1.8SiO_2 samples, respectively

To estimate the cubic phase concentration in all samples, the following index AR was defined.

$$AR[\%] = \frac{A_{(400)c}}{A_{(400)c} + A_{(400)T}} 100\% \tag{1}$$

AR corresponds to the integrated peak area ratio between the tetragonal leucite $A_{(400)T}$ and cubic leucite $A_{(400)C}$.

Figure 4 shows the *AR* index as a function of temperature and a systematic shift of the onset temperature transition toward lower temperature was observed. The onset temperature transition was shifted toward lower temperatures (~200°C) and a broadening of phase transition was observed with the increment in SiO_2 addition. Thus, in control samples, the transition takes place within 50°C, whereas in SiO_2-added samples the range was broadened up to 400°C. Therefore, the SiO_2 addition seems to be effective for partial stabilization of the cubic phase in the tetragonal matrix at low temperature.

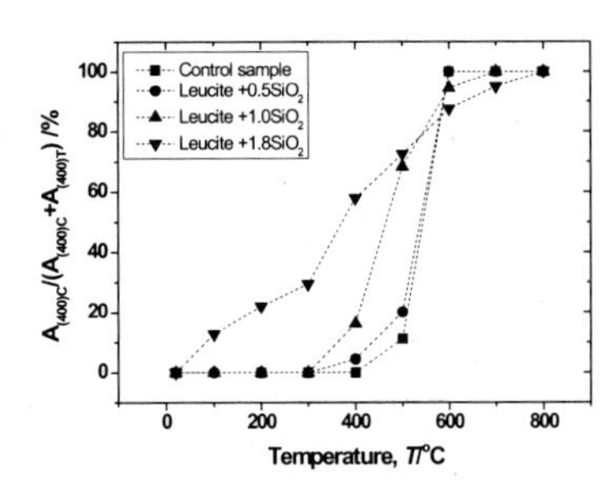

Figure 4 *AR* as a function of temperature. Lines are used as a reference

Figure 5a shows the TEC for control and SiO_2-added leucite samples from room temperature up to 800°C. Regardless of SiO_2 addition, a maximum of TEC as a function of temperature is observed and it tends to shift toward lower temperatures as the SiO_2 addition increases. Figure 5b shows the TEC and *AR* as a function of temperature for a control sample. Apparently, the TEC variation is related to the occurrence of tetragonal–cubic phase transition.

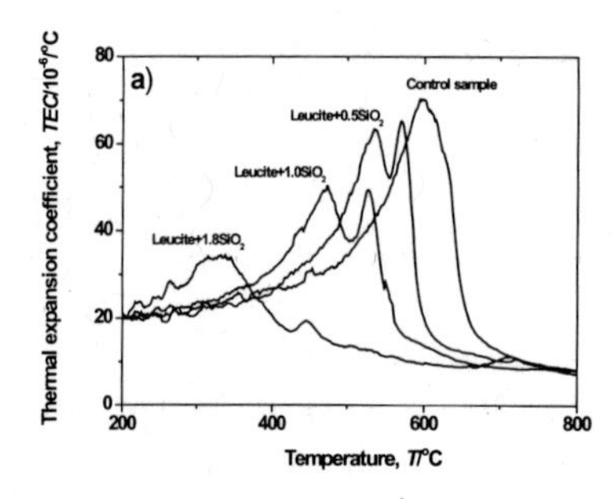

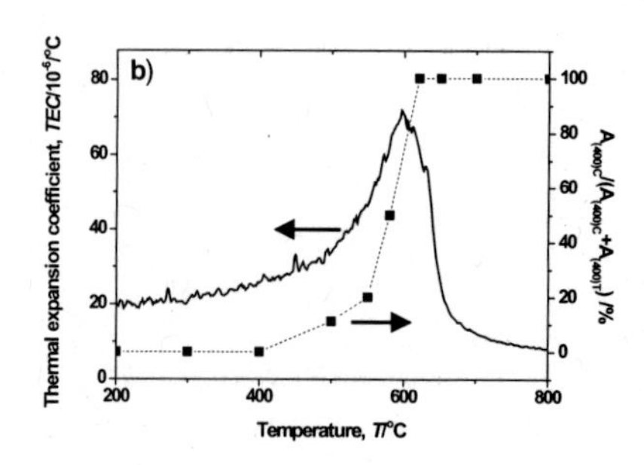

Figure 5a TEC as a function of temperature for control and SiO_2-added samples. Figure 5b TEC and *AR* for Leucite+1.8SiO_2 sample as a function of temperature.

Figure 5a shows that TEC values at temperatures under 300°C and over 650°C are in agreement with the reported values for tetragonal and cubic phases[1,3] whereas there is a large variation of TEC between 300°C and 650°C. Among them, maximum TEC of $70\times10^{-6}\,°C^{-1}$ was observed for the control sample. This abnormally high TEC is the highest one for a conventionally sintered ceramic, but unfortunately it seems to be stable in a narrow range of temperature. Figure 6a also shows that all SiO_2-added samples have a splitting near the maximum TEC. The origin of this splitting is not clear yet. Although DTA analysis (not shown) showed no clear evidence about this phenomenon, it seems that the phase transition occurs through one or two intermediate phases in SiO_2-added samples. In the following discussion, we will treat the phase transition including one intermediate phase (ITP), but the transition through several intermediate phases can be argued analogously.

Figure 5b indicates that the progressive increment of TEC occurs simultaneously with the cubic phase formation. However, transformation from tetragonal to cubic phase is not responsible for this large TEC because of low TEC of cubic phase ($8\times10^{-6}\,°C^{-1}$). Therefore, ITP with extremely high TEC should be considered. Grögel et al. proposed that ITP should be formed close to the tetragonal–cubic phase transformation and it should have a similar crystal structure as the tetragonal $I4_1/a$ phase. Additionally, Newton et al. confirmed the existence of ITP by using DTA analysis and reported a proper tetragonal space group for this intermediate phase ($I4_1/acd$)[6] quite similar with the tetragonal $I4_1/a$ phase, but with extra symmetry conditions. It is believed that the ITP formation might control the maximum TEC in each sample during the phase transition. In addition, the reduction of maximum TEC from $70\times10^{-6}\,°C^{-1}$ to $30\times10^{-6}\,°C^{-1}$ as a function of the SiO_2-addition (Figure 5a) could be explained by the reduction of volume content of ITP.

The cause of the change in the ITP content by means of SiO_2 addition is still not understood; however, it seems that, as Henderson et al. proposed, the distortion caused by Si ions in the leucite framework could be a reasonable mechanism[8]. For example, the c/a lattice ratio reported by Henderson et al. (1.051 at +0.5 mol) was significantly higher than those found in this work (1.046 at +1.8 mol). The difference in c/a may be caused by the amount of SiO_2 addition and/or the presence of iron impurities. Nevertheless, further decrease in the onset temperature and larger broadening of phase transition by higher SiO_2 addition (+1.8 mol) when compared with their work reasonably explains the effect of SiO_2 addition on the phase transformation, indicating that the modification of leucite framework by SiO_2 addition is a responsible mechanism for the change in phase transition and subsequent TEC variation.

CONCLUSIONS

The effect of SiO_2 additions in leucite ceramics on its lattice structure, phase transition, and thermal expansion coefficient were analyzed.

All leucite samples, with various SiO_2 contents, have a maximum TEC during the phase transition, and depending on the SiO_2 addition, both the maximum TEC and its transition temperature can be controlled. The maximum TEC in the control sample is around $70\times10^{-6}\,^{\circ}C^{-1}$, whereas for a SiO_2-added one, it can be reduced up to $30\times10^{-6}\,^{\circ}C^{-1}$, while the onset transition temperature shifts from 600°C to around 200°C. The effect of SiO_2 addition on phase transition and TEC can be understood, as the extra Si ions modify the leucite framework, inducing a partial stabilization of cubic leucite, which modifies the formation of the ITP and subsequent TEC.

REFERENCES

[1]J. Maixner, A. Klouzkova, M. Mrazova and M. Kohoutkova: X-ray analysis in leucite systems. Z. Kristallogr. Suppl., **26**, 531-36 (2007).

[2]T. Sheu, W. O'Brien, S. Rasmussen and T. Tien: Mechanical properties and thermal expansion behaviour in leucite containing materials. J. Mater. Sci., **29**, 125-28 (1994).

[3]T. Ota, M. Takahashi and I. Yamai: High-thermal-expansion polycrystalline leucite ceramic. J. Am. Ceram. Soc., **76**, 2379-81 (1993).

[4]M. Novotna and J. Maixner: X-ray powder diffraction study of leucite crystallization. Z. Kristallogr. Suppl., **23**, 455-59 (2006).

[5]D. Taylor and C. Henderson: The thermal expansion of the leucite group of minerals. Am. Mineralogist, **53**, 1476-89 (1968).

[6]H. Newton, S. Hayward and S. Redfern: Order parameter coupling in leucite: a calorimetric study. Phys. Chem. Minerals, **35**, 11-16 (2008).

[7]T. Grögel, H. Boysen and F. Frey: Phase transition and ordering in leucite. Acta Cryst., A **S40**, C-256 (1984).

[8]C. Henderson: The tetragonal-cubic inversion in leucite solid solutions. Progr. Exp. Petrol., **50**, 50-54 (1981).

[9]J. Rodriguez-Carvajal: Recent advances in magnetic structure determination by neutron powder diffraction. Phys. B: Cond. Matter., **192**, 55-69 (1993).

[10]Powder Diffraction Files, The International Centre for Diffraction, Swarthmore, USA, 2004.

[11]J. Schairer and N. Bowen: Melting relations in the system Na_2O-Al_2O_3-SiO_2 and K_2O-Al_2O_3-SiO_2. Am. J. Sci., **245**, 193 (1947).

Geopolymers

INORGANIC POLYMERS (GEOPOLYMERS) AS ADVANCED MATERIALS

Kenneth J.D. MacKenzie
MacDiarmid Institute for Advanced Materials and Nanotechnology,
Victoria University of Wellington,
Wellington,
New Zealand

ABSTRACT

Geopolymers are alkali-activated inorganic materials, generally aluminosilicates, that harden at ambient temperatures. Although they may have been used as building materials in ancient Egypt and possibly by earlier civilizations, their re-discovery in the mid-20th Century has aroused renewed interest in their uses as environmentally-friendly building materials, fireproof panels and for hazardous waste immobilization and disposal. However, these materials are potentially much more versatile than their relatively simple synthesis chemistry might suggest. The formation of composites with a wide range of inorganic and organic materials opens up the possibility of incorporating new functionalities. Examples to be discussed include the synthesis of bioactive materials by the insertion of calcium-containing compounds and methods for overcoming problems of tissue toxicity associated with the highly alkaline materials. Other examples include the modification of the electrical and mechanical properties of geopolymers by the addition of carbon nanotubes, and the use of carbon and silicon-geopolymer composites as precursors for carbothermal and silicothermal synthesis of oxynitride engineering ceramics. The synthesis of gallogermanate geopolymer analogues with potentially interesting optoelectronic properties has encountered problems that have necessitated the development of alternative new synthesis methods to be outlined. The production of nanoporous geopolymers with aligned pore structures has opened up the possible use of these materials to remediate heat island effects in large cities. These and other possible future directions for inorganic polymers will also briefly be considered.

INTRODUCTION

Inorganic polymers, also known as geopolymers, are aluminosilicates, conventionally prepared by condensation of a solid aluminosilicate such as the dehydroxylated clay mineral metakaolinite with an alkali silicate solution under highly alkaline conditions[1-3]. At ambient temperatures these materials set and harden to an X-ray amorphous product containing solely tetrahedral Al and tetrahedral Si characterised by a broad ^{29}Si MAS NMR resonance at about -92 ppm[2]. The most recent ^{23}Na MAS NMR studies[3] have provided more details about the way in which the charge-balancing alkali ions are incorporated in the structure.

To date, geopolymers have been proposed as environmentally-friendly substitutes for Portland cement[4], as fireproof materials in vehicle body components[4] and to encapsulate hazardous wastes for safe storage and disposal[4]. The present paper extends the geopolymer concept to their use more advanced applications, by reviewing developments in new synthesis methods and discussing some of the new inorganic polymer compounds resulting from these syntheses. Finally, some examples of advanced applications arising from the availability of these new inorganic polymer compounds will be given.

NEW GEOPOLYMER SYNTHESIS METHODS

Conventional Synthesis Without Thermal Dehydroxylation of the Kaolinitic Clay

The conventional synthesis process requires the expenditure of thermal energy to dehydroxylate the clay. For this reason, it would be advantageous if the thermal pre-treatment step could be eliminated. Viable geopolymers cannot be prepared from undehydroxylated kaolinite, but several alternative methods have been investigated for pre-treating the clay to render the aluminium component more reactive[5]. These include high-energy grinding to destroy the clay mineral lattice and treatment with alkali or acid to expose the Al-containing layer[5]. Mechanical pre-treatment (grinding) of the tubular kaolinitic mineral halloysite for 15 min. in a vibratory mill renders the clay X-ray amorphous and converts the Al coordination from solely 6-fold to a mixture of 4, 5 and 6-fold coordination, as determined by ^{27}Al MAS NMR[5]. When subjected to the conventional synthesis process, the ground material forms a viable geopolymer showing all the typical characteristics.

However, grinding itself requires an energy input, making chemical pre-treatment a more energetically attractive option. Pre-treatment of the clay with 1M NaOH converts a significant proportion of the 6-fold coordinated Al to 4-coordinated, and when subjected to the conventional geopolymer synthesis, produces a material that sets; this is, however, not X-ray amorphous, but contains crystalline carbonated phases[5].

Pre-treating the clay with 1M HCl has no effect on the Al coordination and does not produce a viable geopolymer[5]. However, a successful acid synthesis involving reaction of metakaolinite with H_3PO_4 has been reported[6] to produce a product with a crushing strength of 55 MPa and an X-ray amorphous structure containing a proportion of 4-fold coordinated Al (Figure 1). This may simply be an example of a phosphate-bonded material, and, as such, it would be specific to H_3PO_4 and would not represent a more generally applicable acid synthesis method.

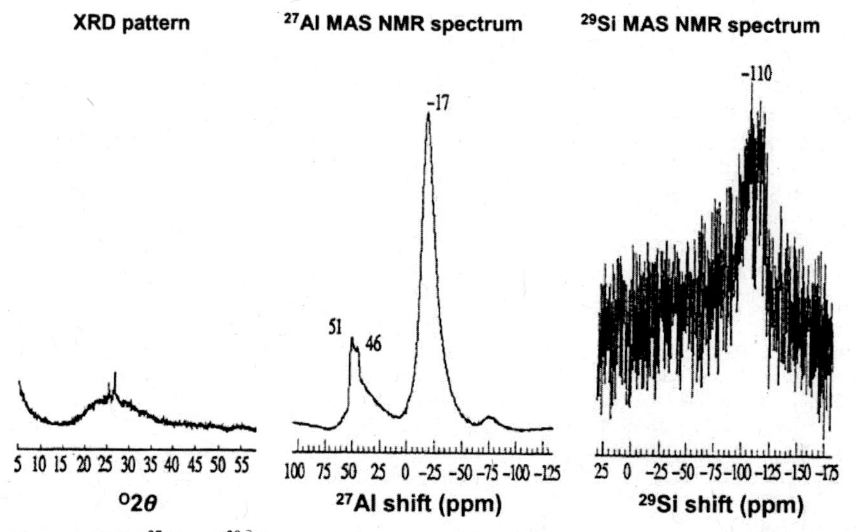

Figure 1. XRD, ^{27}Al and ^{29}Si MAS NMR spectra of phosphate geopolymer, re-drawn from ref. 6.

Sol-Gel Synthesis

In an alternative synthesis method that does not involve a solid aluminosilicate reactant as the Al source, sodium aluminate solution is reacted with alkali silicate, formed *in situ* by reaction of SiO_2 with the appropriate alkali hydroxide[7]. The Na product sets at 40°C with a crushing strength of 26 MPa, is X-ray amorphous and contains solely 4-coordinated Al. Its ^{29}Si MAS NMR spectrum indicates the presence of some unreacted SiO_2 in addition to the typical Si resonance of a geopolymer[7]. This method has been applied to the synthesis of other analogues of aluminosilicate geopolymers such as the galliogermanates, where suitable solid precursors are unavailable.

For example, a potassium galliosilicate inorganic polymer has been synthesised by reacting a solution of $KGaO_2$ with potassium silicate formed *in situ* by reaction of SiO_2 with KOH solution[8]. The product is X-ray amorphous (Figure 2), and is shown by ^{71}Ga MAS NMR to contain Ga in solely 4-fold coordination (166 ppm), by contrast with the Ga_2O_3 from which the $KGaO_2$ was derived[8]. The ^{29}Si MAS NMR spectrum of the product contains a resonance at -81 ppm corresponding to the reaction product, together with some unreacted SiO_2 (-108 ppm) (Figure 2). On heating at 800°C the inorganic polymer is converted to crystalline $KGaSi_2O_6$[8].

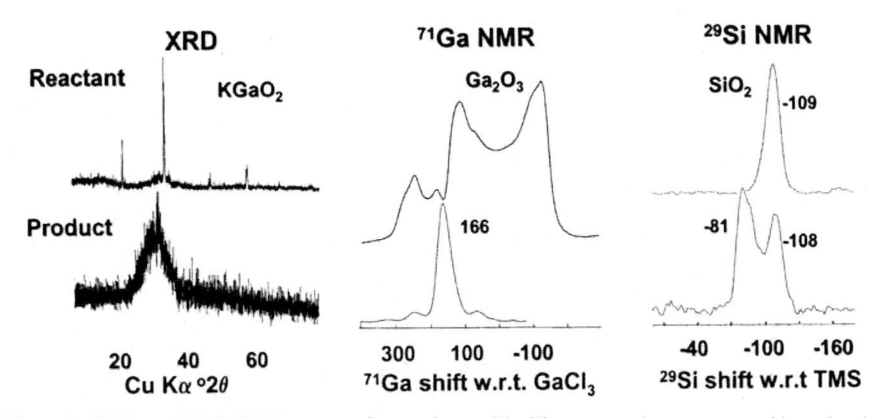

Figure 2. XRD and MAS NMR spectra of potassium galliosilicate geopolymer prepared by sol-gel synthesis, data from ref. 8.

The same method has been used to synthesise a mixed gallium-aluminium silicate inorganic polymer from a mixture of $KGaO_2$ and $KAlO_2$ reacted with SiO_2 in KOH solution[8]. The cured product is a true inorganic polymer, being X-ray amorphous and showing typical ^{27}Al and ^{29}Si MAS NMR spectra (Figure 3). By contrast with the pure Ga compound, the pure Al end member of this series begins to melt at >1100°C, prior to complete crystallisation 800°C[8].

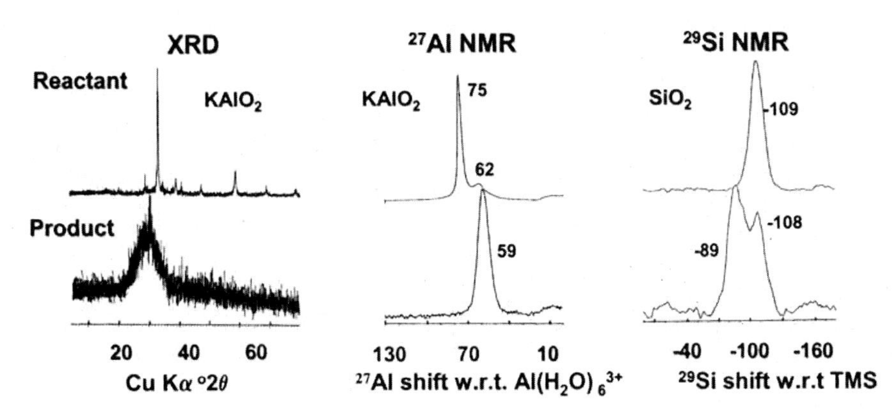

Figure 3. XRD and MAS NMR data for a gallium-aluminium silicate prepared by sol-gel synthesis, data from ref. 8.

Solid-State Synthesis

Kolusek et al., have reported the preparation of a geopolymer-like material by reacting an impure undehydroxylated clay with alkali hydroxides at 550°C[9]. The product shows hydraulic activity and when wetted, sets to a weak (about 1 MPa) material with the Al and Si characteristics of a true geopolymer (Figure 4), but containing crystalline phases[9].

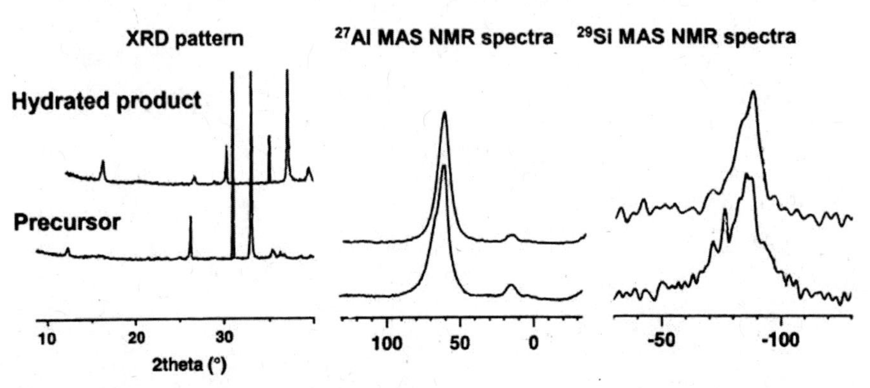

Figure 4. XRD and MAS NMR data for an aluminosilicate geopolymer-like product of solid-state synthesis, data re-drawn from ref. 9.

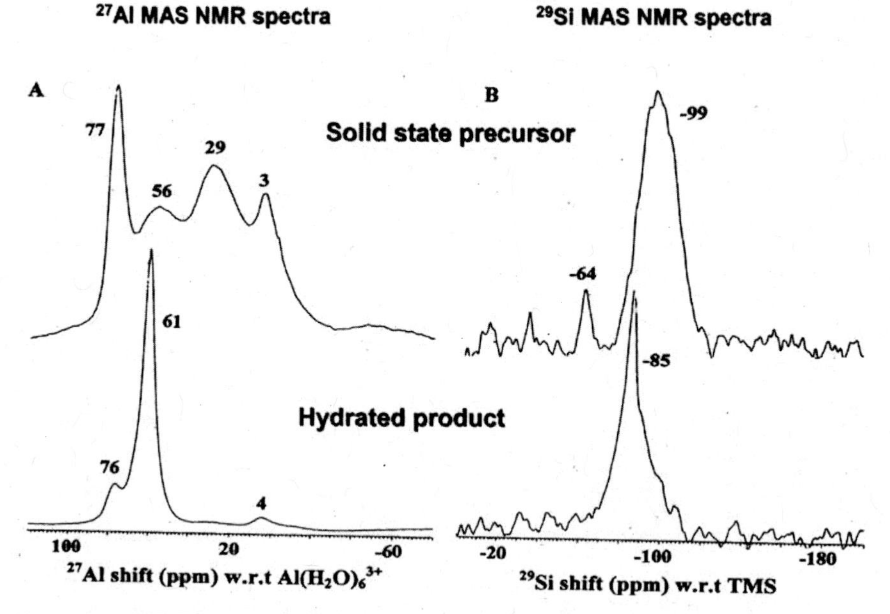

Figure 5. MAS NMR spectra of a lithium geopolymer prepared by solid-state synthesis, data from ref. 10.

This method has been successfully used to synthesise a lithium aluminosilicate geopolymer, which is difficult to prepare by the conventional method because of the poor solubility and weak alkalinity of LiOH solution. In this solid state synthesis[10], halloysite clay was reacted with LiOH at 550°C. When wetted, the product of this solid state reaction sets at 40°C to a compound containing solely 4-fold coordinated Al and a typical ^{29}Si MAS NMR spectrum (Figure 5), but is not fully X-ray amorphous[10]. When heated at 900°C, this geopolymer crystallises to α-eucryptite, $LiAlSiO_4$[10] but if additional quartz is added, heating at 1300°C for 8hr causes the crystallization of β-eucryptite and β-spodumene ($LiAlSi_2O_6$)[10].

The solid-state method has been used to synthesise gallium germanate analogues of the aluminosilicate geopolymers, since the lack of reactivity of both Ga_2O_3 and GeO_2 with KOH militates against sol-gel synthesis of the germanate compounds. Separate solid-state reactions at 550°C of Ga_2O_3 with KOH and GeO_2 with KOH produce the alkali gallate and germanate which are then reacted together at 550°C to form a mixed oxide precursor[11]. When hydrated at 40°C, this precursor sets to form an X-ray amorphous compound. The technical interest in these compounds is in their potentially useful electronic and opto-electronic properties.

NEW APPLICATIONS FOR ADVANCED INORGANIC POLYMER COMPOUNDS

Potential Colour-Change Humidity Indicators

Since the alkalinity of a geopolymer varies reversibly with its dryness, it was expected that the incorporation of a suitable acid-base indicator in an aluminosilicate geopolymer might provide a chromophoric colour-change indicator of the relative humidity of the environment in which the geopolymer was located[12]. Several different acid-base indicators were homogeneously distributed in conventionally-synthesised potassium aluminosilicate geopolymers and exposed to atmospheres of different relative humidity[12]. The high alkalinity of the geopolymer produced intense colours in the samples but not necessarily the normal high pH colour. The most promising indicator for this purpose was thymol blue, which changed from light tan in the dry geopolymer to blue under damp conditions[12]. However, at higher humidity, the colour was found to bleach and fade irreversibly, probably due to the oxidation of the ($-SO_3-$) groups. Proper implementation of this application will require an indicator that is stable to oxidation at high pH and exhibits a suitable colour change in the appropriate pH range.

Bioactive Inorganic Polymer Composites

This application is for a material that bonds to living bone and tissue for the repair of accidental or congenital damage, and requires an inorganic polymer containing the necessary elements for bone formation (Ca and/or P). Calcium-containing inorganic polymers have been produced by the addition of calcium compounds such as amorphous calcium silicate, $Ca(OH)_2$ and $Ca_3(PO_4)_2$ to conventional potassium aluminosilicate geopolymers[13]. The ^{27}Al and ^{29}Si MAS NMR spectra of these composites indicate that they are normal geopolymers, while the environment of the calcium has been shown by natural-abundance 43Ca MAS NMR to be unlike that of the calcium starting materials, and more like calcium silicate hydrate (CSH) gel[13].

The use of inorganic polymers as bioactive materials carries with it two potential problems, namely the risk of cell toxicity due to the high alkalinity and the possible leaching of aluminium from the polymer, with the consequent risk of damage, particularly to brain cells. The latter problem can be addressed by the use of compositions of higher Si:Al ratios, while the problem of leachable alkali can be reduced by heating the inorganic polymer to 600°C, as shown by experiments in which an inorganic polymer containing 10% $Ca(OH)_2$ was exposed to simulated body fluid (SBF)[14]. A sample heated at 600°C for 1hr showed a much reduced release of alkali (Figure 6) and a consequently lower pH of the SBF than the same material unheated[14].

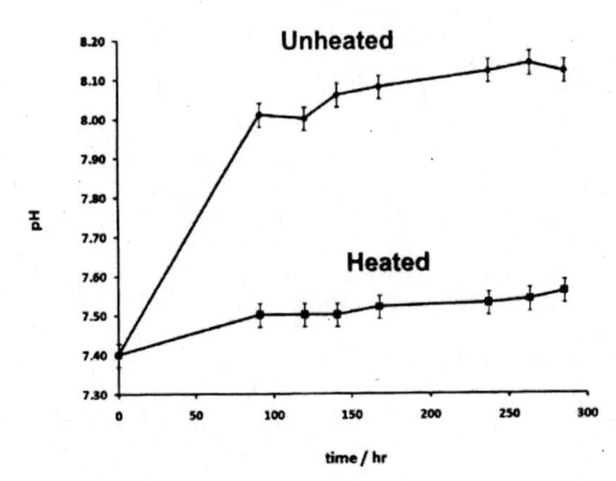

Figure 6. Effect of pre-heating on the release of alkalinity into SBF from a potassium inorganic polymer containing 10% $Ca(OH)_2$, data from ref. 14.

Heating to 650°C to reduce the alkalinity of aluminosilicate geopolymers and increase their porosity has previously been used in samples without added calcium or phosphorus intended for implantation as bioactive materials; in this case the porosity of the material was exploited to allow bone growth by normal biological mechanisms[15]. In other experiments, bone minerals such as hydroxyapatite or tricalcium phosphate were added to standard aluminosilicate geopolymers and successfully implanted in vivo without harmful effects to the surrounding tissue[16]. In *in vitro* experiments with potassium geopolymers containing $Ca(OH)_2$, the Al and Si structure was shown by ^{27}Al and ^{29}Si MAS NMR to be unchanged by soaking in SBF for up to 7 weeks, but bioactivity in the soaked samples was evidenced by the appearance of crystalline hydroxycarbonate apatite and calcite (Figure 7), the latter resulting from carbonation of the geopolymer[14]. The environment of the calcium, monitored by natural-abundance ^{43}Ca MAS NMR (Figure 8), was found to be unchanged both by heating to lower the pH and by soaking in SBF[14].

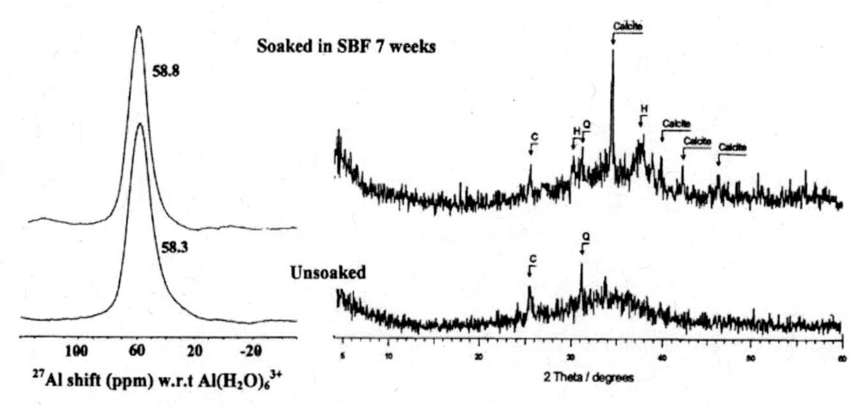

Figure 7. ^{27}Al MAS NMR spectra and XRD traces of potassium inorganic polymer containing 10% $Ca(OH)_2$, before and after soaking in SBF. Q = quartz, C = cristobalite, H = hydroxycarbonate apatite. From ref. 14.

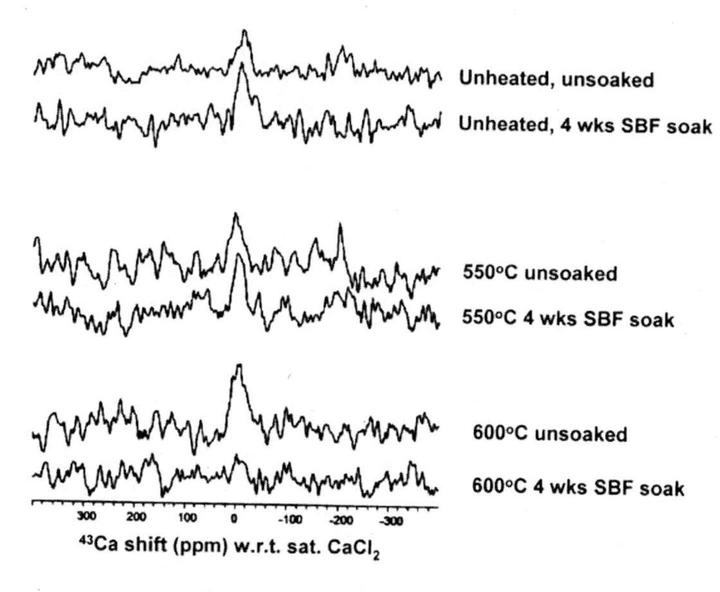

Figure 8. Natural abundance ^{43}Ca MAS NMR spectra of potassium inorganic polymers containng 10% Ca(OH)2, unheated and heated to reduce the alkalinity, before and after exposure to SBF. From ref. 14.

Inorganic Polymer-Carbon Nanotube Composites as New Electronic Materials

The DC electric conductance of potassium aluminosilicate geopolymers has been found to be significantly increased by the addition of single-wall carbon nanotubes (SWCNTs), at all temperatures up to the decomposition temperature of the nanotubes[17]. This increase in conductance was much greater than in composites containing the same amounts of graphite (up to 3 wt%). The SWCNTs were shown by scanning electron microscopy to be homogeneously distributed throughout the composites (Figure 9), with only a small proportion being multiwall nanotubes[17].

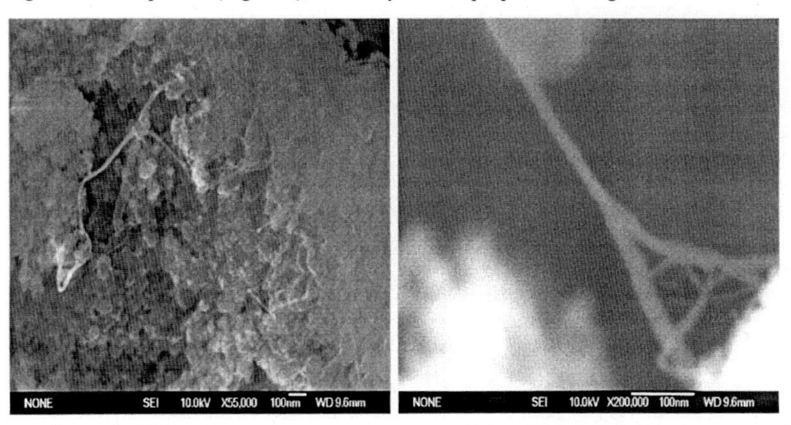

Figure 9. SEMs of potassium geopolymer containing 6% single-wall carbon nanotubes. Micrograph on right shows a multiwall nanotube. From ref. 17.

Since SWCNTs are extremely strong, their presence in the geopolymers was hoped to improved the mechanical strength of the composites, but this was found not to be the case; the strengths did not differ greatly from those of composites containing graphite, and much greater differences in tensile strength were found between Na and K-aluminosilicate-based composites[17].

Geopolymers as Precursors for the Synthesis of SiAlON Engineering Ceramics

SiAlON (silicon aluminium oxynitride) ceramics for engineering applications can be synthesised from aluminosilicates by carbothermal reduction and nitridation (CRN), in which a mixture of the aluminosilicate with carbon is fired in nitrogen at ~1400°C. Composites of potassium aluminosilicate geopolymers with graphite, when fired for 10hr under these conditions, have been found[18] to produce predominantly β-SiAlON (Figure 10) by a reaction such as:

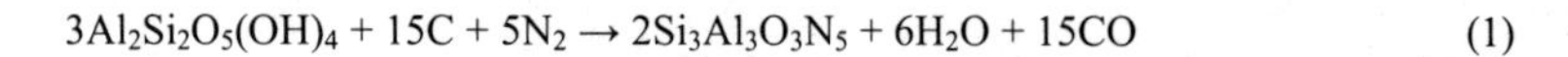

$$3Al_2Si_2O_5(OH)_4 + 15C + 5N_2 \rightarrow 2Si_3Al_3O_3N_5 + 6H_2O + 15CO \quad (1)$$

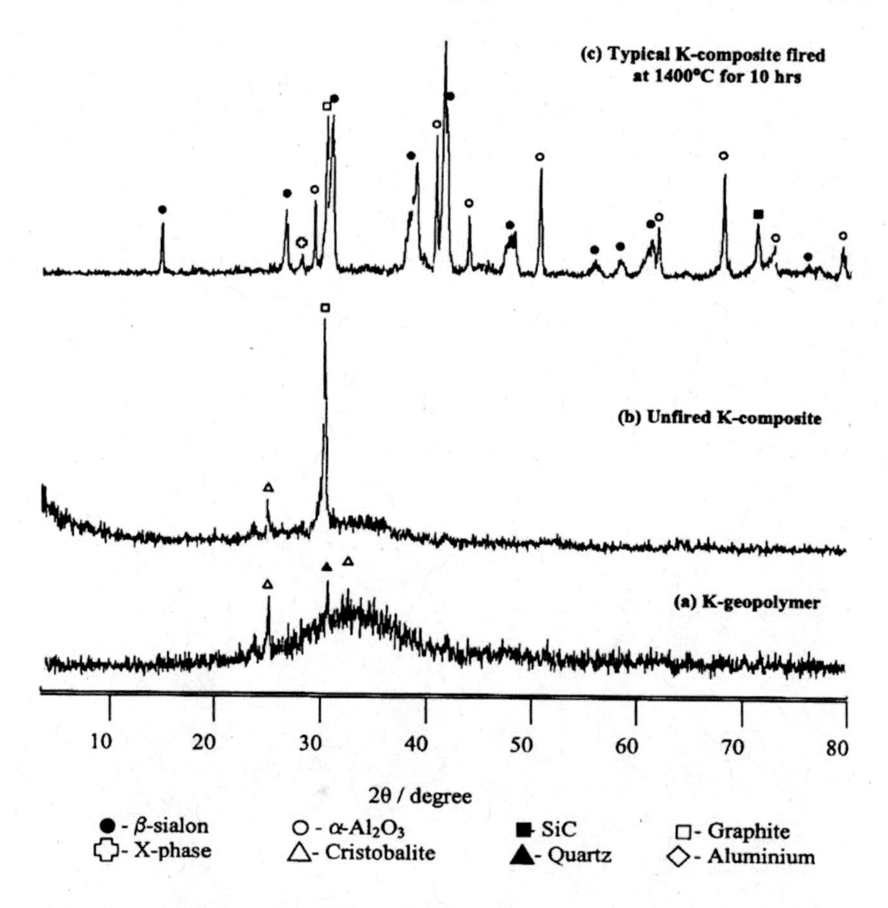

Figure 10. XRD traces of unfired and fired potassium geopolymer/graphite CRN precursor, data from ref. 18.

Other crystalline products can include **O**-SiAlON, Si_3N_4, SiC, α-Al_2O_3 and mullite ($Al_6Si_2O_{13}$), the relative amounts of which depend on the SiO_2/Al_2O_3 ratio, the H_2O/SiO_2 ratio and the K_2O/SiO_2 ratio of the initial geopolymer[18]. The presence of the alkali in the precursor was found not to be a problem, since it is lost during firing and does not affect the melting temperature of the composite[18]. An advantage of geopolymer-carbon composites as SiAlON precursors was expected to be the ability to cast the geopolymer into accurate and detailed shapes which could then be converted to a solid SiAlON component; this was not fully realised, however, since the degree of SiAlON formation in solid geoplymer composites was significantly less than when these were reacted in powdered form (Figure 11), due to the greater ease of nitrogen gas permeation in the latter[18].

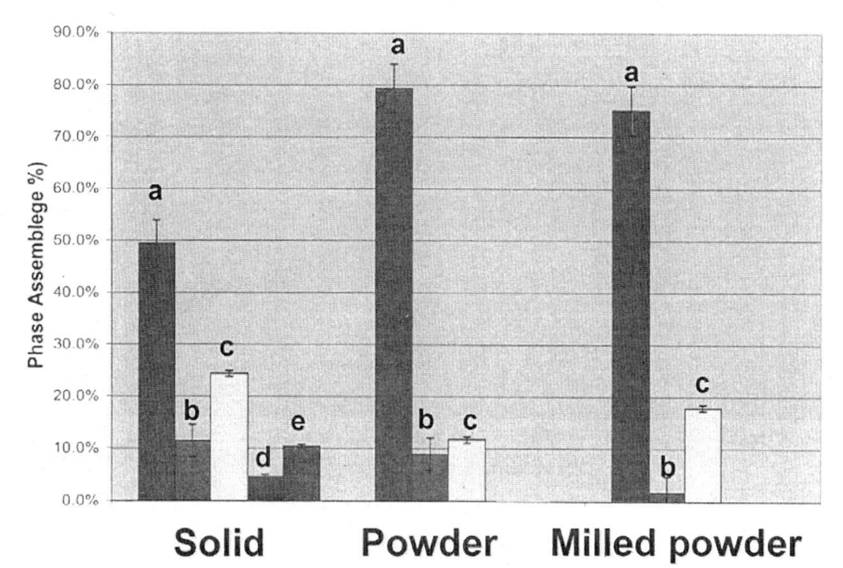

Figure 11. Effect of physical state of the precursor on the degree of CRN sialon formation from potassium geopolymer/graphite composite. Key: a = β-sialon, b = amorphous phases, c = α-alumina, d = graphite, e = SiC. From ref. 18.

Silicothermal reduction and nitridation (SRN) is an alternative synthesis method for SiAlON ceramics, in which the carbon is replaced by elemental silicon as the reducing agent. This produces SiAlONs without the formation of greenhouse gases, by reactions such as:

$$3Al_2Si_2O_5(OH)_4 + 6Al_2O_3 + 6Si + 4N_2 \rightarrow Si_{12}Al_{18}O_{39}N_8 + 6H_2O \quad (2)$$

The alkalinity of geopolymers synthesised by the conventional method militates against the incorporation of elemental Si for the preparation of SRN precursors, making it necessary to use a solid-state synthesis for this purpose. A lithium aluminosilicate geopolymer was prepared by solid-state reaction of halloysite clay with LiOH and elemental Si was added prior to hydration[19]. The choice of a Li precursor was made in the hope of ultimately synthesising Li α-SiAlON[19], but firing in nitrogen at 1150°C was found to form a mixture of β-SiAlON and spodumene ($LiAlSi_2O_6$)[19]. However, these preliminary studies demonstrate the viability of aluminosilicate geopolymer composites as precursors for SiAlON synthesis by both CRN and SRN methods.

Porous Geopolymers as Potential Passive Cooling Elements in Buildings

The temperature of large cities is continually increasing due to "heat island" effects, in which evaporative cooling mechanisms are inhibited by the large impervious city footprint, increase in the generation of heat by increasing human activity, and the increasing attraction and retention of solar heat by large areas of dark-coloured pavements and roads. As a result, the number of 24-hour periods in which a city temperature does not fall below 30°C (so-called "tropical nights") is steadily increasing in cities such as Tokyo.

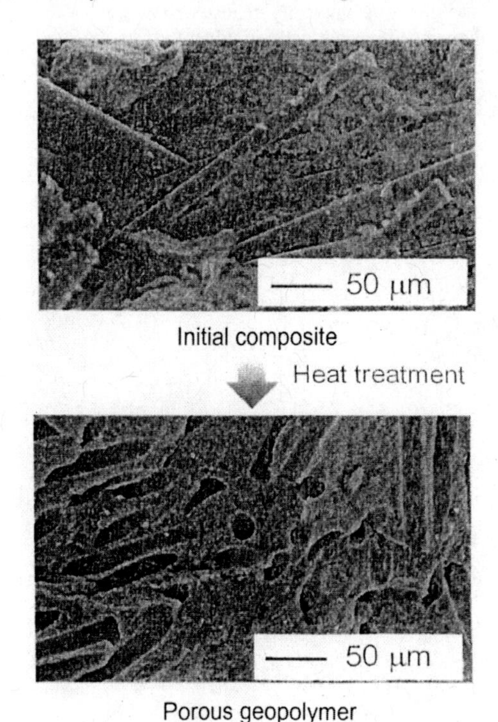

Figure 12. Scanning electron micrographs of a sodium geopolymer containing Nylon 66 pore formers (top) and after heating at 600°C to form aligned pores. Data from ref. 20.

A possible solution to this environmental problem is the use of evaporative cooling in the walls of buildings, utilising porous cladding materials with a high capacity for capillary water lift. A requirement of such materials is that they contain aligned pores of the correct size distribution. Porous geopolymer panels meeting these requirements have been made using Nylon 66 fibres as the sacrificial pore formers, which are removed by heating at 600°C after the panels have cured[20]. Scanning electron micrographs of these materials (Figure 12) show that the resulting pores are well aligned, while their passive cooling properties, monitored by a thermographic camera, are found to be extremely efficient in raising water by capillary action[20]. Comparison of the pore structure of these materials with those of conventional porous ceramics prepared at much higher temperatures suggests that their capillary lift hight should be comparable (~1000mm), while the low processing temperature of the geopolymer equivalents makes them an attractive environmentally-friendly alternative[20].

CONCLUSIONS

These examples show that inorganic polymers are much more than low-technology materials, but have many potential applications as advanced materials. These opportunities are by no means exhausted, and there are many more developments and new applications waiting to be discovered. The full potential of these exciting materials will be realised only if the fundamental science is well understood, but future developments will also require a high degree of creative lateral thinking and a recognition of the advantages and limitations of these materials.

ACKNOWLEDGEMENTS

We are indebted to the MacDiarmid Institute for Advanced Materials and Nanotechnology and the New Zealand Foundation for Research Science and Technology for partial financial support.

REFERENCES

[1]J. Davidovits, Geopolymers: Inorganic New Materials. *J. Thermal Anal.*, **37**, 1633-56 (1991).

[2]V.F.F. Barbosa, K.J.D. MacKenzie and C. Thurmaturgo, Synthesis and Characterisation of Materials Based on Inorganic Polymers of Alumina and Silica: Sodium Polysialate Polymers. *Int. J. Inorg. Mat.*, **2,** 309-317, (2000).

[3]M.R. Rowles, J.V. Hanna, K.J. Pike, M.E. Smith and B.H. O'Connor, ^{29}Si, ^{27}Al, ^{1}H and ^{23}Na MAS NMR Study of the Bonding Character in Aluminosilicate Inorganic polymers. *Appl. Magn. Reson.*, **32**, 663-89 (2007).

[4]J. Davidovits, Geopolymer Chemistry and Applications, Institut Geopolymere, St. Quentin, France (2008).

[5]K.J.D. MacKenzie, D.R.M. Brew, R.A. Fletcher and R. Vagana, Formation of Aluminosilicate Geopolymers from 1:1 Layer-lattice Minerals Pre-treated by Various Methods: A Comparative Study. *J. Mat. Sci.*, **42,** 4667-74 (2007).

[6]D. Cao, D. Su, B. Lu and Y. Yang, Synthesis and Structure Characterization of Geopolymeric Material Based on Metakaolinite and Phosphoric Acid. *J. Chinese Ceram. Soc.*, **33**, 1385-9, (2005).

[7]D.R.M. Brew and K.J.D. MacKenzie, Geopolymer Synthesis Using Silica Fume and Sodium Aluminate. *J. Mat. Sci.*, **42,** 3990-3 (2007).

[8]K.J.D. MacKenzie, A. Durant and H. Maekawa, To be published.

[9]D. Kolusek, J. Brus, M. Urbanova, J. Andertova, V. Hulinsky and J. Vorel, Preparation, Structure and Hydrothermal Stability of Alternative (Sodium Silicate-Free) Geopolymers. *J. Mat. Sci.*, **42**, 9267-75 (2007).

[10]K.J.D. MacKenzie and S.J. O'Connor, To be published.

[11]A.Durant, S.J. O'Connor and K.J.D. MacKenzie, To be published.

[12]K.J.D. MacKenzie and B. O'Leary, Inorganic Polymers (Geopolymers) Containing Acid-Base Indicators as Possible Colour-Change Humidity Indicators. *Mat. Lett.*, **63,** 230-2 (2008).

[13]K.J.D. MacKenzie, M.E. Smith and A. Wong, A Multinuclear MAS NMR Study of Calcium-Containing Inorganic Polymers. *J. Mat. Chem.*, **17**, 5090-6 (2007).

[14]K.J.D. MacKenzie, N. Rahner and A. Wong, To be published.

[15]H. Oudadesse, A.C. Derrien, M. Mami, S. Martin, G. Cathlineau and L. Yahia, Aluminosilicates and Biphasic HA-TCP Composites: Studies of Properties for Bony Filling. *Biomed. Mat.*, **2**, 59-64 (2007).

[16]S. Martin, A.C. Derrien, H. Oudadesse, D. Chauvel-Lebret and G. Cathlineau, Implantation of Aluminosilicate/Calcium Phosphate Materials; Influence on Bone Formation in Rabbit Tibias. *Eur. Cells and Mats.*, **9**, 71-2 (2005).

[17]K.J.D. MacKenzie and M.J. Bolton, Electrical and Mechanical Properties of Aluminosilicate Inorganic Polymer Composites with Carbon Nanotubes. *J. Mat. Sci.*, (Submitted).

[18]K.J.D. MacKenzie and B. O'Leary, To be published.

[19]K.J.D. MacKenzie and J. Barnes. To be published.

[20]K. Okada, A. Ooyama, T. Isobe, Y. Kameshima, A. Nakajima and K.J.D. MacKenzie, Water Retention of Porous Geopolymers for Use in Cooling Applications. *J. Eur. Ceram. Soc.*, (In press).

PROPERTIES AND PERFORMANCE OF SI-RICH GEOPOLYMER BINDER SYSTEMS

Kwesi Sagoe–Crentsil
CSIRO Materials Science and Engineering
PO Box 56, Highett, Victoria 3190, Australia

ABSTRACT

This paper examines specific roles of various constituent oxides on the hydrolysis and condensation reactions that underpin the properties and performance of high silica geopolymer binder systems. Geopolymer systems formulated to high Si/Al ratios provide an ideal system for this form of analysis given their potential for mainstream engineering applications. For this study, specific emphasis was placed on the roles of silica and alkali species present in the feedstock material and their impact on mechanical properties such as early strength development.

It is observed, that high silica geopolymers with $SiO_2/Al_2O_3 > 15$ can be synthesized as compared to conventional geopolymers which generally have $SiO_2/Al_2O_3 = 2\text{-}4$. Prior to curing, such high-Si mixtures display a more viscous consistency than conventional geopolymers and have a lower pH after setting. The relative high initial strength gains of the systems are complemented with good bonding characteristics. The overall performance trends of these silica-rich systems are explored and discussed in this Paper.

INTRODUCTION

Considerable research effort has been expended over the past two decades on synthesis mechanisms governing geopolymer binders owing to the potential performance advantages associated with this class of materials. Based on the remarkable contributions of Davidovits[1], subsequent research effort by several groups[2-4] has consolidated fundamental understanding of these systems largely focussing on conventional binders characterised by Si/Al < 3. However, there has also been corresponding interest in geopolymer systems with Si/Al ≥ 3 an d the reaction mechanisms of speciation governing these systems[1,5].

For generic geopolymer systems the prevailing chemical reaction processes involving Al_2O_3 and SiO_2 constituents are largely controlled by the stability of the respective speciated phases. These phases in-turn dictate microstructural characteristics as well as properties and performance of the end-product. In this regard, the dissolution of metakaolin, fly ash and other clays under alkaline conditions was studied by Phair et al.[6] and others[7]. X-ray diffraction (XRD) analysis shows geopolymers to be largely amorphous[8,9] although there is published evidence of occurrence of nanocrystalline particles, within the geopolymer[10,11] matrix structure. The reaction between metakaolin and alkaline silicate solutions has also been investigated by thermal analysis (DTA and TGA)[9,10], FTIR and NMR[10]. Correspondingly, in the alkaline aqueous solutions of geopolymers, aluminium is present mostly as monomeric aluminate $(Al(OH)^{4-}$ ions [12,13].

This paper draws on experimental and theoretical studies to examine the relationships between chemical formulation, microstructure and mechanical properties of high silica geopolymer systems with $SiO_2/Al_2O_3 > 15$ as an alternative to conventional geopolymers having $SiO_2/Al_2O_3 = 2\text{-}4$. Analysis of the compressive strength development characteristics and system pH provides a basic index to assess the interrelationships of key oxide components and their implications on geopolymer performance.

EXPERIMENTAL

Several geopolymer formulations were prepared with varying proportions of ingredients selected to allow the effect of alkali and silica contents to be assessed; supplementary silica was provided with the use of either colloidal silica (Ludox HS-40) and/or amorphous silica fume (Australian Fused Materials Pty Ltd). Commercial grade pearl sodium hydroxide pellets was used for all mixtures. The alkali silicate solution was supplied by PQ Australia and had the following composition: sodium silicate (8.9 wt% Na_2O, 28.7 wt% SiO_2, and 62.5 wt% H_2O). For example, 10.0g metakaolin, 71.7g AFM silica fume, 56.85g sodium silicate, 10.29g sodium hydroxide was used the formulation given as $5.0Na_2O.35SiO_2.Al_2O_3.50H_2O$.

High silica geopolymers were made by adding a large proportion of silica fume to conventional metakaolin geopolymer formulations. In preparing geopolymer systems with silica contents from SiO_2/Al_2O_3 = 15-56, more water had to be added to the higher silica formulations to make them mixable. In general the high silica geopolymer systems were made with very high alkali concentrations i.e. Na_2O/Al_2O_3 = 1.5-6.5 compared to 1.0 ratio for conventional geopolymers.

Sample preparation involved initial mixing of alkali hydroxide, alkali silicate solution and cooled, followed by the metakaolin. After mixing, samples were immediately poured into preheated moulds which were then sealed. Samples were cured at 85°C for 2 hours, after which they were demoulded and cooled in a refrigerator to arrest reaction kinetics. Compressive strength tests were performed on 25.4 mm cubes in an Instron (Model 5585H) compression testing machine at a load rate of 20 MPa/min. A minimum of three samples were tested for each mix formulation. Scanning electron microscopy (SEM) and energy dispersive x-ray analysis were carried out on carbon coated fracture surfaces of geopolymer samples using a Philips XL31 series microscope operated at 15 KeV and fitted with EDX Link ISIS manufactured by Oxford Instruments.

RESULTS

Characteristics of high Silica geopolymer systems

Figure 1 shows an SEM micrograph of a high silica geopolymer having a very dense microstructure; comparatively denser than conventional geopolymers. The dense microstructures suggest that high silica geopolymers could good mechanical properties. This has been noted by Davidovits[1] who used NMR to show that the silica was present in cross-linked chains rather than a full 3D network as proposed for conventional geopolymers. The dense microstructure is partly attributable to the gelling characteristics of silica-rich geopolymer systems during the mixing and setting stages. On a macroscopic scale little change on initial mixing of the reactants is evident. This is typical gelation behaviour, in that the change occurs at the gel point (where viscosity starts to increase) caused by the silicate oligomers reaching sufficient size to overlap one another and span the liquid phase.

Figure 1. SEM image of a high silica geopolymer with SiO_2/Al_2O_3 = 34.

Water Content

It was observed that the more water present, the lower the strength which is in agreement with results of conventional geopolymers. The pH after setting also dropped slightly with increased water. These results indicate that the water content should be minimised, however, in practice the water content of these high silica formulations is governed by the viscosity required for efficient processing and subsequent casting. The distribution of water in the final product may also effect diffusion through the structure.

A typical geopolymer composition of $Na_2O.4SiO_2.Al_2O_3.10\text{-}20H_2O$ can have water loadings of about 31-47 wt% water. In the initial mixture this water is added as an alkali silicate solution. During the dissolution phase of the synthesis, this water is involved in the hydrolysis reaction:

$$\text{-R-O-R-} + H_2O \rightarrow 2(\text{-R-OH}),\ (R = \text{Si or Al})$$

Assuming $H_2O/Al_2O_3 = 10$, in the above mentioned formulation 70% of the water would be required for complete hydrolysis reaction. This significant reduction in the free water would affect the viscosity of the mixture, however it may not occur to the full extent as condensation reactions are presumed to occur simultaneously. Essentially higher water content geopolymers generally set slowest. This is partly because more time is available for mobility of ions in the system as the solids dissolve. A porosity increase is to be expected, as there will be more water-filled pores in the product with increased water content akin to high water/binder ratio in hydraulic cement systems. The practical importance of the viscosity of geopolymer formulations has been commented upon previously by other authors [14,15].

Silicon and aluminium content

Figure 2 shows the strength development characteristics of geopolymer systems with varying SiO_2/Al_2O_3 content ranging from SiO_2/Al_2O_3 = 15-56. The compressive strength was a maximum at around $SiO_2/Al_2O_3 = 32$. As noted previously, the higher silica content formulations also displayed excess water demand.

In general the Si/Al ratio in geopolymer systems has an important influence on the structure and mechanical properties. Other researchers such as Palomo et al.[16] showed that Al solubility is generally much higher than Si under alkaline conditions, thus aluminate anions are likely to react with silicate species from alkaline silicate solutions. A higher Al content assists in condensation occurring more feasibly, and tends to lead to a denser network structure due to the removal of more hydroxyl groups.

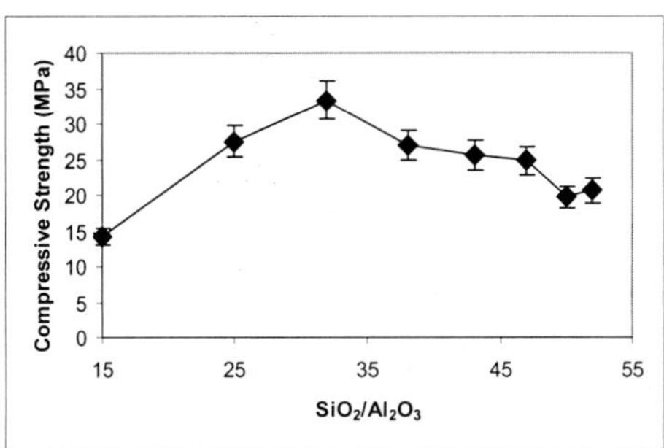

Figure 2. Relationship between compressive strength and SiO_2/Al_2O_3 ratio.

Considering the fact that aluminium component of geopolymer mixtures tends to dissolve more easily than the silicon component[17], it is plausible that more $Al(OH)_4^{4-}$ species and relatively less Si species are available for condensation in the system with low SiO_2/Al_2O_3 ratios. Therefore condensation is more likely to occur between aluminate and silicate species producing poly (sialate) polymer structures. With increasing Si content, more silicate species are available for condensation and reaction between silicate species, resulting oligomeric silicates, becomes dominant. And further condensation between oligomeric silicates and aluminates results in rigid 3D network of poly (sialate-siloxo), and poly (sialate-disiloxo) 3-D rigid polymeric structures. The heat-cured high silica content samples did not display obvious elastic properties in contrast to findings of Fletcher et al.[5] in their ambient-cured samples for which increasing elastic behaviour was observed for samples with SiO_2/Al_2O_3 >24.

Alkali content

There appears to be a very clear correlation between the alkali content and the compressive strength as shown in Fig 3. The pH of the cured material also rises with alkali content as would be expected. Indeed the high silica geopolymers also require very high alkali concentrations Na_2O/Al_2O_3 = 1.5-6.5 compared to a ratio of ~1.0 for conventional geopolymers. Furthermore, increasing the alkali content decreases the viscosity of the formulation. However for practical applications higher alkali dosages may be limited by handling and plant equipment issues.

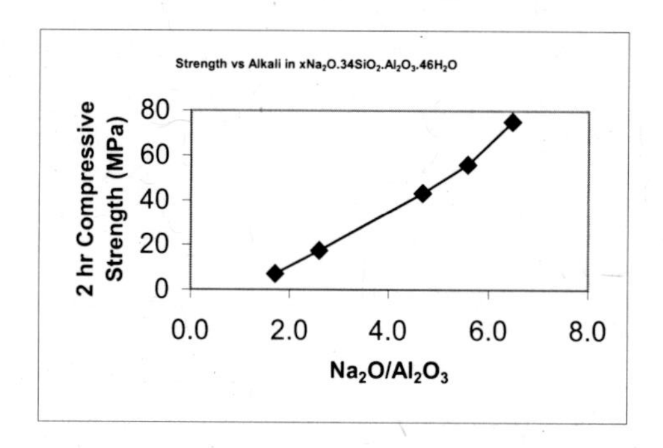

Figure 3. Relationship between Na_2O/Al_2O_3 and compressive strength of Geopolymer with $SiO_2/Al_2O_3 = 34$.

In theory, the alkalinity of the activator governs the rate of condensation between aluminate and silicate species, besides such factors as temperature and the nature of the feedstock as shown previously by calorimetric results[17]. The highly alkaline environment with an initial pH regime of around 13.9 makes silicate and aluminosilicate oligomeric ions less stable than at pH<12[18]. Phair et al.[18] found that geopolymer condensation was favoured in conditions of high pH in which most of the dissolved silicon was monomeric. By comparison, zeolitic reactions which have similar chemistries to geopolymers are also formed in conditions favouring monomeric or small oligomeric silicate ions. However the speciation of aluminosilicate solutions at high pH and high silica concentration is still evolving[19].

The pH after curing for samples with varying SiO_2/Al_2O_3 ratios cured for 2 hrs/85°C showed a trend of lowering pH with increase in SiO_2/Al_2O_3 ratios. The silica content however remains the main variable, in these samples although the water and alkali contents also varied between samples.

The observed trend inverse between pH and silica content of the mixtures suggests pH reduction primarily arises from silica dissolution reactions.

Curing conditions

The strength development over time at 50°C and 85°C are shown in Fig 4. The setting is much faster at 85°C and the 24 hr strength is much higher. The ultimate strength at long times has however not been tested. At 50°C the geopolymer was still soft at 3 hours and under ambient conditions the formulation remains soft for several days.

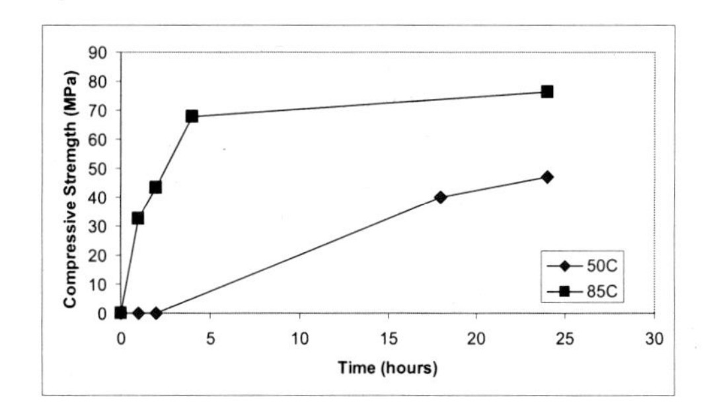

Figure 4. Effect of curing temperature (85°C and 50°C) and time on compressive strength of high-Si Geopolymer with $SiO_2/Al_2O_3 = 34$.

DISCUSSION

The prospects of synthesising high silicate geopolymer derivatives, generally referred to as poly(sialate-disiloxo) systems remain good, particularly with the selection of reactive Si constituents that contribute to overall target silicate dosage of the binder. Traditionally, there appears to be a limit of Si/Al < 3 favoured in the development of higher strength geopolymers, mainly determined by the selection and availability of suitable feedstock materials. However, as indicated in this study, it is apparent higher SiO_2/Al_2O_3 molar ratio geopolymers with improved strength characteristics are plausible. Evidently, the initial molar contents of Na_2O, Al_2O_3 and SiO_2 play a key role in controlling the chemistry and properties of these systems[11]. In particular, increasing Na_2O is shown to be critical for processing of Si rich formulations as it induces initial dissolution of the amorphous silica feedstock material. The high SiO_2 contents also tend to favour rather dense and notably less brittle microstructures.

The Al reaction remains a critical factor in the performance of these Si rich systems. Given that aluminate anions for the reaction are solely derived from the dissolution of mineral oxides under alkaline conditions, monomeric $[Al(OH)_4]^-$ ions are probably the only aluminate species existing under high alkaline conditions[18]. On the other hand, silicate species come from both soluble alkaline silicates and the dissolution of mineral oxides. In the specific case of metakaolin systems, the silicate species from the dissolution of particles are difficult to predict because the hydrolysis process of amorphous silica is kinetically dependent on various factors, such as the reactivity of the particles, temperature, time, and the concentration and pH value of alkaline silicate solutions.

Higher Al composition also suggests that condensation occurs more feasibly, and leads to a denser network structure due to the removal of more hydroxyl groups. Therefore, as expected, geopolymers composed of a higher Al component may be brittle, and have high hardness. Moreover, geopolymers with a low Al component will be better binders since more hydroxyl groups exist in the structure, resulting in improved characteristics primarily from existing hydrogen bonds. These observations suggest that the condensation process in these systems occurs in two stages: (a) quick condensation between aluminate and silicate species; followed by (b) a slow condensation stage solely involving silicate species[4].

CONCLUSIONS

1. It is observed that the high silica geopolymer formulations investigated achieve near optimal strengths at $SiO_2/Al_2O_3 \approx 35$ and show characteristic low porosity and dense, fine grained microstructures.
2. Relatively higher alkalinity characterises silica-rich systems with observed alkalinity levels equivalent to $Na_2O/Al_2O_3 = 4.5\text{-}5.5$; compared to conventional geopolymers with nominal alkali contents $Na_2O/Al_2O_3 = 1.0$.
3. Water demand of high silica geopolymer systems appear to be comparatively high, however observed effects of oxide components correlate well with known trends with respect to geopolymer mix composition, microstructure and strength.
4. The potential exists for developing very high performance high silica $SiO_2/Al_2O_3 \approx 35$ binder systems, although issues with complete Si dissolution or solubility need to be addressed.

REFERENCES

1. J. Davidovits, Geopolymer chemistry and properties. In: Davidovits, J, Orlinski, J. (Eds.), Proceedings of the 1st International Conference on Geopolymer '88, vol. 1, Compiegne, France, 1–3 June, 25–48(1988)
2. P.V. Krivenko, G.Yu Kovalchuk, Directed synthesis of alkaline aluminosilicate minerals in a geocement matrix. J. Mater. Sci., **42,** 2944–2952(2007).
3. H. Xu, and Van J.S.J. van Deventer, The geopolymerisation of aluminosilicate minerals. International Journal of Mineral Processing, **59,** (3), 247–266(2000).
4. K. Sagoe-Crentsil and L. Weng "Dissolution processes, hydrolysis and condensation reactions during Geopolymer synthesis: Part II - High Si/Al ratio systems" Journal of Materials Science, Vol. 42 #9, pp 3007-3014. (2007)
5. R.A. Fletcher, K.J.D MacKenzie, C. L. Nicholson and S. Shimada. "The composition range of aluminosilicate geopolymers" Journal . Eur. Ceram. Soc., **25**, 1471-77. (2005).
6. J.W. Phair, and J.S.J. Van Deventer, Effect of silicate activator pH on the leaching and material characteristics of waste-based inorganic polymers. Miner. Eng. **14** 289–304(2001)
7. A. Fernandez-Jimenez and A. Palomo, Composition and microstructure of alkali activated fly ash binder: Effect of activator Cem. Concr. Res., **35**, 1984-1992(2005)
8. H. Rahier, W. Simons, B. van Miele, and M. Biesemans, Low temperature synthesised aluminosilicate glasses. Part III: Influence of the composition of the silicate solution on production, structure and properties, J. Mater. Sci. **32,** 2237-2247 (1997).
9. M. Rowles, and B. O'Connor, Chemical optimisation of the compressive strength of aluminosilicate geopolymers synthesised by sodium silicate activation of metakaolinite. J Mat. Chem. **13.** 1161-1165, (2003).
10. H. Rahier, B. Van Mele, M. Biesemans, J. Wastiels and X. Wu, "Low-Temperature Synthesized Aluminosilicate Glasses. Part I: Low-Temperature Reaction Stoichiometry and Structure of a Model Compound," J. Mater. Sci., 31[1] 71-79 (1996)
11. A. Palomo, M. T. Blanco-Varela, M. L. Granizo, F. Puertas, T. Vazquez and M. W. Grutzeck, "Chemical Stability of Cementitious Materials Based on Metakaolin," Cem. Concr. Res., 29[7] 997-1004 (1999)

12. B. B. Sabir, S. Wild and J. Bai, "Metakaolin and Calcined Clays as Pozzolans for Concrete: A Review," Cem. Concr. Compos., 23[6] 441-454 (2001)
13. V.F.F. Barbosa, K.J.D. MacKenzie, and C. Thaumaturgo, Synthesis and characterisation of materials based on inorganic polymers of alumina and silica: sodium polysialate polymers, International Journal of Inorganic Materials, 2, 309-317(2000).
14. P. De Silva, K. Sagoe-Crenstil, and V. Sirivivatnanon, Kinetics of geopolymerization: Role of Al_2O_3 and SiO_2. Cem. Concr. Res. 37 (4), 512–518. (2007)
15. P. Duxson, J.L. Provis, G.C. Lukey, S.W. Mallicoat, W.M. Kriven, and J.S.J. van Deventer, Understanding the relationship between geopolymer composition, microstructure and mechanical properties, Colloids and Surfaces A: Physicochem. Eng. Aspects, 269, 47-58, (2005).
16. A. Palomo, M.W. Grutzeck, and M.T. Blanco, Alkali-activated fly ashes, A cement for the future, Cem. Concr. Res., 29, 1323-1329(1999).
17. M. Steveson, and K. Sagoe–Crentsil, Relationship between composition, structure and strength of inorganic polymers Part I – Metakaolin-derived inorganic polymers J. Mater. Sci., 40, 2023-2036, (2005). Relationship between composition, structure and strength of inorganic polymers Part II – Flyash-derived inorganic polymers J. Mater. Sci., 40, 4247-4259(2005)
18. J.W. Phair, S.J. van Deventer, and J.D. Smith, Mechanism of polysialation in the incorporation of zirconia into fly ash-based geopolymers, Ind. Eng. Chem. Res., 39, 2925-2934(2000)
19. T.W. Swaddle, J. Salerno, and P.A. Tregloan, Aqueous aluminates, silicates and aluminosilicates, Chemical Society Reviews, 319-325(1994)

COLD SETTING INORGANIC NETWORKS INCLUDING PHOSPHATES

Christian Kaps and Marc Hohmann
Bauhaus-University Weimar
Chair of Building Chemistry
D-99423 Weimar, Germany, Coudraystrasse 13C

ABSTRACT

The aim of the experimental investigations is to find out possibilities for compositional and structural modifications of inorganic network binders (silicates and aluminosilicates) in respect to new application fields by the chemical reaction of different phosphates. The preparations have been made in two ways: (1) Addition of cyclo-phosphates to water-glass solutions and (2) Reactions of phosphoric acid with convenient types of reactive oxides like Fe_2O_3 and $Al_2O_3 \cdot 2SiO_2$. Aspects of the setting, structure and compositions of the networks are discussed. The results demonstrate glass-like binder materials and improved durabilities for special applications in construction components under strong chemical attack.

INTRODUCTION

Cold setting inorganic binders are in the field of vision for construction materials. Recent research is focused on exploitation of secondary raw-materials, savings of energy or carbon dioxide reduction and on the development of unconventional properties for new applications of construction materials. The roots of these tendencies go back to the sixties and seventies of the foregoing century. Glukhovsky[1] investigates cement systems using the chemical activation by alkalis for slag raw-materials and described these binders as "soil silicates". Davidovits[2,3] initiated systematic research in respect to the chemical reactions and technologies for generation and application of amorphous aluminosilicates with a polymeric character. He created the well-known term "Geopolymer" and described the network structure. As typical network components or connecting oxo-coordination tetrahedra act $SiO_{4/2}$ or $O_{3/2}SiO^-$ and $Al^-O_{4/2}{}^*$ in charge compensation with the alkaline cations Na^+ and/or K^+.

Also including the water-glass binder, the general formation of these inorganic polymer binders appears in a similar way, starting with any at least silicate containing solid from an industrial or natural origin.

I. Melting and rapid cooling or thermal activation
   (→ defect-rich or amorphous silicatic materials)

II. Solution process in water or alkaline solutions
   (→ reactive suspension or solutions with monomers / oligomers with OH-groups, available for condensation)

III. Setting process by polycondensation enabling condensation with escape of water molecules
   (→ inorganic networks and more or less alkaline by-products)

In dependence on the binder composition, a strong attack with water can reactivate a part of the alkaline by-products and a partial dissolution is observable. Understandably these materials show no resistance against acidic solutions. However, the civil engineering searchs for construction materials with a sufficient durability under acidic attack (floors in chemical laboratories).

With respect to this effort, the aim of the investigation can be described as an attempt to use the chemical component phosphate for the generation of stable binders on basis of network structures.

RAW MATERIALS, REACTIONS AND CHEMICAL BONDING

Geopolymer networks consisting of aluminosilicates are often formed from metakaolin reacting with water-glass solutions with a defined amount of alkali hydroxide. In recent years however a lot of inexpensive secondary raw-materials have also been tested and used for the generation of network binders. Some of the investigated starting materials are compiled in the survey of Figure 1.

Single oxides in a reactive state (highly dispered, partially dehydroxylated)	Industrial products (from a high-temperature process)	Natural products (thermal activated under hydrothermal conditions)
• metakaolin • silica • alumina • halloysite	• fly ash • brick powder ←** • blast-furnace slags*** • clinker phases***	• clay (2- and 3-layer minerals) • loam

Figure 1. Starting materials for aluminosilicate network binders

One advantage of the aluminosilicate networks seems to be that these binder materials maintain their binder action also in the high-temperature range. This is in contrast to a conventional binder like gysum, lime or cement, corresponding to the chemistry of these materials with evaporable or decomposable components. In this connection it should be also remarked, that a sufficient content of free reactive alkaline earth oxide (especially CaO) in the starting mixtures of geopolymers causes the generation of alternative C-S-H phases[4] in the process of setting (see *** in Fig. 1). C-S-H phases contain water as a relevant constituent and decompose at higher temperatures.

In the last years an useful relation between geopolymers and ceramic technology has been developed. On one hand, several brick powders are an effective component in mixtures for generation of aluminosilicate networks (a "geopolymer" way) and on the other hand, cold setting mixtures with thermal activated clay show a helpful stabilisation of ceramic green bodies in the drying process (a "ceramic" way[5]). In analogy to the latter, loam building components as ecological products can be stabilized by geopolymers[6] in respect to mechanical load and weathering.

The formation of the networks is characterized by chain growth and interlinking of the dissolved monomers or oligomers in a sol-gel transformation process during the setting (see step III. in the INTRODUCTION). In general the basic condensation reaction is also characterized by acid-base relation of the two monomer species (different acidity of OH-groups). These reactions can be described for a water-glass binder (eq. 1) and aluminosilicate network (eq. 2) in the following manner.

$$(HO)_2O^-Si-OH + HO-Si(OH)_3 \rightarrow (HO)_2O^-Si-O-Si(OH)_3 + H_2O\uparrow \quad (1)$$

$$(HO)_2O^-Si-OH + HO-Al^-(OH)_3 \rightarrow (HO)_2O^-Si-O-Al^-(OH)_3 + H_2O\uparrow \quad (2)$$

The remaining OH-groups are available for further condensation (escape of water) and for carry on the network growth by binding of chemical species. The structure of the formed network is a more or less an amorphous arrangement of the monomer and oligomer units. The sequence of the condensation steps can be regarded as independent and governed by the local meeting up of condensation-able OH-groups in a random process. In this sense the formation of inorganic networks is characterized by a chain growth and interlinking of an organic polymer and seems very different from an electrostatic ally dominated precipitation process of crystalline solids (salt-like)[7].

The negative charges on the non-bridging oxygen (NBO) and on the aluminate are compensated by the presence of alkali ions from the solution process (see step II. in the INTRODUCTION). The bridging oxygen (BO, uncharged) and the bondings in the network former silicate and aluminate are important for formation tendency and stability of the inorganic networks. Some aspects of the element-oxygen bonding (E-O) are compiled in Table 1.

Table 1. Comparison of the E-O bonding in respect to networks

	Bonding E - O	Difference of electro-negativ-ities	Covalency (covalent part of the E - O bonding	Monomer species	
Aluminosilicate networks	Si - O Al - O	1.76 2.03	0.46 0.35	$Si(OH)_4$, $Al^-(OH)_4$	$SiO^-(OH)_3$
Alkali metals in oxo-coordination	(Li, Na, K)	2.5 – 2.6	≈ 0.20	-	
Semi-metals in oxo-coordination	(B, Ge, As)	1.3 – 1.5	≈ 0.55 – 0.65	-	
Transition metals in oxo-coordination	(Ti, Fe, Zn)	1.3 – 1.8	≈ 0.45 – 0.65	-	
Condensed phosphates	P - O	1.44	0.60	$OP(OH)_3$,	$OPO^-(OH)_2$

It is understandable that a suffient covalency of the E-O bonding must be regarded as a precondition for the stability of a network with bridging oxygens. Obviously, covalencies between 0.35 and 0.46 enable stable aluminosilicate networks. Low covalencies of about 0.20 consistently cause a dominant electrostatic binding of alkali ions on the charged non-bridging oxygen of a network. Semi-metals and transition metals demonstrate in oxo-coordinations convenient or a little too high covalencies. Borate units have been discussed in networks by Nichelson et al.[8] and Perera et al.[9] described ferric units introduced in geopolymers. In view of the wide variety of known condensed phosphate structures[10], phosphate units are also attractive components for mixed inorganic networks.

It should be possible to bind phosphate units by condensation reactions in a simplified description corresponding to equation 3.

$$(HO)_2E-OH+HO-PO(OH)_2 \rightarrow (HO)_2E-O-PO(OH)_2+H_2O \qquad (3)$$

The phosphate units possess an advantageous oxo-coordination number of four and can form an oxygen bridging with three OH-groups to the network, at least.

We tried to examine the introduction of phosphate in inorganic networks on two ways. The first one was the reaction of water-glass solutions with different cyclo-phosphates and the second one consisted in the reaction of phosphoric acid with convenient solid oxides.

NETWORK BINDER FROM WATER-GLASS SOLUTIONS AND CYCLO-PHOSPHATES

Preparation of the binders and structural aspects

The binders have been prepared by addition of the two cyclo-phosphates to the water-glass solution (WGS), mixing and intense homogenization in two preparation rows (a and b) up to an amount of 30 wt.%. Table 2 contains the information about the starting materials used.

Table 2. Preparation materials

- Water-glass solution (WGS, Sigma Aldrich, ρ=1.39 g/cm^3, SiO_2/Na_2O=3.41)
- Added cyclo-phosphates (ACP, Sigma Aldrich)
  - Preparation series a: Sodium - cyclo - triphosphate $Na_3(PO_3)_3$
  - Preparation series b: Aluminium - cyclo - tetraphosphate $Al(PO_3)_3 = Al_4(P_4O_{12})_3$
- Optimized quartz sand filler (OQF, Quarzwerke Frechen, H 33 and W 3 mass ratio: WGS/OQF = 2/1)

The sodium-cyclo-phosphate showed a weak slow reaction with the water-glass solution. The aluminium-cyclo-phosphate caused a pronounced setting process[11].

The phase contents of the set binder were analysed after 28 days (60 % r.h., 25°C) by X-ray diffraction (XRD, 3003 – OED Seifert, Rietfeld refinement, Auto Quan ® with ZnO standard). Table 3 gives the comparison of the analysed contents in the case of 20 wt.% ACP addition.

Table 3. Phase content of the binder after XRD analysis

Binder from preparation a with 20 wt.% $Na_3(PO_3)_3$	Binder from preparation b with 20 wt.% $Al(PO_3)_3$
63.3 wt.% amorphous network	76.9 wt.% amorphous network
28.2 wt.% $Na_4P_2O_7$ } crystalline	10.0 wt.% $Al(PO_3)_3$ } crystalline
	8.0 wt.% $Na_4P_2O_7$ } crystalline
8.5 wt.% Na_2HPO_4 } crystalline	9.6 wt.% Na_2HPO_4 } crystalline
	1.5 wt.% sodalite

On a comprehensible way, the identified components of the network binders result from the three dominant reaction steps[12]:

- Opening of the phosphate-cycles
- Uptake of the alkali ions by the phosphate fragments
- Polycondensation of the silicate network

Caused by an addition of 20 wt.% $Na_3(PO_3)_3$ the amorphous network constitutes about 2/3 of the binder mass. Estimations of the balance of components show that the network is a silicatic one. In preparation series a the increasing addition of $Na_3(PO_3)_3$ enlarges the contents of crystalline by-products $Na_4P_2O_7$ and Na_2HPO_4. The reaction of WGS with $Na_3(PO_3)_3$ can be described by equation 4.

$$2(Na_2O - xSiO_2 - yH_2O) + Na_3(PO_3)_3 \rightarrow 0.5Na_2O \bullet 2xSiO_3 \bullet (2y - 0.5)H_2O + Na_4P_2O_7 + Na_2HPO_4 \quad (4)$$

In preparation series b the addition of 20 wt.% $Al(PO_3)_3$ constitutes an amount of 3/4 of the binder mass in fact. There is a surplus of added $Al(PO_3)_3$. Corresponding to the above three steps of reaction, the amount of crystalline by-products $Na_4P_2O_7$ and Na_2HPO_4 increases also with an increase in added $Al(PO_3)_3$ in this preparation series b of note is the observation of the crystalline aluminosilicate, sodalite. This fact can be valued as a hint for introduction of aluminate in the formed network. Evaluations of the balance of components show a distinct dominance of the silicate for the amorphous phase (Reference[13]).

For more detailed information Magic Angle Spinning - Nuclear Magnetic Resonance measurements (MAS-NMR, Avance Bruker,14.1 T ^{27}Al (Larmor frequency 156.34 MHz, spinning speed 12.5 kHz), ^{29}Si (119.9 MHz, 4.0 kHz) and ^{31}P (242.88 MHz, 12.5 kHz); CP ^{1}H - ^{29}Si and ^{1}H - ^{31}P) were carried out. The spectra of a set WGS and of the binders with 20 wt. % $Na_3(PO_3)_3$ (series a) and $Al(PO_3)_3$ (series b) are represented in Figure 2.

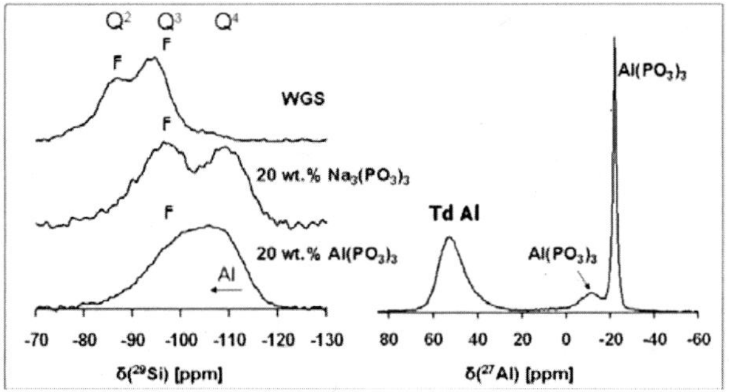

Figure 2. ^{29}Si and ^{27}Al MAS-NMR spectra of the binders (F Position of a CP-resonance, ^{1}H - ^{29}Si)

The ^{29}Si spectra demonstrate a higher connectivity of the network (Q^4 groups in the range from -100 to -110 ppm) after the addition of 20 wt.% $Na_3(PO_3)_3$ and $Al(PO_3)_3$ in comparison with that of a set water-glass solution (WGS). Apart from the resonance of the added $Al(PO_3)_3$ (at -21 ppm) the ^{27}Al spectrum shows a broad peak around +55 ppm. This fact confirms the introduction of aluminate in the network (Td Al means $AlO_4Q^4(Si)$)[14].

As expected in the ^{31}P spectra of Figure 3 the resonances of the crystalline by-products $Na_4P_2O_7$ (+2 ppm) and Na_2HPO_4 (+6 ppm) appear. Nevertheless, the elevated background in the from +10 to -25 ppm can be valued as a hint for amorphous phosphate arrangements[15,16] in the case of preparation with 20 wt.% $Al(PO_3)_3$. The type and the degree of binding of these phosphate components in the aluminosilicate network remains uncertain.

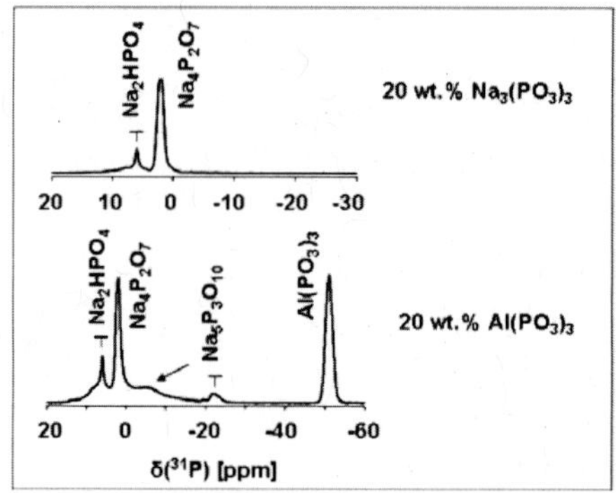

Figure 3. ^{31}P MAS-NMR spectra of the binders (┬ Position of a CP-resonance, ^{1}H -^{31}P)

Properties of materials with quartz sand filler

The network binder prepared from water-glass solution (WGS) under addition of cyclo-phosphates (ACP) were used in mixtures with an optimized quartz sand filler (OQF, see Table 2) to generate prismatic samples (1 x 1 x 6 cm^3) for characterisation of the mechanical stability and chemical durability. Figure 4 shows the compressive strength before and after chemical attack in dependence on the addition of $Al(PO_3)_3$ (Preparation series b).

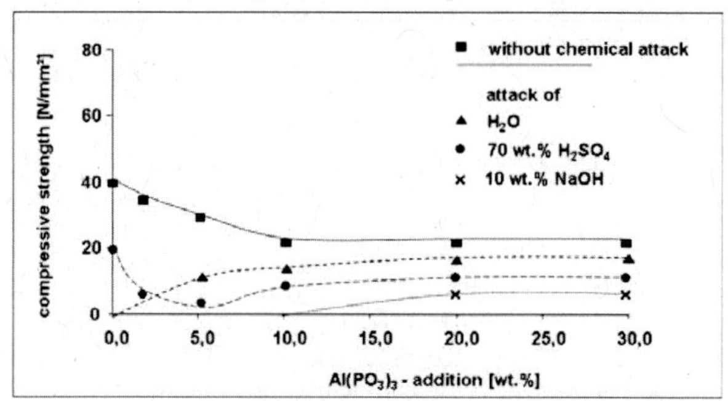

Figure 4. Compressive strength of prisms with network binder from preparation series b, influence of chemical attack (sample size: 1 x 1 x 3 cm^3, loaded area: 3 cm^3)

In Figure 4 a transition from a silicate network to an aluminosilicate network modified by phosphate units is represented with rising additions of $Al(PO_3)_3$. The unexposed prisms exhibit a slow lessening of the compressive strength in the series b, but in every case the values are larger than 22 N/mm^2. However, after an attack of water (H_2O, 6 h, 100°C) and also 70 wt.% H_2SO_4 (6 h, 100°C) the phosphate-modified aluminosilicate networks demonstrate an advantageous increase of chemical durability. The compressive strengths are at least 11 N/mm^2.

NETWORK BINDER FROM PHOSPHORIC ACID AND REACTIVE OXIDES

Preparations and reaction behaviour

In an alternative way compared to preparation with strong basic water-glass solutions we started the investigation of the possibility for phosphate introduction in inorganic networks from strong acidic solutions of the phosphoric acid. The setting processes were managed by two reactive oxides: a water-containing iron oxide and a water-free metakaolin. Table 4 contains the parameters of the raw materials.

Table 4. Starting materials

- Phosphoric acid ($H_3PO_{4\ conc.}$; 85 wt.%, ρ= 1.69 g/cm³)
- Added reactive oxides:
  - Iron oxid (Waste product, amorphous; average oxidation number of iron: + 2,99 , by cerimetric titration,
    Fe_2O_3; about 50 wt.% Fe_2O_3 and 10 wt.% SiO_2, more than 25 wt.% H_2O)
  - Metakaolin (MetaStar 402, Imerys Minerals Limited, England
    $Al_2O_3 \cdot 2SiO_2$, amorphous)
- Quartz sand filler (Quarzwerke Frechen, $0<d<0.5$ mm)

With respect to a network formation with phosphate as an immediate constituent, a setting reaction can be expected from a solution process of the iron oxide and a hydrolysis of the iron cations in a simplified form corresponding to equation 5.

$$(HO)_2Fe-OH + HO-PO(OH)_2 \rightarrow (HO)_2Fe-O-PO(OH)_2 + H_2O\uparrow \quad (5)$$

In Table 5 the relevant mixtures and additives of the preparation are compiled.

Table 5. Preparation conditions

Slurry mixtures	Additives	pH start value	Preparation type
Fe_2O_3 : H_2O : H_3PO_4	without	< 1.0	A
mass ratios	about 1 wt.% Fe powder	< 1.0	B
1 : 1 : 2	K_2CO_3 for pH increase	≈ 1.5	C
$Al_2O_3 \cdot 2SiO_2$: H_2O : H_3PO_4 mass ratios 1 : 1 : 2	without	< 1.0	D

The addition of a trace of iron powder was carried out in the preparation type B for a better solution process of the iron oxide by a partial reduction of Fe^{3+} to Fe^{2+} corresponding to the observation of Wagh[18]. A higher value of pH by a K_2CO_3 addition (Preparation type C) should reinforce the probability for a more developed hydrolysis of the iron cations to realize condensationable OH-groups. This is a well-known process for the formation of charged iron hydroxide gels (Rähse[19]).

Finally, the metakaolin in the slurry mixture of the preparation type D should give information about the action of aluminate in the network forming process.

The preparation A leads to a clear sol-gel process. In a few minutes there are tacky products with the characteristic feature of the suitability for "drawn out fibres". The setting process is slow and needs several days or weeks. Solidified binder materials seem to be glass-like. This fact is also confirmed by a corresponding micrograph of the binder morphology in Figure 5.

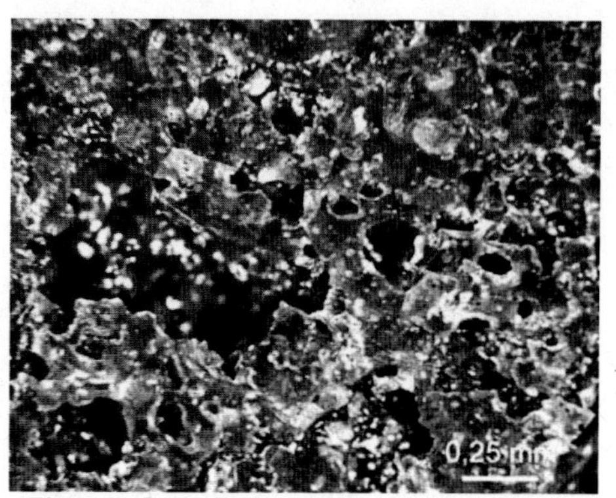

Figure 5. Optical micrograph of a type A binder

In the preparation B, the small amount of iron powder causes an acceleration of the setting in the time range from hours to a few days. The preparation of the type C at a higher pH-value gives a mixture with a very rapid setting in minutes or a few hours. The slurry mixtures with metakaolin of the preparation type D show also a moderate setting rate, which is comparable to that of the type A and exhibits a noticeable dependence on temperature. Moreover, the solidified binders demonstrate, after a few week, a remarkable mechanical stability.

Characterisation of binders and materials with quartz sand filler

For comparison of structural aspects the set binders have been characterized by X-ray diffraction (XRD). Figure 6 proves for the type A binder, an amorphous structure to a large extent, and for the B type binder pronounced crystalline features by sharp Bragg reflexes.

Thus, the XRD analysis confirms the formation of an amorphous network consisting of iron and phosphorus oxo-coordination units from the preparation way A. The broad "bump" around $2\Theta \approx 28°$ corresponds to a value $\sin\Theta / \lambda = 0.157$ and amounts with it a little higher than that of set SiO_2 gels ($\sin\Theta / \lambda = 0.125$, after Scholze[20]). Inhomogeneities in this amorphous binder can not exclude completely. In contrast, the diffraction pattern of the binder B shows first the dominance of a crystalline iron hydrogen phosphate hydrate ($Fe_3H_9(PO_4)_6 \cdot 6H_2O$, 44-812) as the result of accelerated setting, including a crystallization process, induced by a trace of metallic iron. Under comparable conditions Wagh and Jeong[21] also observed the formation of iron hydrogen phosphates and discussed the meaning of the reduction process of Fe^{3+} by the action of elemental iron.

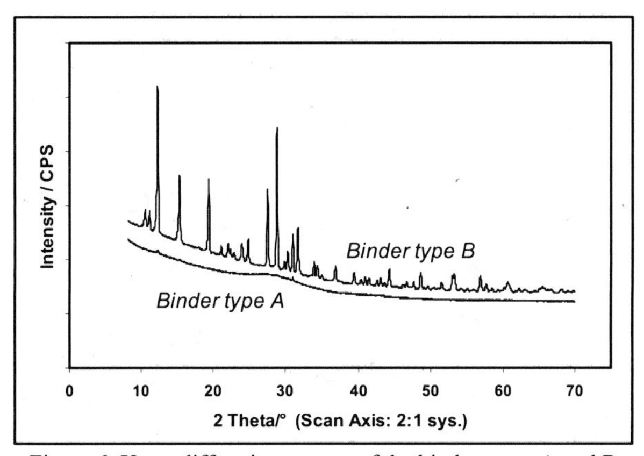

Figure 6. X-ray diffraction pattern of the binders type A and B

In Figure 7 the diffraction pattern exhibits features of an amorphous binder with crystalline inclusions for the setted binder from the preparation method D. The crystalline phase formed is at least quartz. Obviously, the selective solubility of the aluminate component of the metakaolin in the acidic range seems to be important for the generation of an amorphous network. It is well known that clay minerals can be an useful Al^{3+} source in contact with strong acids. For instance acid-activated bentonites[22] are generated by a partial solution of the octahedral aluminate layer with strong acids and staying of disordered silicates as reactive component. On the other hand Wagh et al.[23] prepared chemically bonded phosphate ceramics from different alumina-based raw-materials with phosphoric acid and achieved a sufficient solubility by increase of the process temperature in to the range from 100 to 150° C, including a crystallization of berlinite ($AlPO_4$).

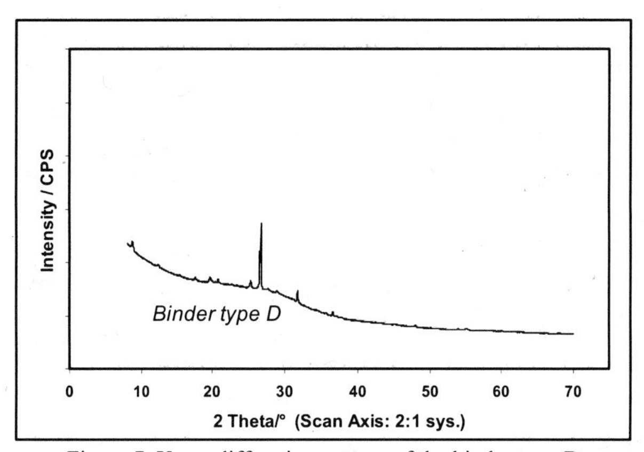

Figure 7. X-ray diffraction pattern of the binder type D

For the characterization of the binder behaviour the four types of slurries were mixed with a quartz sand filler in a mass ratio of slurry/ sand = 1/ 1.25. The fresh suspensions have been poured in round teflon® moulds (d = 2.0 cm). Cylindrical samples from the four preparation ways are shown in Figure 8.

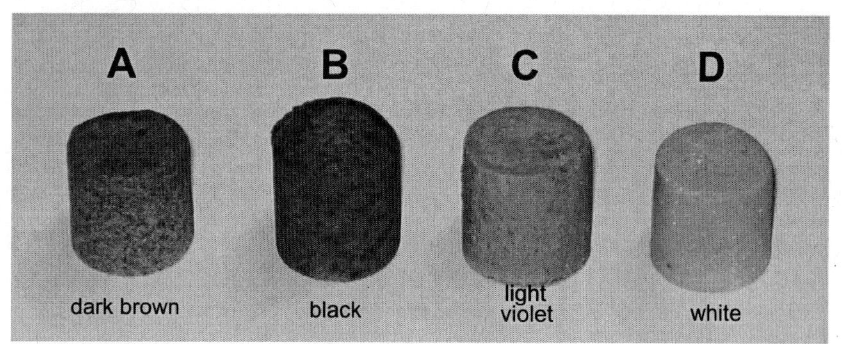

Figure 8. Solidified samples with quartz sand filler, bonded by different binders (A, B, C and D)

In the presence of the component sand the composite materials appear in different colours. The black colour of the sample B can be caused by the coexistence of Fe^{3+} and Fe^{2+} ions in the form of developed magnetite pigments. On the way of preparation B, hydrogen gas is generated by the redox-reaction ($Fe + H_3PO_4 \rightarrow Fe^{2+} + HPO_4^{2-} + H_2$). Because of that the top plane of the cylinder B is remarkably bent. The white samples of the D type show an excellent mechanical stability. This composite material demonstrates compressive strength up to 70 N/mm² and shows no deterioration after a strong attack of boiling water (6 h). This result reveals that metakaolin is also a suitable raw-material for generation of an attractive network binder by reactions with phosphoric acid in the strong acidic pH-range. Just like in the alkaline solution process of metakaolin, the sol-gel transformation leads preferably to an amorphous network, obviously consisting of aluminate and phosphate units (References[24, 25]).

Further systematic investigations are necessary for more detailed information about phosphate as a constituent of an inorganic network.

CONCLUSIONS

Cold setting inorganic networks are in the research focus of building chemistry because these materials show an advantageous binder behaviour and can be produced under environmentally friendly conditions. The amorphous networks formed keep up their binder action in the high-temperature range, which is in contrast to the conventional binders like gypsum, lime or cement.

Silicate (water-glass) and aluminosilicate (geopolymer) networks are generated in alkaline suspensions, but the setted binders have no durability in water and acids, respectively. To overcome these disadvantages, investigations using phosphates as additives or network formers were carried out for the development of network binders with exploitable properties.

On a first pathway, cyclo-phosphates as additives in water-glass solutions lead advantageously to enlarged connectivities of the networks. It is assumed that the reaction starts with an alkaline opening of the phosphate-cycles and goes via uptake of alkali ions by the phosphate fragments to an induced polycondensation of an amorphous network.

The addition of 20 wt.% $Na_3(PO_3)_3$ realizes a silicate network (63.3 wt.% of the binder) and the crystalline by-products $Na_4P_2O_7$ and Na_2HPO_4 are formed. The addition of 20 wt.% $Al_4(P_4O_{12})_3$ generates an aluminosilicate network (76.9 wt.% of the binder) including the same by-products and unreacted aluminium cyclo-phosphate is detectable. There is also a weak hint of an unordered phosphate component.

Composite materials, prepared with this aluminosilicate network binder and quartz sand, demonstrate a remarkable increase of the chemical durability against water and 70 wt.% H_2SO_4 (both at 100 °C, 6 h) with rising addition of $Al_4(P_4O_{12})_3$ up to 30 wt.%. With respect to the useful chemical durability and the compressive strength of about 22 N/mm^2 these composite materials have a large application potential in the fields of waste-water pipes and collection units.

On a second pathway, reactive oxides like Fe_2O_3 or $Al_2O_3 \bullet 2SiO_2$ give, after a solution process in phosphoric acid, amorphous networks with binder action by a slow setting.

In the case of Fe_2O_3, an addition of a trace of iron initiates an acceleration of the setting process and a remarkable crystallization of the binder (iron hydrogen phosphates). In the case of $Al_2O_3 \bullet 2SiO_2$, the metakaolin leads to a very strong binder. It is assumed that these amorphous networks consist preferably of aluminosilicate units.

Composite materials of these aluminosilicate binders with a quartz sand filler show an excellent water resistance (100 °C, 6 h) and demonstrate high compressive strength up to 70 N/mm^2 for robust construction components.

The experimental results clearly exhibit the requirement for further detailed investigations on the action of phosphates in inorganic networks as an attractive challenge.

FOOTNOTES

* The symbols $XO_{4/2}$ or $O_{3/2}XO^-$ mean oxo-coordinations (tetrahedra) with four bridging oxygen or with three bridging oxygen and one non-bridging oxygen.

** Brick powder is generated by burning of clay minerals. Both consist of similar chemical components.

REFERENCES

[1] V. D. Glukhovsky, Soil silicates, Their properties, technology and manufacturing and fields of application, Doc. Tech. Sc. Degree Thesis, Civil Engineering Institute, Kiev, 1965

[2] J. Davidovits, Solid phase synthesis of a mineral block polymer by low temperatures polycondensation of aluminosilicate polymers, IUPAC Int. Symp. on Macromolecules, Stockholm, 1976, Topic III

[3] J.Davidovits, Geopolymer Chemistry and Applications – Institute Geopolymer, France, Saint-Quentin, 2008

[4] A. Polomo and A. Fernandez –Jimenez, Nature of Alkali Alumosilicate Polymers, in Int. Conf. - Alkali Activated materials, Prague 2007, p.509-522

[5] Ch. Kaps, A. Buchwald and M. Hohmann, Geopolymers in Ceramic Building Materials in 2nd International Congress on Ceramics “Global Roadmap for Ceramics”, Verona 2008, ed. by Bellosi and G.N. Babini, ISTEC-CNR p. 723-733

[6] Ch. Kaps, A. Buchwald, M. Hohmann: Geopolymer as inorganic additive to stabilize clay containing materials; in Geopolymers, green chemistry and sustainable development solutions, ed. J. Davidovits, St. Quentin, France, , 2005, 167-172

[7] Ch. Kaps, Polycondensation in waterglass solution a geopolymer generation? in Geopolymer Binders Interdependence of composition, structure and properties, ed. by A. Buchwald, K,. Dombrowski and M. Weil, Workshop Proc., Weimar, 2006, p. 87-96

[8] L. Nichelson, B. J. Murray, A. Fletscher, D. R. M. Brew, K. J. D. Mackenzie and M. Schmücker, Novel Geopolymer Materials Constaining Borate Structural Units, in Geopolymer, Green Chemistry and Sustainable Development Solutions, ed. by J. Davidovits, Saint-Quentin, 2005, p. 31-33

[9] D. S. Perera, E. R. Vance, Y. Zhang, J. Davis and P. Yee, Specification Studies of Fe, Mn, Ca and Ti and Dissolution of a Metakaolinite - based Geopolymer with Si/Al ~ 2, in Geopolymer, Green chemistry and Sustainable Development Solutions, ed. by J. Davidovits, Saint-Quentin, 2005, p. 57-59

[10] A. Durif, Crystal Chemistry of Condensed Phosphates, Planum Press; New York and London, 1995
[11] H.-D. Zellmann, Metaphosphat – modifizierte Silikatbinder als Basis säurebeständiger Beschichtungsmaterialien, Diss., Weimar, 2008
[12] H.-D. Zellmann and Ch. Kaps, Chemically modified Water Glass Binders for Acid-resistance Molars, *J. Amer. Ceram. Soc.*, **89,** 1369-1372 (2006).
[13] A. Buchwald, H.-D. Zellmann and Ch. Kaps, Condensation of aluminosilicate gels - modul system for geopolymer binders (in progress)
[14] K. J. D. Mackenzie and M. E. Smith, Multinuclear Solid-State NMR of Inorganic Materials, Pergamon Press, New York and London, 2002, p. 436 and 273, resp.
[15] R. Dupree, D. Holland, M. G. Mortuza, J. A. Collins and M. W. G. Lochyer, Magic Angle Spuming NMR of Alkali Phospho-Alumino-Silicate Glasses, *J. Non-Cryst. Solids,* **112,** 111-119 (1989).
[16] M. J. TOPLIS and T. Schaller, A ^{31}P MAS NMR Study of Glasses in the System xNa_2O-(1-x)Al_2O_3-$2SiO_2$-yP_2O_5, *J. Non-Cryst. Solids,* **224,** 57-68 (1998).
[17] H.-D. Zellmann and Ch. Kaps, Chemisch resistente Silikatmörtel basierend auf Natriumsilikatlösungen und aluminiumphosphat, GDCh-Monographie Bd. 35, Bauchemie, S. 111-118
[18] A. S. Wagh, Chemically Bonded Phosphate Ceramics, Elsevier, Amsterdam, Boston, 2004, p. 136
[19] W. Rähse, Z. *Anorg. Allg. Chemie*, **438**, 222-232 (1978).
[20] H. Scholze, Glas-Natur, Struktur und Eigenschaften, Springer Verlag, Berlin, 1988, p. 91
[21] A. S. Wagh and S. Y. Jeong, Chemically Bonded Phosphate Ceramics: III Reduction Mechanism and Its Application to Iron Phosphate Ceramics, *J. Am. Ceram. Soc.,* **86,**1850-55 (2003).
[22] K. Jasmund and G. Lagaly, Tonminerale and Tone, Steinkopff Verlag, Darmstadt, 1993, p. 208 and 363
[23] A. S. Wagh, S. Grover and S. Y. Jeong, Chemically Bonded Phosphate Ceramics II, Warm-Temperature Process for Alumina Ceramics, *J. Am. Ceram. Soc.,* **86,** 1845- 1849 (2003.)
[24] R. Giskow, J. Lind and E. Schmidt, The Variety of Phosphates for Refractory and Technical Applications by the Example of Aluminium Phosphates, *Ber. DKG,* **81,** 5, E 27-E32 (2004).
[25] R. Kniep, Orthophosphate im Dreistoffsystem Al_2O_3-P_2O_5-H_2O, *Angew. Chem.,* **98,** 520-529 (1986).

PROPERTIES OF PHOSPHORUS-CONTAINING GEOPOLYMER MATRIX AND FIBER-REINFORCED COMPOSITE

Oleg Bortnovsky, Petr Bezucha
Research Institute of Inorganic Chemistry,
Revoluční 84, Ústí nad Labem, Czech Republic

Jiří Dědeček, Zdeněk Sobalík
J. Heyrovsky Institute of Physical Chemistry, Academy of Sciences of the Czech Republic,
Dolejškova 3, Prague, Czech Republic

Věra Vodičková, Dora Kroisová,
Technical University Liberec
Hálkova 6, Liberec, Czech Republic

Pavel Roubíček
Czech Kaolin Company, Inc.
Nové Strašecí 1171, Czech Republic

Martina Urbanová
J. Heyrovsky Institute of Physical Chemistry, Academy of Sciences of the Czech Republic,
Dolejškova 3, Praha, Czech Republic
Institute of Macromolecular Chemistry, Academy of Sciences of the Czech Republic,
Heyrovského sq. 2, Prague, Czech Republic

ABSTRACT

In this work properties of geopolymer matrices based on silica fumed prepared with different alumina sources and phosphoric acid were discussed. To evaluate the pot-life of geopolymer resin at room temperature, the viscosity of geopolymer resins was coherently measured. Matrices were analyzed by means of XRD, FTIR ATR, SEM-EDS, ^{27}Al, ^{29}Si and ^{31}P MAS NMR. Thermo-dimensional stability of fiber-reinforced composites was measured by dilatometry. Water resistance was tested in boiling water; chemical composition and NMR spectra of treated geopolymer matrices were analyzed.

The nature of the alumina source influenced the mechanical and thermomechanical properties of the matrix. Although the phosphorus added enhanced the dimensional stability of geopolymer matrix/composites at high temperature, more than 50% of phosphorus was soluble and washed out of the matrix during boiling in water. The rest of phosphorus was probably bound to geopolymer or adhered to silica fumed particles. Still, the amounts of soluble SiO_2 and Al_2O_3 were rather low and the total Si/Al ratio in the matrix remained constant.

Properties of fiber-reinforced composites with 40-65 wt % of fiber were influenced more by fiber used than by the matrix. In contrast to an E-glass composite, which is brittle, a basalt fiber composite exhibits properties near to carbon fiber composites, except for a lower elastic modulus. Material mechanical parameters independent of sample size were calculated from flexural tests with various span-to-sample height ratios. In contrast to composites reinforced with fiber rovings, composites reinforced with fabric fiber were more brittle and less anisotropic, which was reflected in material mechanical parameters.

INTRODUCTION

Application of geopolymer for preparation of fiber-reinforced composites has been well known for almost 20 years, since the first Davidovits patent was filed[1]. There are some well-known advantages of these composites: including rather simple production conditions comparable with conditions of plastic-based composites, together with no flammability and high heat and fire resistance[1-5].

However, some drawbacks of using geopolymer resin for impregnation of different types of fiber still exist. For effective impregnation of fabric or fiber rovings containing single filaments with diameter from 7 to 25 μm, resin with low viscosity and maximum particle size lower than the filament diameter should be used. Therefore a geopolymer resin based on classical metakaolin and similar materials, containing rather large particles and marked with high viscosity, can be hardly used for effective fiber impregnation, or very high pressure would have to be applied to penetrate the resin between single filaments[6]. Application of silica-based geopolymer with nanosized amorphous silica as a main component could solve to do with particle size. Moreover only the special type of silica with low surface allows the achievement of low viscosity resin with high solid content. However, the high-silica geopolymers are generally not dimensionally stable at high temperature and rather unstable in water[7]. The use of hardeners on the base of aluminum, zinc or iron phosphate, similar to commercial hardeners and accelerators for water glass, provides the dimensional and water stability[7]. Unfortunately the pot-life of the resin and workability greatly decrease so that prolonged fiber impregnation at temperatures of 15–25 °C, which is the normal case for pultrusion technology, is problematic. Application of phosphorus additives in geopolymer silica-alumina matrices is rare and described predominantly in the patents [8,9]. In the patent [8] the addition of soluble phosphate notably phosphoric acid to geopolymer resin was applied.

In this work, the properties of high-silica geopolymer resins with various alumina sources and with phosphoric acid additives, the structure of resulting geopolymer matrices, and mechanical and thermomechanical properties of basalt, carbon and E-glass fiber-reinforced geopolymer composites were studied. For preparation of reinforced composite, different types of fiber rovings and simple vacuum-bagging procedures without extra-pressure were applied. Utilization of fiber rovings allows simplification and mechanization of fiber impregnation on the laboratory scale, while also approaching real conditions of pultrusion technology. To compare properties of fiber rovings composites with generally prepared fabric composites, various types of basalt and E-glass fabric were used under the same preparation conditions.

EXPERIMENTAL

Geopolymer resins were prepared according to the simplified procedure described in the patent[8]. As a silica source, thermal silica (**D_{50}** 0.62 μm, **D_{90}** 3.24 μm, Saint-Gobain, France) containing 93.8 wt.% of SiO_2 and 2.9 wt. % of Al_2O_3 was used. Thermal silica was blended with 48.5 wt. % KOH and mixed for 30 minutes, and then phosphoric acid which was diluted with water 1:1 by weight was admixed. Finally an alumina source, chosen from alumina (**D_{50}** 0.50 μm, **D_{90}** 12.32 μm, PFR, Groupe Alcan - Aluminium Pechiney), aluminum hydroxide (**D_{50}** 1.9 μm, **D_{90}** 3.18 μm, Matrinal OL107, Albermal Corp.), kaolin (**D_{50}** 8.00 μm, **D_{90}** 17.26 μm, KKAF, LB MINERALS, Ltd.) and metakaolin (**D_{50}** 4.06 μm, **D_{90}** 10.36 μm, calcined shale, Czech Kaolin Company, Inc.), was added. Although measured with a MALVERN 2 600 Sizer, the particle size of kaolin is rather high. According to SEM observation this size corresponds to agglomerates of 1-3 μm kaolin plates. For comparison, geopolymer resin without alumina source was also prepared. The amount of alumina source added was calculated to achieve similar total ratio K/Al without the consideration of incorporating Al from these additives into the geopolymer structure. Main calculated total molar ratios in geopolymer resins are depicted in Table I.

Table I.Calculated Chemical Composition of Geopolymer Matrices

Matrix	Al source	Si/Al	K/Al	K/Si	P/Al	K/P	H_2O/K
Q-none	None	28.2	7.5	0.26	1.77	4.2	5.2
Q-alumina	Al_2O_3	9.4	2.5	0.26	0.58	4.2	5.2
Q-hydroxide	$Al(OH)_3$	9.7	2.5	0.26	0.58	4.2	5.2
Q-kaolin	Kaolin	9.7	2.4	0.24	0.58	4.2	5.2
Q-MK	Metakaolin	9.8	2.4	0.24	0.58	4.2	5.2

Roving fiber geopolymer composites were prepared by a three-stage procedure. At first a roving was impregnated with geopolymer resin in a lab-scale home-made equipment showed on Fig. 1. Roving fiber was wetted with resin in the bath, then the rest of the resin removed between two moving rollers, and finally with a rubber scratcher. Impregnated fiber with suitable resin content was taken up on the reel. By finally cutting the coil to the reel length, approx. 20 pieces of pre-preg with the same length were obtained. Next, the pre-pregs were rollered to achieve the desired width, and stratified layer by layer into a silicon rubber mold. In the case of basalt roving with 2520 tex 12 layers of pre-preg, to prepare a sample with dimensions aprox. 5×5×100 mm, 7 layers for 2×7 ×100 mm were used. In the case of carbon roving with 800 tex 20, 12 layers was needed to prepare 5×5×100 mm and 2×7×100 mm samples, respectively. The mold with pre-preg was then placed into a sealed polyethylene bag under vacuum, and left at room temperature for 1 hour, followed by 5 hours at 85°C in the oven. After cooling, the composite coupons were dried for 5 hours at 85°C. During all preparation procedures samples were weighed to calculate the fiber content. Composite samples were finally calcined at 200, 400, 550 and 700°C for 3 hours with a ramp rate of 10°C min^{-1}.

Figure 1. Apparatus for roving fiber impregnation.

Fabric composites were prepared under the same conditions as for the roving composites. Pre-pregs were prepared by manual impregnation of fabric with a roller. For preparation of composite plates, 12 to 16 pieces of pre-preg fabric were used to achieve approx. 2 mm in height. E-glass fabric plane and serge weaves with density 163 $g.m^{-2}$ for both, as well as basalt fabric plane and satin weaves with densities 180 and 220 $g.m^{-2}$, respectively, were used.

The flexural tests with samples 5×5×100 mm were conducted over a simply supported span of 76 mm (the span to height ratio was approx. 16) with a center-point load on the testing machine TMZ-

3U electronic with maximum load of sensor 1000 N. The loading was in a direction parallel to the pre-preg layer in the composite to prevent delamination of samples. Flexural strength was calculated from maximum loadings, flexural modulus was calculated from the linear part of the loading deflection curve. In the next case of roving composites 2×7×100 mm and fabric composites 2×12×100 mm, the loading was in a direction perpendicular to the pre-preg layer with spans 38 and 76 mm (span to height ratios approx. 18 and 37) to calculate the material Young modulus, shear modulus and material flexural strength.

Dilatation of roving basalt composites having dimensions 5×5×50 mm were measured twice at the temperature range 50–700°C in a NETZSCH DIL 402 PC dilatometer using a quartz calibration standard, with the ramp rate of $10°C.min^{-1}$.

Preparation of neat resin samples for characterization of structure parameters was done under the same conditions, except evacuation. For next analysis except SEM-EDS all samples was ground. SEM-EDAX analysis was done on the fresh fracture surface of the sample.

Stability of geopolymer matrix in water was studied by immersing of geopolymer disks into distilled water and boiling for 1 hour. The weight loss after drying at 85°C was then estimated. The geopolymer sample after boiling in water was ground and characterized with MAS NMR and XRF analysis.

Dynamic viscosity of geopolymer resins was measured intermittently (every 15 minutes) with rotation viscosimeter Reotest 2 at 20°C. Between subsequent measurements resin was not stirred

SEM analyses were obtained using the scanning electron microscope VEGA TESCAN II EasyProbe with integrated EDS microanalyser. The fresh fracture surface of the sample was coated with a 10 nm thick layer of Au/Pd alloy. The EDS analyses and mappings were performed using an EasySEM software interface for measurement, image processing and analysis. Application of the EDAX method for chemical analysis enabled us to distinguish between the geopolymer matrix and unreacted particles of raw material and to check the homogeneity of individual elemental distribution.

FTIR-Attenuated Total Reflectance (FTIR-ATR) spectroscopy was used to analyze networking of the geopolymer structure. Ground samples were placed onto the ATR silicon crystal and clamped to obtain good contact. The IR spectra were recorded using an Nexus 670 FT-IR Thermo Nicolet spectrometer with DTGS detector and Smart 7 MIRacle ATR (Thermo Nicolet) adapter with Si plate..

^{29}Si, ^{27}Al and ^{31}P MAS NMR experiments were carried out using a Bruker Avance 500 MHz (11.7 T) Wide Bore spectrometer with 4 mm o.d. ZrO_2 rotors with a rotation speed of 5 kHz for ^{29}Si MAS NMR and 12 kHz for ^{27}Al MAS NMR and with 2.5 mm o.d. ZrO_2 rotors with a rotation speed of 25 kHz for ^{31}P MAS NMR. A ^{29}Si MAS NMR high-power decoupling experiment with a $\pi/6$ (1.7 μs) excitation pulse and a relaxation delay of 30 s was employed to collect a single pulse spectrum and a 50 % ramp cross polarization (CP) pulse sequence were applied to collect a single pulse and cross polarization spectra. The ^{29}Si NMR chemical shifts were referred to octasilsesquioxane ($((CH_3)_3SiO)_8Si_8O_{12}$, Q8M8). High-power decoupling pulse sequences with $\pi/12$ (0.7 μs) excitation pulse were employed to collect ^{27}Al MAS NMR single pulse spectra. The ^{27}Al NMR observed chemical shift was referred to an aqueous solution of $Al(NO_3)_3$. A ^{31}P MAS NMR high-power decoupling experiment with a $\pi/2$ (4.0 μs) excitation pulse and a relaxation delay of 10 s was employed to collect a single pulse spectrum and a 50 % ramp cross polarization (CP) pulse sequence were applied to collect a single pulse and cross polarization spectra. The ^{31}P NMR chemical shifts were referred to Na_2HPO_4.

RESULTS

The time course of dynamic viscosity of geopolymer resins with different types of Al sources is shown on Figure 2. It is obvious that there is almost no difference among various compositions except for the Q-alumina resin, which exhibited a more rapid increase of viscosity, although it contained the lowest amount of additional Al sources (note the similar total Si/Al ratio). Resin Q-hydroxide, Q-MK and Q-kaolin held their low viscosity (less than 500 mPa.s) for a long time and easily created drops of fluid up to 6 hours at 20°C.

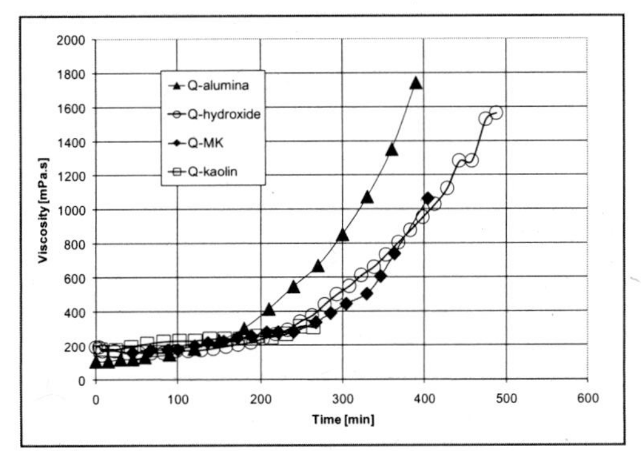

Figure 2. Dynamic viscosity of various geopolymer resins at 20°C

XRD diffraction pattern of Q-none sample is shown on Figure 3. It is seen that only the ZrO_2crystal phases from raw materials were observed, together with potassium phosphate KH_2PO_4 and probably potassium pyrophosphate which were formed from potassium hydroxide and phosphoric acid. Any other new phases were not observed. Boiling in water led to a decrease in intensity of the diffraction lines corresponding to KH_2PO_4.

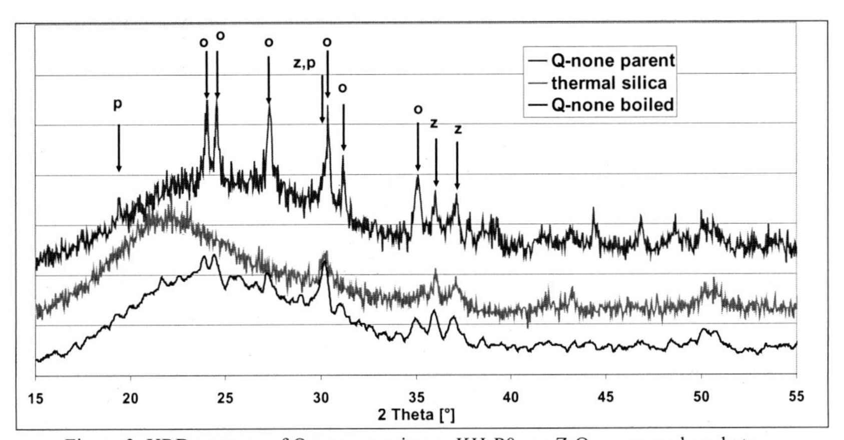

Figure 3. XRD patterns of Q-none matrix, o= KH_2P0_4, z=ZrO_2, p=pyrophosphate

FTIR-ATR spectra of geopolymer matrices with additional Al sources are depicted in Figure 4. In case of the Q-kaolin sample with predominantly octahedrally coordinated Al (see below) the two main rather sharp T-O-T vibrations at 1027 and 1006 cm^{-1} were observed, in contrast with broad vibration centered at 1043 cm^{-1} having shoulder at 1006 cm^{-1} for the Q-MK sample. The bands 940 and 926 cm^{-1} were observed only in P-containing geopolymer matrix and could represent P-O-P bridging and PO_4^{3-} anion respectively. In case of samples made from pure Al sources (alumina and aluminum hydroxide) only a shift of main T-O-T band to 1020 cm^{-1} was observed. All spectra exhibited the shoulder at 1118 cm^{-1}. For the sample Q-MK no P was measured to distinguish the presence of P and position of main T-O-T bands.

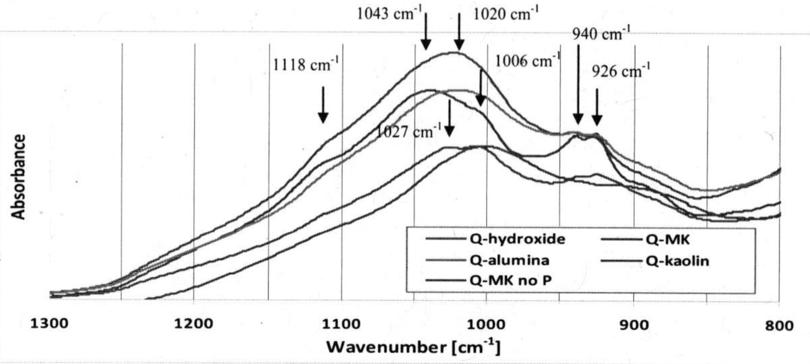

Figure 4. FTIR-ATR spectra of geopolymer matrices.

SEM image of matrix containing additional Al sources with some EDS mapping of individual elements were depicted on Figures 5-7. There is clear evidence of spherical, unreacted particles of thermal silica presented, in accordance with references [7,10]. On the other hand it is clearly seen from Al mapping that in all matrices (not all Al mapping included) unreacted particles of Al sources were observed. On the other hand the distribution of other elements: K, Si and P except the places with unreacted thermal silica and Al sources particles is rather homogeneous. It can be stressed that P in contrast with K covered the unreacted particles of thermal silica. The ratios of individual elements calculated from EDS measurement (area approx. 150 × 150 μm) were shown in Table II. Chemical composition of the Q-none and Q-none/boiled matrices was done with the XRF technique. It can be concluded that the distribution of all main elements is rather homogeneous in this area. However, there are high inhomogeneity of matrix in size up to tens of μm (see Fig. 5-7).

Table II. Chemical Composition of Geopolymer Matrices by EDS, Area 150 x 150 μm

Matrix abbr.	Si/Al	K/Al	K/P	K/Si	P/Al
Q-none	23.44	5.85	4.82	0.25	1.21
Q-none/boiling	25.19	4.24	7.54	0.17	0.56
Q-alumina	9.13	1.56	4.44	0.17	0.35
Q-hydroxide	8.94	2.06	3.38	0.23	0.61
Q-kaolin	10.54	2.18	4.14	0.21	0.53
Q-MK	10.37	2.40	4.49	0.23	0.59

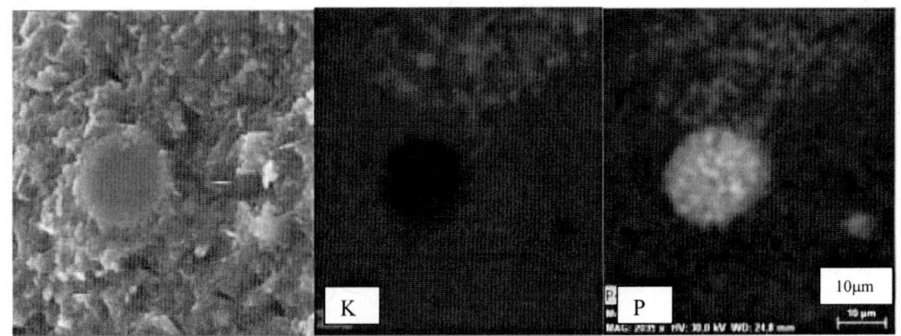

Figure 5. SEM and EDS mapping of an individual element of Q-kaolin matrix.

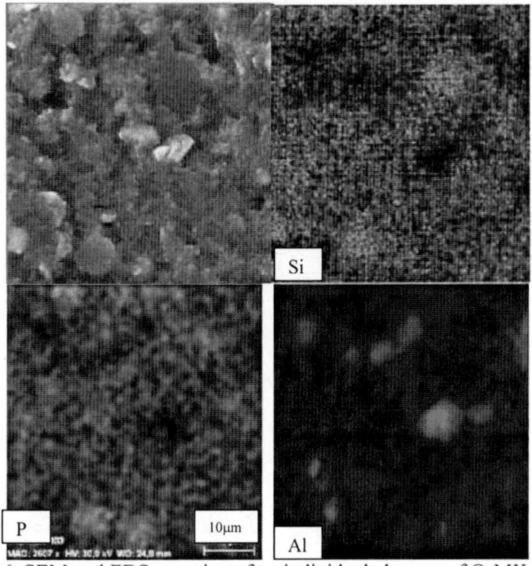

Figure 6. SEM and EDS mapping of an individual element of Q-MK matrix.

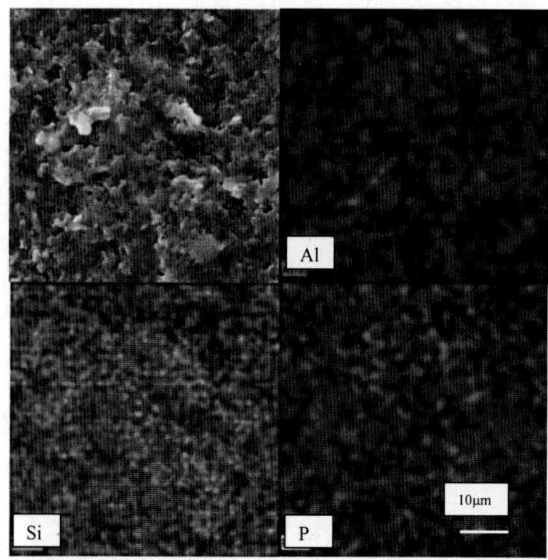

Figure 7. SEM and EDS mapping of an individual element of Q-hydroxide matrix.

^{31}P MAS NMR spectra of P-containing geopolymers

The ^{31}P MAS NMR single pulse spectrum of Q-none and its simulation are shown in the Figure 8A, the effect of Al source on the ^{31}P MAS NMR spectra of geopolymers is depicted in the Figure 8B and the effect of boiling on the ^{31}P MAS NMR spectrum of Q-none geopolymer is illustrated in Figure 8C. All spectra exhibit resonances only in the range 0–10 ppm. The changes of the ^{31}P MAS NMR spectra with the presence of Al sources before as well as after boiling enable us to distinguish three resonances in the spectra. Presence of three resonances - narrow ones at 0.9 and 4.5 ppm and a broad one at 5.0 ppm was confirmed by the spectra simulations. The simulation of the spectra enabled us to quantify the concentration of individual phosphorus containing species in the samples, see Table III.

The absence of ^{31}P NMR signal with negative chemical shift value enables us to exclude the formation of alumophosphate analogues of zeolites as well as other types of alumophosphate species in the samples, as reported in ref [11]. The narrow signal at 0.9 ppm reflects the presence of pyrophosphate in samples, see ref. [12]. The low width (400 Hz) of the 0.9 ppm resonance indicates a well-defined form of pyrophosphate prevails in all the samples investigated. The resonance at 4.5 ppm can be attributed to KH_2PO_4 with reported chemical shift of 4.3 ppm [12]. The narrow band (width 250 Hz) indicates a well-defined crystalline form of KH_2PO_4.

The resonances at 5.0 ppm was not yet reported for the potassium containing phosphorus compound. Nevertheless, the chemical shifts at 5.0 ppm indicates formation of some Q^0 species containing exclusively one isolated P atom[11]. The presence of K_3PO_4, $K_2HPO_4.3H_2O$ and KH_2PO_4 having chemical shift values 13, 2.3 and 4.3-4.5 ppm can be excluded[12]. KH_2PO_4 exhibits resonance at 4.3 ppm. This suggests the possibility that an other atom than potassium can occupy the position of the next nearest neighbor of the P atom. Species with chemical shifts 5.0 ppm represent predomination P species in samples. This indicates that only Si atoms can be taken in account to occupy next nearest neighbor position of the P atom. Thus, a resonance at 5.0 ppm can be suggested to reflect

$(SiO_3)K_xH_yPO_4$ group with Si atom belonging to the silicate network. A second possibility is that the resonance at 5.0 ppm represents KH_2PO_4 molecules with significantly changed geometry of the P atom, due to an interaction of unknown nature.

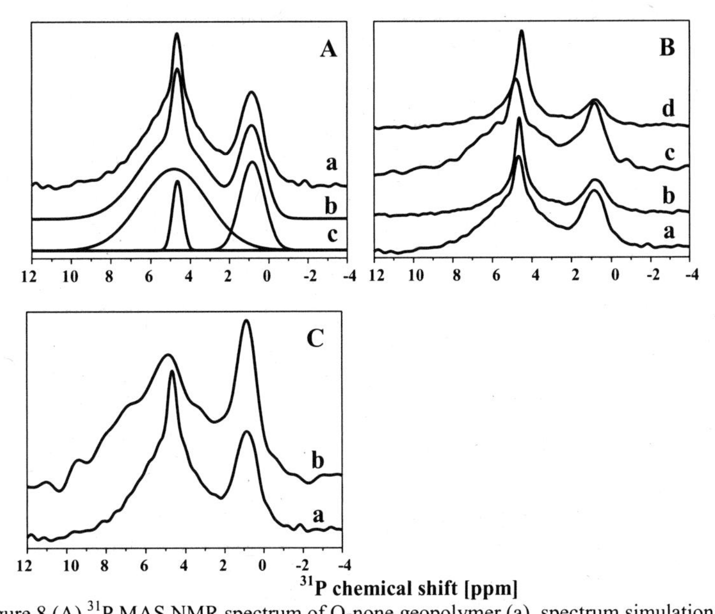

Figure 8 (A) ^{31}P MAS NMR spectrum of Q-none geopolymer (a), spectrum simulation (b) and individual Gaussian bands of the spectral simulation (c); (B) ^{31}P MAS NMR spectra of Q-none (a), Q-alumina (b), Q-hydroxide (c) and Q-kaolin (d) geopolymers; (C) ^{31}P MAS NMR spectra of boiled (a) and parent Q-none geopolymer (b).

Table III. Attribution of Resonances of ^{31}P MAS NMR Spectra of P-containing Geopolymers

Matrix	Relative Concentration (% rel.)		
	Resonance 0.9 ppm Pyrophosphate	Resonance 4.5 ppm KH_2PO_4	Resonance 5.0 ppm
Q-none	26	9	65
Q-none/boiled	21	4	75
Q-alumina	27	20	53
Q-hydroxide	24	8	68
Q-kaolin	22	25	53

^{27}Al MAS NMR spectra of P-containing geopolymers

The effect of Al sources on the ^{27}Al MAS NMR single pulse spectra of P-containing geopolymers is shown in the Figure 9A and the effect of boiling on the ^{27}Al MAS NMR spectrum of Q-none geopolymer is illustrated in Figure 9B. In the case of the geopolymer without additional Al sources, tetrahedral Al of the AlO_4 tetrahedral network in the silicate framework, characterized by the resonance at 56.3 ppm, predominates in the sample[13]. Octahedral or pentacoordinated extra-network Al atoms with bands between 10 and 35 ppm[13] represent 20–30 % of Al atoms. The boiling of geopolymer resulted in the disappearance of the octahedral signal. Thus, an exclusively network with tetrahedral Al is present in the boiled geopolymer sample. In the case of P-containing geopolymers prepared using additional Al sources, resonances corresponding to the Al atoms in these Al sources predominate or represent substantial fraction in the ^{27}Al MAS NMR spectra. ^{27}Al MAS NMR single pulse spectra of sources used are concluded in the Figure 9C. Kaolin is marked with a narrow band at 3.6 ppm, Al_2O_3 with a narrow band at 14 ppm, $Al(OH)_3$ with a band at 9 ppm with a shoulder at 2.7 ppm, and metakaolin (shale calcined at 750°C) with a complex spectrum with resonances at 3.9, 14, 30 and 53 ppm. The simulation of the ^{27}Al MAS NMR spectra of geopolymers (not shown in Figures) enabled us to quantify concentration of tetrahedral Al atoms in samples, as summarized in Table IV.

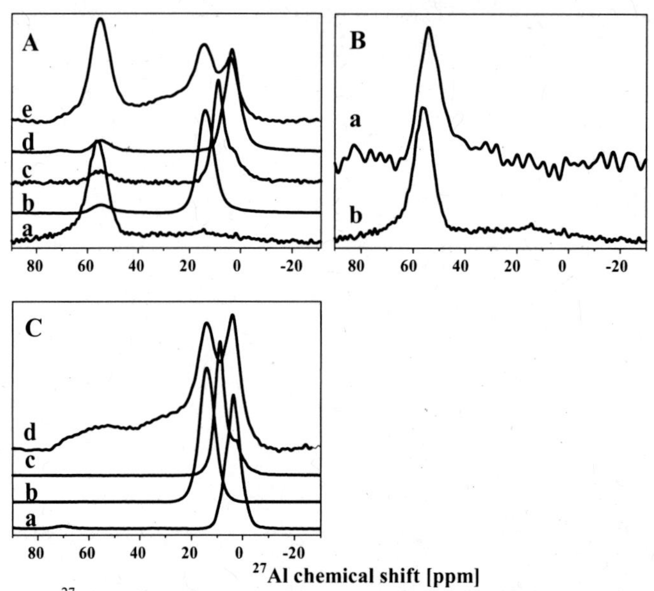

Figure 9(A) ^{27}Al MAS NMR spectra of Q-none (a), Q-alumina (b), Q-hydroxide (c), Q-kaolin (d) and Q-MK (e) geopolymers; (B) ^{27}Al MAS NMR spectra of boiled (a) and parent Q-none geopolymer (b); (C) ^{27}Al MAS NMR spectra of kaolin (a), Al_2O_3 (b), $Al(OH)_3$ (c) and metakaolin (d).

Table IV. Relative Concentration of Tetrahedral Al from ^{27}Al MAS NMR and Si/Al^{total} vs. Si/Al^{Td} Ratios

Sample	Tetrahedral Al (rel. %)	Si/Al^{total}	Si/Al^{Td}
Q-none	70-80	23.4 (XRF)	32
Q-none/boiling	100	25.2 (XRF)	25
Q-alumina	10	9.1 (EDS)	100
Q-hydroxide	10	8.9 (EDS)	98
Q-kaolin	10	10.5 (EDS)	114
Q-MK	25	10.4 (EDS)	45

^{29}Si MAS NMR spectra of P-containing geopolymers

The ^{29}Si MAS NMR single pulse spectra of P-containing geopolymer and the illustration of the effect of Al sources on the ^{29}Si MAS NMR spectra as well as the cross-polarization spectrum of the Q-none sample are given in the Figure 10A, simulation of the ^{29}Si MAS NMR spectrum of Q-none sample is given in the Figure 10B. Quantitative results of the spectral analysis are concluded in the Table V. ^{29}Si MAS NMR spectrum of the P-containing sample exhibits four components. Resonance at -96.3 ppm reflects Q^3 silicones – Si(3Si,OH) tetrahedral atoms, while resonance at -88.3 ppm can be attributed both to the Q^3 and Q^2 Si atoms. The terminal SiOH atoms represent 14 % of Si atoms. Resonance at -109.3 ppm reflects also Q^4 – Si(4Si) atoms of the geopolymer network. Resonance at -97.9 ppm can be attributed to an other type of the Q^4 Si atom. However, to the Si(4Si) atoms corresponds resonances with chemical shift value significantly lower than -100 ppm. Thus, Si(3Si,1P) atoms can be attributed to the resonance at -97.9 ppm. Attribution of this resonance to the Si(3Si,1Al) can be excluded due to the low concentration of Al atoms in the geopolymer network. For the details on the interpretation of Si NMR spectra see ref. [14].

Table V. Quantitative Analysis of ^{29}Si MAS NMR Spectra of Q-none Matrix

Resonance (ppm)	Concentration (rel. %)
-109.3	66
-97.9	20
-96.3	7
-88.3	7

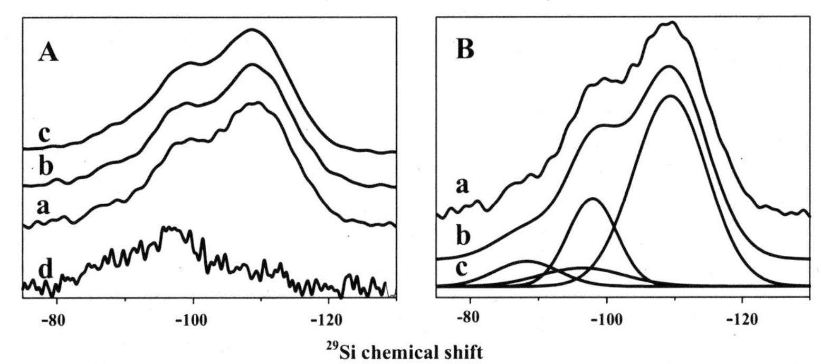

Figure 10(A) ^{29}Si MAS NMR spectra of Q-none (a), Q-alumina (b) and Q-MK geopolymers (c) and ^{29}Si CP MAS NMR spectrum of Q-none sample (d); (B) ^{29}Si MAS NMR spectrum of Q-none geopolymer (a), spectrum simulation (b) and individual Gaussian bands of the spectra simulation (c).

Mechanical properties of fiber-reinforced geopolymer composites

Mechanical properties of basalt roving reinforced composites with dimensions 5×5×100 mm after curing and drying, and after calcination at different temperatures having a span 76 mm with crosshead speed 2 mm/min are shown in Tables VI and VII. Only average values of flexural strength and Youngs modulus are included. In all cases, the variation coefficient did not exceed 10 % of average values. All composite samples did not change in length, and very small volume shrinkage from 0.8 % at 200°C to 2 % at 700°C was observed after heating up even to 700°C and following cooling.

Table VI. Flexural Strength of Basalt Roving Reinforced Geopolymer Composites

Matrix	Fiber content	85°C	200°C	400°C	550°C	700°C
	wt. %	MPa				
Q-none	66	110		76		58
Q-alumina	64	129		117		61
Q-hydroxide	64	124		103		50
Q-kaolin	60	285	229	225	166	55
Q-MK	62	279		195		37

Mechanical properties of roving and fabric reinforced composites were measured on samples 2×7×100 mm and 2×12×100 mm respectively. with crosshead speed of 1 mm/min and with 2 different span to height ratios, which allow one to calculate the material Youngs modulus (E), shear modulus (G), and flexural strength (σ_m) from linear regression of estimated E^* and σ_m^* against $(H/L)^2$ (reciprocal span : height ratio) values - see Table VIII.

Table VII. Youngs Modulus of Basalt Roving Rreinforced Geopolymer Composites

Matrix	Fiber content	85°C	200°C	400°C	550°C	700°C
	wt. %	GPa				
Q-none	66	33.1		32.4		41.2
Q-alumina	64	38.9		40.4		41.2
Q-hydroxide	64	39.1		34.9		33.3
Q-kaolin	60	41.9	43.0	45.9	49.4	48.1
Q-MK	62	43.7		41.7		38.0

Table VIII. Calculated Material Parameters of Reinforced Geopolymer Composites

Matrix	Fiber	Fiber content	E linreg	G linreg	E/G	σ_m linreg	$d(1/\sigma_m^*)/$ $/d(H/L)^2$
		wt.%	GPa	GPa		MPa	
Q-kaolin	E glass plane	64	31	0.53	58	63	1.546
Q-kaolin	E glass serge	56	34	0.49	69	88	1.402
Q-kaolin	Basalt plane	54	34	1.74	19	115	0.797
Q-kaolin	Basalt satin	55	27	1.30	21	117	0.643
Q-alumina*	Basalt 2520tex	61	41	0.29	140	274	0.378
Q-hydroxide	Basalt 2520tex	66	43	0.16	262	201	0.505
Q-hydroxide*	Basalt 2520tex	63	41	0.29	140	171	0.545
Q-kaolin	Basalt 2520tex	62	50	0.30	169	341	0.270
Q-kaolin	Carbon 800tex	41	95	1.60	57	585	0.157

A fictitious Youngs modulus E^*, calculated for each height : span ratio according to the equation (1), which is valid for isotropic materials:

$$E^* = \frac{F \times L^3}{4 \times s \times B \times H^3}, \quad (1)$$

where F is the applied bending load, L span, H height, B width of the rectangular cross section profile, s (central) flexion, is a quantity linearly dependent on $(H/L)^2$. The true modulus E is only the reciprocal value of the intercept in the linear regression ($1/E^*$) vs. $(H/L)^2$.

The material shear modulus G was calculated by the equation (2):

$$G = \alpha \Big/ \frac{d(1/E^*)}{d(H/L)^2}, \quad (2)$$

where α is a correction factor, theoretically equaling 1.178, the variable in the denominator is the slope in this regression.

Quite similarly, a fictitious flexural strength σ_m^*, calculated by the equation (3), which is valid for isotropic materials

$$\sigma_m^* = \frac{3}{2}\frac{F_m \times L}{B \times H^2}, \quad (3)$$

where F_m is the maximum bending load, is a dimension dependent quantity. It appeared that the linearity of ($1/\sigma_m^*$) is reasonable with quite the same independent variable $(H/L)^2$ as above. The true material property σ_m (here for simplification still called flexural strength) is thus even here obtainable as the reciprocal intercept of the regression. The picture for both regressions is shown on Figures 11.

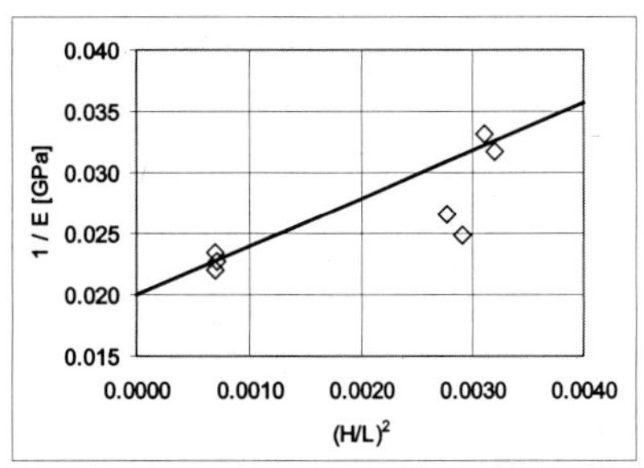

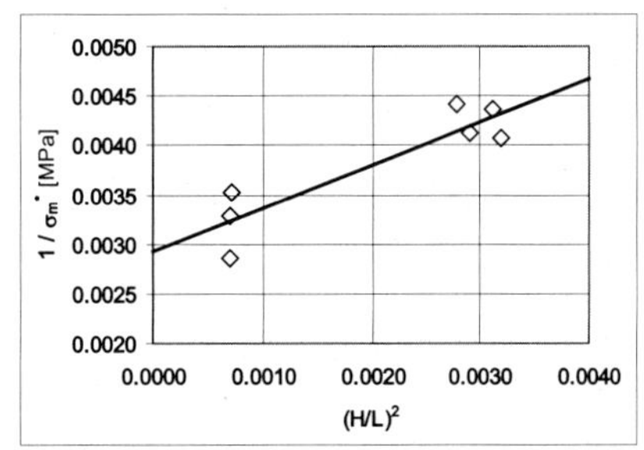

Figure 11. Regression curves for calculation of material parameters of composites: E, G, and σ_m.

Dilatation of roving basalt composites was measured twice to compare the effect of matrix transformation at first and second heating cycle. The results are depicted on Figure 12 with sample order according to their expansion at 700°C for each heating cycle. At the first heating all composites exhibited rather slow expansion from 50 to 400°C, from 400 to 600°C composites were rather stable or slightly shrunk (Q-none). From approx. 580°C a great expansion was observed for all composites. The lowest total expansion was observed in case Q-none and Q-kaolin composites. At the second heating, composites exhibited up to approx. 650°C in linear expansion, which increased from 650°C due to the softening of basalt fiber. Q-kaolin composites exhibited the lowest expansion.

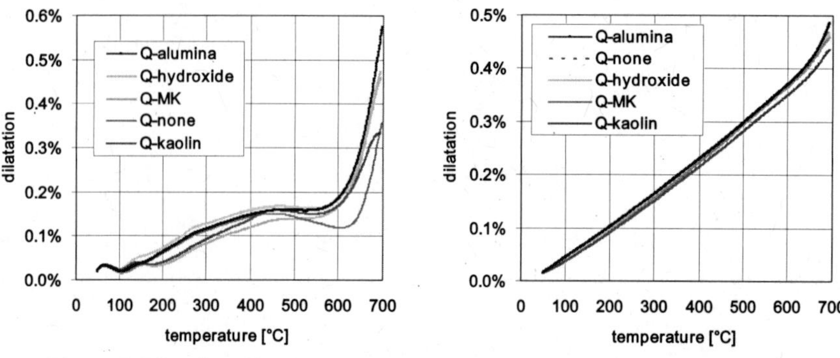

Figure 12. Dilatation of basalt roving composites: left – first heating, right – second heating.

DISSCUSION

The properties of geopolymer resins with different types of Al sources were almost the same. The exception in behavior of Q-alumina resin is difficult to explain.

XRD patterns showed the absence of newly formed crystalline phases except for potassium phosphate in geopolymer matrices, which confirmed that at high Si/Al ratio even with using potassium hydroxide as an activator no crystallization of alumosilicates during geopolymerization at temperatures as high as 85°C occurred. From FTIR-ATR spectra, the presence of unreacted kaolin and the formation of potassium phosphate were clearly observed. Presence of another Al source led to a shift of main the T-O-T vibration to lower wavenumbers.

According to SEM and EDS measurements the geopolymer matrices with additional aluminum sources were highly inhomogeneous in size by tens of μm. The clear evidence for unreacted particles of thermal silica and particles of unreacted Al sources with different sizes was found. The latter was confirmed by ^{27}Al MAS NMR spectroscopy which showed the presence of unreacted Al sources, where chemical shifts exactly matched the chemical shifts of original Al sources (see Figures 9A and 9C). Therefore, almost no dissolution of Al sources occured obviously during geopolymerization even in the case of reactive metakaolin. Moreover, from comparison of the ratio of total silicon by EDS/XRF analysis to tetrahedral (Td) aluminum by ^{27}Al MAS NMR spectroscopy (Si/Al^{Td}) to total silicon to aluminum ratio by EDS/XRF analysis (Si/Al^{total}) it is clearly seen that only in the case of the Q-none sample and mostly in the case of Q-none/boiled sample were these values comparable (see Table IV). Less Al transformed to tetrahedral Al in the case of the Q-MK sample with $Si/Al^{Td} = 45$ in comparison with $Si/Al^{total} = 10.4$ and much less in the case of other samples when Si/Al^{Td} is only 1/10 of Si/Al^{total}. It seems that not only does dissolution of additional Al sources occur, but also the addition of these Al sources reduces dissolution of thermal silica and therefore less tetrahedral Al is formed.

The surprising result of EDS analysis was that phosphorus covered the surface of unreacted particle of thermal silica, which generally makes a lot of OH group easily dehydrate at temperature above 150°C [10,15]. According to ^{29}Si MAS NMR only 14 % of Si atoms in Q-none matrix bear OH groups. Also ^{29}Si MAS NMR and ^{31}P MAS NMR offers information about possible neighboring of P and Si atoms. Both these bonds can be formed in a newly formed geopolymer phase due to rather high concentration of P among unreacted particles of thermal silica and potassium phosphate. In both cases it leads to the reduction of total amount of OH groups. High matrix expansion at high temperatures occurs due to dehydroxylation of OH groups. The main part of P is not firmly bound to the geopolymer framework or adhered to thermal silica particle due to its easy dissolution during boiling, as seen in XRD diffraction, ^{31}P MAS NMR and chemical analysis evidence. Increasing dissolution of this P with addition of Al sources, leading to the decrease of amount of dissolved thermal silica, probably indicates that fewer P atoms are bound to the geopolymer network. However, in all cases the appropriate amount of P added prevented expansion of geopolymer matrix and kept the mechanical properties (flexural strength) at elevated temperatures sufficiently high.

It should be stressed that the mechanical properties of basalt roving composites were the highest in the case of Q-kaolin and Q-MK matrix, because a lower degree of geopolymerization was observed for these matrices in comparison with Q-none matrix. Addition of lower weight parts of alumina and aluminum hydroxide, although with smaller particles, in comparison with metakaolin and even with kaolin did not bring up the considerable increase in flexural strength in the case of 5×5×100 mm samples. Although the relative decrease in strength at elevate temperature was comparable for different matrices or rather lower in the case of Q-none, Q-alumina and Q-hydroxide matrices, only in the case of Q-MK and mostly Q-kaolin matrices were the absolute values of remaining strength very high. It is surprising with regards to a rather rapid decrease of mechanical properties of basalt fiber even at 400°C [16]. At 700°C the flexural strength of all basalt roving composites is almost the same, but the Youngs modulus increases, except Q-MK matrix. At this temperature, the mechanical properties of basalt fiber are almost immeasurable; therefore geopolymer matrix must protect basalt fiber at least to some extent. Considerable decrease of flexural strength of Q-MK and Q-kaolin composites at 700°C corresponds to a high expansion of composites between 580 and 700°C, due to the softening of basalt fiber (see Figure 12). Therefore the maximum working temperature for basalt fiber-reinforced composites is approx. 550°C, where more than 50 % of original strength remains.

Higher amounts of additional sources of Al serves as a filler, but it can also contribute to increasing the strength. Nevertheless, according to ref. [1], smaller filler particles should be used. To compare the effect of total amount of Al filler in the resin, the increased amount of alumina and aluminum hydroxide equal to the weight of kaolin was added, when composite samples 2×7×100 mm were prepared (see the samples in Table VIII marked with asterisk). Only in the case of Q-alumina* composite, was the flexural strength fairly close to that of corresponding Q-kaolin 2×7×100 mm composite. At the same time, the flexural strength of Q-hydroxide and Q-hydroxide*composites with the smallest particle size of the Al source was still lower than the strength of Q-kaolin, and moreover without apparent effect of the additional amount of aluminum hydroxide. Anyway, the highest values of flexural strength and Youngs modulus were estimated at Q-kaolin composite.

From comparison of mechanical properties of the roving and fabric composites it is clear that, fabric composites were more brittle and less anisotropic than roving composites, with almost 3-times lower flexural strength and 5-times higher shear modulus. Note that fabric composites contained lower amounts of fiber. It is also seen that, E-glass fabric composites were more brittle than basalt ones.

From comparison of mechanical properties of thin (800 tex) carbon and thick (2520 tex) basalt roving composites it follows that higher flexural strength and higher Youngs modulus appeared with carbon composites. Both composites contained approximately the same volume fraction of fiber (the density of basalt is slightly higher than 50 % that of carbon). In the case of thick carbon 1600 tex roving was used, which is comparable with 2520 tex basalt roving, the flexural strength of these composites is comparable [16]. Therefore, basalt roving geopolymer composites exhibited a great potential in achieving mechanical properties retained even at rather high temperatures up to 550°C.

CONCLUSIONS

It was concluded that the geopolymer resins modified with phosphoric acid showed low viscosity and long pot life, suitable to performing the impregnation of fibers. On the other hand a majority of phosphorus added to geopolymer resins did not become part of the geopolymer matrix but rather soluble filler within the matrix. Three types of phosphorus species were identified: KH_2PO_4 easily dissolved in water, larger pyrophosphate species, which were not easily dissolved probably due to a steric problem inside the network or due to the lower relative solubility in comparison with KH_2PO_4, and probably phosphorus anion fragments bound to silicon atoms in the geopolymer network or adhered to unreacted thermal silica particles. The existence of the latter P entities was probably the reason why this geopolymer matrix does not expand at high temperatures, due to Si-OH group extinction.

Additional Al sources led to a decrease of polymerization degree moreover with less dissolution of thermal silica mostly in the case of Al sources with octahedral aluminum: alumina, aluminum hydroxide and kaolin, and to a lower extent with metakaolin, where also tetrahedral Al was present. The geopolymer matrix was rather homogeneous in size being hundreds of μms, but was highly inhomogeneous in the size range of tens μms due to the presence of undissolved particles of thermal silica and Al sources. The additional Al sources therefore played rather the role of microfiller inside the geopolymer matrix.

Higher flexural strength was exhibited by the basalt roving composite with a matrix containing kaolin having the largest particle among other Al sources up to even 550°C. This fact was difficult to explain considering almost no dissolution of octahedral Al in the kaolin, in comparison with other octahedral Al which prevailed in the aluminum sources.

Application of basalt roving fiber for reinforcing geopolymer matrix was comparable, with respect to flexural strength, with carbon roving fiber, but basalt composites exhibited half the Youngs modulus of carbon composites. In comparison with basalt roving composites, the basalt fabric composites and moreover E-glass fabric composites were more fragile and less anisotropic. Therefore the combination of roving and fabric reinforced composites, known from plastic-based ones produced by the pultrusion technique, can be recommended.

Application of two or more various span to height ratios allowed the calculation of the material flexural mechanical parameters independent of sample size and inclusion the shear parameters.

REFERENCES

[1]N. Davidovits, M. Davidovics and J. Davidovits, Ceramic-ceramic Composite Materials and their Manufacture, US patent 4,888,311 (1989).

[2]J. Davidovits and M. Davidovics, Geopolymer: Ultra-high Temperature Tooling Material for Manufacture of Advanced Composites, Proc. 36th International SAMPE Symposium 1939-49 (1991).

[3]R.E. Lyon, A.J. Foden, P.N. Balaguru, J. Davidovits and M. Davidovics, Properties of Geopolymer Matrix-carbon Fiber Ccomposites, *Fire and Materials*, **21** 67-73 (1997).

[4]J.A. Hammel, P.N. Balaguru and R.E. Lion, Influence of Reinforcement Types on the Flexural Properties of Geopolymer Composites, Geopolymer'99 Proceeding, 155-64 (1999).

[5]C.G. Papakonstantinou, P.N. Balaguru and R.E. Lyon, Comparative Study of High Temperature Composites, *Composites Part B: Engineering*, **32** (8), 637-49 (2001).

[6]J.L. Bell, D.C. Comrie, M. Gordon, and W.M. Kriven, Graphite Fiber Reinforced Geopolymer Molds for Near Net Shape Casting of Molten Diferrous Silicide, in GGC 2005: International Workshop On Geopolymers And Geopolymer Concrete. Perth, Australia: Curtin University of Technology, (2005).

[7]J. Davidovits and R. Davidovits, Ready to Use Liquid Geopolymer Resin and Method for Production Thereof, PCT WO 03087008 (2003).

[8]A.M. Mazany, J.W. Robinson and C.L. Cartwright, Inorganic Matrix Compositions, Composites and Pprocess of Making the Same, US patent 6,899,837 (2005).

[9]A.M. Mazany, J.W. Robinson and C.L. Cartwright, Inorganic Matrix Composition and Composites Incorporating the Matrix Composition, US patent 6,969,422 (2005).

[10]J.Davidovits, Geopolymer Chemistry and Application, Institut Geopolymere, Saint-Quentin, 268-75 (2008).
[11] K. J. D. MacKenzie, M. E. Smith, Multinuclear Solid-State NMR of Inorganic Materials, Pergamon, Amsterdam, 433-42 (2002).
[12]G.L. Turner, K.A. Smith and R.J. Kirkpatrick, Structure and cation effect on ^{31}P NMR Chemical Shifts and Chemical Shift Anisotropies of Ortho-phosphate, *Journal of Magnetic Resonance*, **70** (3), 408-15 (1986).
[13] K. J. D. MacKenzie, M. E. Smith, Multinuclear Solid-State NMR of Inorganic Materials, Pergamon, Amsterdam, 272-73 (2002).
[14] K. J. D. MacKenzie, M. E. Smith, Multinuclear Solid-State NMR of Inorganic Materials, Pergamon, Amsterdam, 205-8 (2002).
[15]R.A. Fletcher, K.J.D. MacKenzie, C.L. Nicholson and S. Shimada, The Composition Range of Aluminosilicate Geopolymers, *Journal of the European Ceramic Society*, **25** 1471-77 (2005).
[16]T.D.Hung, D. Kroisová, O. Bortnovsky, P. Louda and N.T.Xiem, Preliminary Abilities of Thermal Sustainment of Composites Based on Geopolymer Matrices, International Conference on Vacuum and Plasma Surface Engineering, Hejnice – Liberec, Czech Republic (2008).

ACKNOWLEDGEMENT

This work was supported by the Ministry of Industry and Trade of the Czech Republic (# FT-TA4/068).

FORMATION OF AN IRON-BASED INORGANIC POLYMER (GEOPOLYMER)

Jonathan L. Bell[*] and Waltraud M. Kriven[*†]

*Department of Materials Science and Engineering, University of Illinois at Urbana-Champaign, Urbana, IL 61801
*Member, The American Ceramic Society
†Fellow, The American Ceramic Society

ABSTRACT
High surface area (132 m^2/g) $Fe_2O_3•2SiO_2$ powder was synthetically made using an organic steric entrapment method. The resultant powder was poorly crystalline and contained small grains of magnetite and/or maghemite, and a minor amount of hematite. An iron-based inorganic polymer analogue ($K_2O•Fe_2O_3•4SiO_2•13H_2O$) was fabricated by mixing the synthetic $Fe_2O_3•2SiO_2$ powder into potassium silicate solution and curing at 50°C for 24 h. The resultant material was a water-soluble, rubbery gel. However, by aging this gel for 361 days at room temperature in a sealed container, the crystalline phases slowly dissolved, and the gel further hardened and became partially insoluble in water. The water insoluble component was amorphous and of the composition $0.48K_2O•1.38Fe_2O_3•4SiO_2$ as determined from SEM-EDS analysis. X-ray analysis showed a broad amorphous peak centered near 25° 2θ, indicative of short range ordering. The microstructure consisted of 10-20 nm sized precipitates, similar to conventional Al-based geopolymers. X-ray pair distribution function (PDF) results suggest that the atomic structure was more disordered than equivalent Al-based geopolymers, and that iron was present in predominately octahedral coordination.

INTRODUCTION

Geopolymers are a class of inorganic, aluminosilicate polymers, which harden at low temperature in a relatively short amount of time.[1] They are typically made by mixing metakaolin ($Al_2O_3•2SiO_2$) or fly-ash powders into alkaline silicate solution. Although geopolymers are generally based on aluminosilicate chemistry, it has recently been shown that alternative metals such as nickel or magnesium can be added to silicate solution to form gel-based amorphous precipitates.[2-4] The metal can be introduced in the form of nitrate solutions or as an oxide powder, and this offers a low-cost way to fabricate novel ceramic precursors. It may be possible to utilize a similar approach to produce an iron-based geopolymer analogue.

Preparation of an iron-based geopolymer has special significance that extends beyond the desire to produce novel ceramic precursors. This significance is due to multiple factors: (1) the possibility of using iron as a substitute for framework Al, as has been done in many zeolites, (2) interest in using geopolymer as an adhesive for steel, and (3) because iron is a common impurity in fly-ash. A variety of zeolites including ZSM-5, Y, beta, mordenite, and LTL have been synthesized with Fe^{3+} incorporated into a tetrahedral lattice framework instead of Al^{3+}.[5-9] Geopolymers and zeolites are both fabricated hydrothermally at high pH, with concentrated alkali, and can be made using similar source materials.[1] Therefore, it may be possible to prepare a highly reactive iron-source that can be used to form an iron-based geopolymer analogue. Despite this potential, the role of iron in geopolymerization has not been adequately explored.

In prior investigations concerning the behavior of iron in alkali silicate solution, iron was only added as a minor component.[10, 11] Jimenez *et al.*[10] found that iron added as crystalline Fe_2O_3 powder was relatively unreactive in highly alkaline solution. However, iron located within the glassy fly-ash was reactive enough to be incorporated into the hydration product of the final geopolymer matrix.

Although this study points out the potential for iron to play a role in geopolymer formation, subsequent effort was not made to determine the crystallographic details of iron in this system.

Perera *et al.*[11] added iron to an aluminosilicate based geopolymer using either ferric nitrate solution or freshly precipitated ferric hydroxide. It was determined that the iron was present in octahedral sites, either as isolated ions in the geopolymer matrix, or as unreacted oxyhydroxide aggregates. In this study, typical aluminosilicate chemistries were used (i.e. metakaolin added to sodium silicate solution) and the iron was added as a minor component ($\leq$5 wt%). Under these conditions, the highly reactive aluminosilicate geopolymers set fairly quickly and may have not allowed sufficient time for iron to participate in the reaction process.

The behavior of iron in a conventional aluminosilicate geopolymer has also been partially examined in studies where geopolymer was used as an adhesive for steel. In these prior investigations,[12, 13] it was suggested from electron microscopy results that a chemical bond was not established between the metallic iron surface and the geopolymer. However, little information was provided regarding the expected role iron would play in chemical bonding. A better understating of how iron reacts with highly basic, silicate solution could lead to improved surface treatments and stronger bonding.

The objective of this investigation was to determine the synthesis viability and structure of a pure iron-geopolymer analogue (i.e. analogous to conventional $K_2O{\bullet}Al_2O_3{\bullet}4SiO_2{\bullet}13H_2O$ geopolymer). To this end, a $K_2O{\bullet}Fe_2O_3{\bullet}4SiO_2{\bullet}13H_2O$ geopolymer analogue was prepared in a similar fashion as conventional Al-based geopolymers; by adding high surface area $Fe_2O_3{\bullet}2SiO_2$ powder to potassium silicate solution. The precursor $Fe_2O_3{\bullet}2SiO_2$ powder was intended to be compositionally similar to metakaolin ($Al_2O_3{\bullet}2SiO_2$), which is commonly used in geopolymer synthesis. The resultant material was examined using X-ray diffraction (XRD), X-ray pair distribution function (PDF) analysis, scanning electron microscopy (SEM), and energy dispersive X-ray analysis (EDS).

EXPERIMENTAL PROCEDURES

High surface area $Fe_2O_3{\bullet}2SiO_2$ powder preparation

High surface area $Fe_2O_3{\bullet}2SiO_2$ powder was made by the organic steric entrapment method.[14] A 50 wt. % solution of 98 % ferric nitrate nonahydrate (Sigma Aldrich, Pittsburgh, PA) was mixed under low heat with a 24.5 wt. % suspension of Ludox SK colloidal silica (Grace Division, Columbia MD) in a Pyrex glass beaker. After one hour of stirring, a 5 wt. % solution of Celvol polyvinyl alcohol (PVA) (Celanese Chemicals, Dallas TX) was added to the mixture to achieve a final molar ratio of $Fe_2O_3{\bullet}2SiO_2{\bullet}7PVA$. The mixture was magnetically stirred at 180 rpm at 150°C for 12 h until the solution achieved a foam consistency, which was subsequently pulverized using a mortar and pestle.

The resulting powder was calcined at 600°C for 1 hour in a commercial furnace (Model Isotemp 650, Fisher Scientific, Pittsburgh, PA) at 10°C/min ramp rates. After calcination, the powder color changed from a yellow-red to deep red. The powder was then ball milled at 68 rpm in a 500 ml Nalgene cylindrical plastic container. The container was filled to ~66% of its overall volume with 200 ml of 5 mm sized diameter, monodisperse alumina balls, 20 g of the $Fe_2O_3{\bullet}2SiO_2$ powder, and isopropanol. After milling, the powder was dried and sieved to < 44 μm.

Geopolymer Synthesis

Iron based geopolymer of the composition $K_2O{\bullet}Fe_2O_3{\bullet}4SiO_2{\bullet}13H_2O$ was made by adding the high surface area $Fe_2O_3{\bullet}2SiO_2$ powder to $K_2O{\bullet}2SiO_2{\bullet}13H_2O$ potassium silicate solution. The geopolymer was hand-mixed, cast into sealed 50 ml polypropylene tubes (Corning Premium Quality Centrifuge Tubes, Corning, NY), and cured at 50°C for 24 h. Details regarding sample preparation and labeling are given in Table I. Samples GP50-7 and GP50-361 were found to dissolve in water, although an insoluble component was left behind for GP50-361. In order to isolate the insoluble component, GP50-361 was centrifuged using a centrifugal mixer (Model Marathon 10K, Fisher Scientific, Pittsburgh, PA) operated at 8000 rpm for 3 min.

Table I – Precursor Powder and Geopolymer Preparation Condition

Sample name	Description	Calcination / Curing conditions*	Seal time*
Precursor Powder	Precursor $2SiO_2 \cdot Fe_2O_3$ powder	600 °C 1h to remove polymer	N/A
GP50-7	$4SiO_2 \cdot Fe_2O_3 \cdot K_2O \cdot 13H_2O$ geopolymer	50 °C for 24 h	7 days
GP50-361	$4SiO_2 \cdot Fe_2O_3 \cdot K_2O \cdot 13H_2O$ geopolymer	50 °C for 24 h	361 days

*The seal time is how long the sample was sealed in the polypropylene container at room temperature prior to being tested.

Analysis

The $Fe_2O_3 \cdot 2SiO_2$ precursor powder (sample PP) was analyzed to determine its density, specific surface area, and particle size characteristics. The density was measured using a helium-based pycnometer (Model 1330, Micromeritics, Norcross, GA). After being filled with powder, the sample chamber was purged 50 times prior to analysis to ensure removal of atmospheric gases. A total of 10 measurements were acquired for each sample. The surface area was measured by nitrogen absorption (Model ASAP 2400, Micromeritics, Norcross GA) using a best fit of seven datum points according to the BET method.[15] Particle size analysis was performed with a Horiba Particle Size Distribution Analyzer (Model CAPA-700, Horiba International Corporation, CA) using Sedisperse A-12 (Micromeritics Instrument Corporation, Norcross GA) as the dispersant.

The geopolymer and precursor powder samples were sieved to < 44 μm (325-mesh) and analyzed using a Rigaku XRD (step size 0.02, rate 0.5° 2θ/min, operating voltage of 45 kV and 20 mA, from 5° to 70° 2θ, Cu Kα radiation - Model D Max II, Rigaku, Danvers, MA). Analysis of the XRD data was performed using JADE 7 software (Minerals Data Inc., Livermore, CA). Energy dispersive X-ray analysis (EDS) work was carried out using a JEOL Low Vacuum SEM (Model 6060LV, JEOL USA Inc., Peabody, MA). The Si/Fe molar of the precursor powder was determined by examining flat pellet-pressed samples at 1000x, 20 kV, and a 10 mm working distance. Copper was used to calibrate the energy. For the geopolymer samples, a minimum of 10 spots were analyzed at 1000x and 20KeV on flat fracture surfaces. High resolution SEM micrographs were taken using a Hitachi SEM (Model S-2700, Hitachi High Technologies America, Schaumburg, IL) on fracture surfaces.

Synchrotron X-ray data was collected at the Argonne National Laboratory, Advanced synchrotron Photon Source, Beamline 11-IDB at ~90.3 keV (λ= 0.1372 Å). A MAR image plate-camera (IP) (Model MAR-345, Marresearch, Norderstedt, Germany) was used to collect diffraction data beyond Q = 30 Å^{-1}, although a Q_{max} = 29 Å^{-1} was used as a cutoff due to data corruption near the IP edges. More details regarding this setup can be found in the work of Chupas *et al.*[16] All sample powders were pressed into polyimide tubes (2.84 mm OD, Accellent Cardiology, Trenton GA) and examined in transmission geometry. A total of five frames were collected for each sample, with exposure times ranging from 90-120 s in order to ensure adequate high Q scattering statistics. The sample to detector distance, detector tilt, angle of rotation, and zero position were determined using a CeO_2 standard (Standard 674a, NIST, Gaithersburg, MD). All 2D image plate data was corrected for polarization and converted to a 1D scattering pattern via the program Fit2D.[17] Raw 1D data was further corrected for container and background scattering, image plate geometry, Compton scattering, and energy dependence using the program "PDFgetX2" in order to generate a properly normalized and corrected total scattering function, Q[S(Q)-1].[16, 18] Finally, the PDF function, G(r), was generated by taking a Fourier transform of the corrected and normalized data using PDFgetX2. Experimental patterns were compared to a calculated model using the program PDFFIT.[19]

RESULTS AND DISCUSSION

Fe_2O_3•$2SiO_2$ Precursor Powder Preparation

As shown in this Table II, use of the organic steric entrapment method successfully produced a high surface area powder with a composition very close to the desired molar ratio of Si/Fe = 1. The precursor powder was X-ray amorphous prior to being calcined, but crystallization of magnetite (Fe_3O_4 - PDF card 01-075-0449) and/or maghemite ($Fe_{2.67}O_4$ PDF card 01-075-0449) and a minor amount of hematite (Fe_2O_3 – PDF card 01-089-8103) was observed after calcination at 600°C (Fig. 1). It was difficult to distinguish between magnetite and maghemite because of their similar X-ray pattern and the line broadening as shown in Fig 1. The presence of magnetite and/or maghemite caused the powder to stick to the magnetic stir bar.

Table II – Fe_2O_3•$2SiO_2$ Precursor Powder Characteristics

Specific Surface Area	Average Particle size	Density	Si/Fe molar ratio
132 m^2/g	1.39 μm ± 2.14 μm	3.1724 g/cm^3	1.02

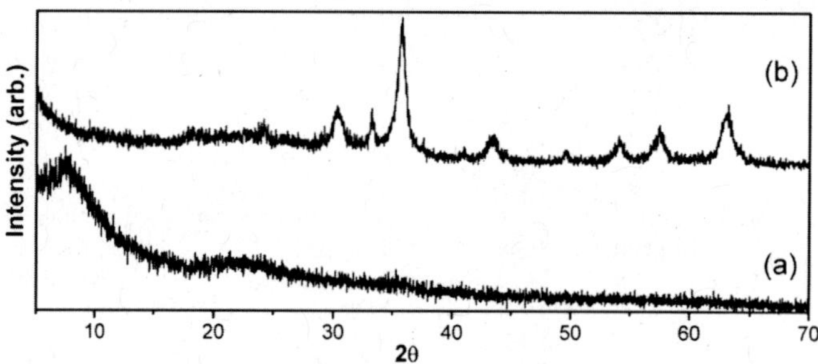

Figure 1. XRD results for Fe_2O_3•$2SiO_2$ powder (a) before calcination and (b) after being calcined at 600°C for 1 h.

An amorphous precursor powder was desired to facilitate iron dissolution and geopolymer formation. However, when using the PVA method, a minimum calcination of 600°C was necessary to ensure PVA polymer removal.[20] Shown in Fig. 1(a), the low angle peak near 10° 2θ was due to the presence of PVA, but was removed after the powder was heated to 600°C (Fig. 1(b)). The broadness of the peaks in Fig. 1(b) indicates that the calcined powder was poorly crystalline. In addition, calcination of the precursor powder caused its color change from a yellow-red to a deep red color.

Crystallization of both hematite and magnetite/maghemite suggests that the resultant precursor powder was of mixed oxidation state. Poorly crystalline maghemite was previously shown to crystallize from a mixture of $Fe(NO_3)_3$ and ethylene glycol, when heated to 300°C under oxygen-poor conditions.[21] Hematite was produced from the same source materials, but in an oxygen-rich environment.[21] In PVA solution, it is expected that that some of the cations will be bound by hydroxyl groups, while others will be sterically entrapped within polymer chains. As water is evaporated, the polymer will shrink which further reduces cation mobility and precipitation.[14] Fe cations which were not strongly trapped in the PVA would be more exposed to the air during sintering, thus favoring hematite formation. More strongly bound ions are more likely to form magnetite or maghemite depending on the nature of PVA and the exact calcination conditions. The fact that primarily magnetite and/or maghemite were formed suggests that oxygen was not readily available.

Geopolymer Formation

Iron based geopolymer of the composition $K_2O \bullet Fe_2O_3 \bullet 4SiO_2 \bullet 13H_2O$ was made by mixing the high surface area powder precursor into potassium silicate solution and curing in a sealed container. The resultant mixture displayed an exothermic reaction shortly after mixing most likely due to silica dissolution. Sample GP50-7, which was cured at 50°C for 24 h and aged at room temperature for 7 days in a sealed container, attained a rubbery texture and was water-soluble. However, by aging this sample for 361 days (i.e. sample GP50-361), the resultant geopolymer further hardened and became partially water soluble.

The slow reaction process is likely due to multiple factors including (1) hindered reaction kinetics and diffusivity within the viscous gel, (2) inadequate or sluggish dissolution of precursor powder, and (3) a limited potential of iron to form a potassium-based aluminosilicate inorganic polymer. Hematite is not expected to be very soluble in concentrated alkali hydroxide solution, although it dissolution is dependent on the alkali used such that solubility is highest in NaOH, followed by KOH and LiOH.[22] It is expected that as maghemite, magnetite and hematite dissolve, Fe^{2+} and Fe^{3+} ions would be released and would then be free to interact with the silica gel. Iron is versatile and can exist in multiple oxidation and coordination states in the final material. In zeolite systems, the more dilute conditions favors the diffusion of iron and other species, and a more thermodynamically stable state can be reached. The use of water-soluble iron sources such as iron nitrate and potassium ferrate in hydrothermal conditions have been shown to be an effective way to produce iron zeolites in which iron is located in tetrahedral coordination.[6, 23]

The precursor powder used in this study, which was fabricated as a powder in order to make the processing similar to that of conventional geopolymers, was far less reactive in potassium silicate compared to equivalent high surface-area $Al_2O_3 \bullet 2SiO_2$ precursors. In prior studies,[24, 25] synthetic high surface-area $Al_2O_3 \bullet 2SiO_2$ precursors were found to be highly reactive in alkaline solution and often led to uncontrolled, flash setting. The aluminosilicate system was highly reactive even when the $Al_2O_3 \bullet 2SiO_2$ precursors were over-calcined and contained crystalline mullite ($3Al_2O_3 \bullet 2SiO_2$). Therefore, the ability of iron-silicate to form an inorganic polymer was far less than the conventional aluminosilicate based system under typical preparation conditions.

XRD results for the geopolymer samples are shown in Fig. 2. For the GP50-7 sample (Fig. 2(a)), the relative proportions of magnetite and hematite remained similar to that present in the initial precursor powder. This indicated that the crystalline phases were stable under the alkali conditions after being cured at 50°C for 24 h. However, a broad peak near 28° 2θ was also formed and is similar to what has been observed in aluminosilicate systems.[26] Because this sample was completely water soluble, it is believed that this broad peak was due to the silica gel phase formed. Heating of sample GP50-7 to ≥ 100°C resulted in the further growth of crystalline phases, and the formation of a porous gel.

For sample GP50-361, a 361 day age at room temperature was sufficient to dissolve the crystalline phases although the diffuse peak near 28° 2θ remained (Fig. 2(b)). Heating of GP50-361 to 100°C for 48 h did not result in crystallization of any new iron phases. This indicated that the iron was located in a more stable environment and was not freely able to oxidize or diffuse to the extent required to create new phases. Nevertheless, GP50-361 was partially soluble in water. X-ray analysis of the insoluble part of GP50-361 (Fig. 2(c)) revealed it had an amorphous structure similar to that shown for GP50-361. However, the peak centered at 28° 2θ broadened and shifted toward lower values. This is consistent with prior observations for Fe substitution of Al in zeolites, which have been shown to produce crystals with a higher d-spacing (i.e. increased unit cell parameters) compared to their Al analogues.[6, 23, 27]

SEM micrographs for GP50-7 and the insoluble part of GP50-361 are shown in Fig 3. The micrograph of GP50-7 was predominantly featureless and resembled a gel structure (Fig. 3(a)).

However, the insoluble component in Fig. 3(b) had a ~10-20 nm spherical precipitates similar to what is commonly seen in aluminosilicate geopolymer systems.[28] The entire surface of sample GP50-361 appeared to be homogenously comprised of the precipitate structure shown in Fig. 3(b).

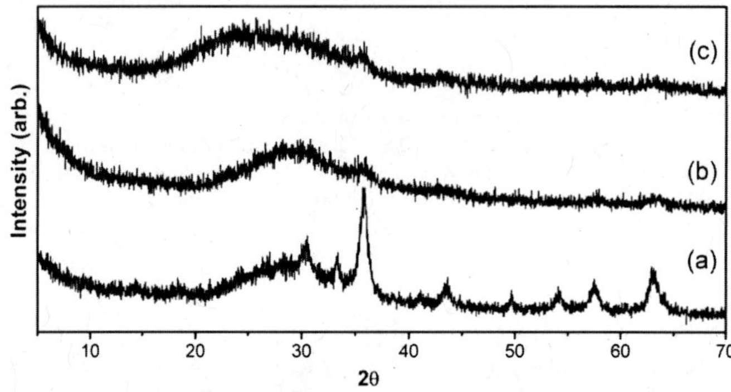

Figure 2. XRD results for $K_2O\bullet Fe_2O_3\bullet 4SiO_2\bullet 13H_2O$ geopolymer for samples (a) GP50-7, (b) GP50-361, and (c) the water insoluble part of GP50-361. Please refer to Table I for sample descriptions.

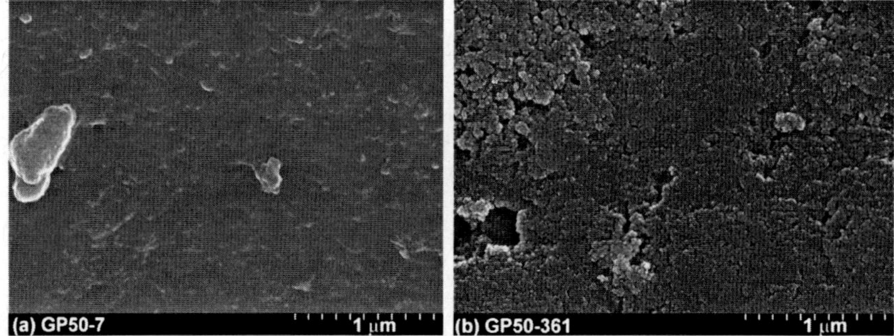

Figure 3. SEM fracture surfaces of (a) GP50-7 and (b) the water insoluble component of GP50-361.

The insoluble part of sample GP50-361 had a composition of $0.48K_2O\bullet 1.38Fe_2O_3\bullet 4SiO_2$ as determined by SEM-EDS, which was iron-rich and potassium poor compared to initial composition ($K_2O\bullet Fe_2O_3\bullet 4SiO_2\bullet 13H_2O$). Iron is able to exist in a multiple oxidation states. Fe^{3+} has been shown to occupy various sites within zeolite frameworks including tetrahedral framework sites, exchangeable cationic sites, and extraframework hydrates oxides.[29] Fe^{2+} has been found in zeolite cages and extraframework sites.[30] In the iron-geopolymer, Fe^{2+} could potentially be located in geopolymer tetrahedral cavities, and thus take over some of the charge balancing role from K^+ ions. This would account for the observed surplus of Fe and deficiency in K.

In order to further explore the local bonding structure of iron in sample GP50-361, further analysis was deemed necessary. Initial attempts to use Mössbauer spectroscopy were unsuccessful due to insufficient signal resolution. It was decided to use an alternative method known as the X-ray pair distribution function (PDF) analysis. The PDF method can be used to examine both the short (2 - 5 Å) and medium range order (5 - 20 Å) in amorphous and nanocrystalline materials by utilizing both Bragg as well as diffuse scattering.[31, 32] The combination of high energy X-rays at third generation synchrotron sources and advanced detectors has allowed various materials to be quickly and accurately analyzed with a high real space resolution.[33, 34] The X-ray PDF method involves collection of total scattering X-ray data, which is corrected for experimental effects such as background and detector geometry, and is subsequently Fourier transformed to generate the PDF function, $G(r)$. The PDF gives the probability of finding any two atoms as a function of interatomic distance.[35]

In an iron-based geopolymer, Fe^{3+} in tetrahedral coordination has a larger ionic radius (0.63 Å) compared to tetrahedrally coordinated Al^{3+} (0.53 Å), which makes Fe-O bonds clearly distinguishable from Si-O bonds. This difference facilitates analysis of bond lengths, bond length distributions, and coordination number calculations using PDF analysis. In conventional geopolymers, it is much more difficult to distinguish between Al-O and Si-O bonds because of their similar bond lengths.[36]

PDF results for precursor powder and GP50-361, plotted in terms of the scattering vector Q ($Q = 4\pi\sin\theta / \lambda$), are shown in Figs. 5-6. The reduced structure function F(Q) for the precursor powder had sharp peaks at low Q while GP50-361 was comprised of broad peaks for all Q-values (Fig. 4). This was consistent with prior X-ray data and was indicative of the presence and lack of long range order for the precursor powder and GP50-361 respectively. Compared to the X-ray data in Fig. 2, the short to medium-range order was amplified in F(Q) and can be seen by the broad oscillations at high Q.

Subsequent Fourier sine transformation of F(Q) yielded the PDF and provides a real space representation of the short to medium-range atomic order. Because of the presence crystalline magnetite/maghemite, ordering in the PDF of the precursor powder extended beyond $r = 20$ Å^{-1}. The PDF of GP50-361 indicated that its atomic structure was highly disordered and only contained well-resolved features below $r = 5$ Å^{-1}. This further supports the notion that all crystalline phases present in the precursor powder have been completely dissolved over the 361 day period, and were no longer ordered even on a fine nanoscale. The atomic structure of GP50-361 appeared to be more disordered that that of Al-based geopolymers of an analogous composition, which have been shown to have short to medium range ordering extending out to ~8 Å^{-1}.[37] The wave-like fluctuations beyond $r = 5$ Å^{-1} are artificial and were due primarily to finite termination of the PDF data.[38]

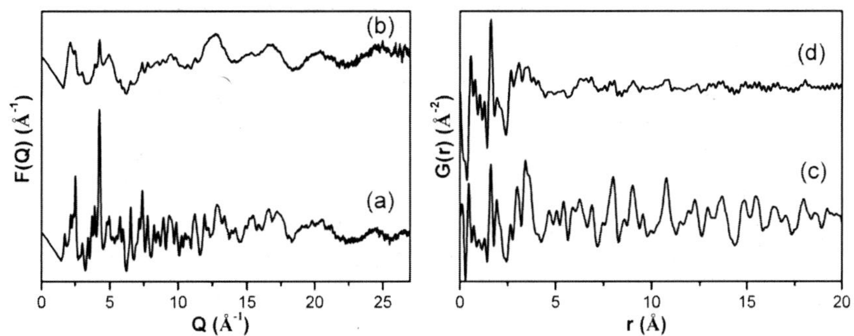

Figure 4. Reduced structure functions F(Q) for (a) the precursor powder (b) GP50-361 and the corresponding PDF data for (c) the precursor powder and (d) GP50-361.

The PDF for GP50-361 is compared to PDF results for the precursor powder, a magnetite model, conventional $K_2O{\bullet}Al_2O_3{\bullet}4SiO_2{\bullet}11H_2O$ geopolymer, and iron-substituted leucite ($KFeSi_2O_2$) in Fig. 5((a)-(d)). As shown in Fig. 5(a), both the precursor and GP50-361 had a large peak near 1.6 Å due to Si-O bonds, and a second peak at approximately 2.0 Å, due to Fe-O pairs. Precursor powder exhibited more well-defined peaks at 3.0 and 3.5 Å, in accordance with first and second nearest neighbor Fe-Fe bonds in magnetite. These peaks were absent in GP50-361 and were indicative of a more disordered Fe bonding environment.

A comparison of GP50-361 to the magnetite structure (Fig. 5(b)) further reinforced the notion that the Fe in GP30-361 was located in a disordered environment. Magnetite has an inverse spinel structure[39] and contains iron in multiple oxidation and coordination states. It is comprised of one tetrahedrally coordinated Fe^{3+} and Fe^{2+} with the remaining Fe^{3+} in octahedral coordination.[40] First nearest neighbor, average Fe-O bonding distances are listed in Table III. Because of the large range of Fe-O bond lengths possible, the first peak in the PDF of magnetite is very broad and centered at r = 2.0 Å. GP50-361 also exhibited a similar broad peak at r = 2 Å, but appears to have a doublet. However, the location of the higher r peak (centered at r = 2.2 Å) varied with the Q_{max} cutoff value and was therefore due to finite termination errors rather than the GP30-361 atomic structure.

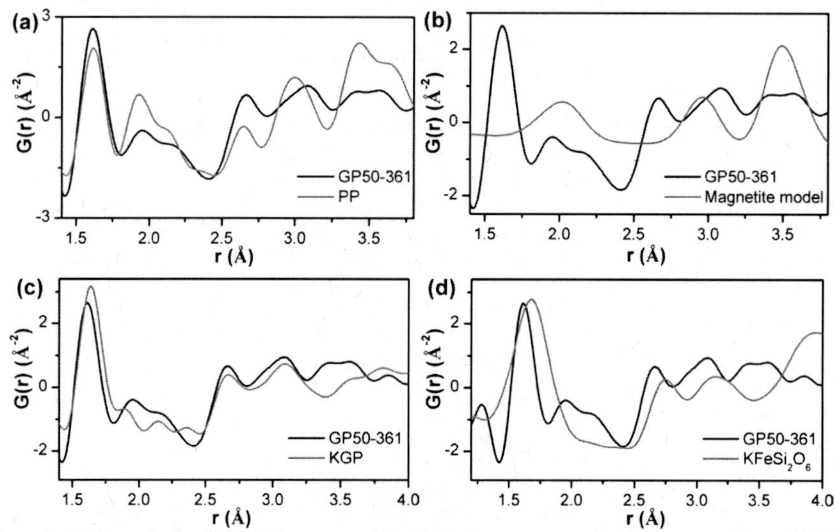

Figure 5. PDF results for GP50-361 compared to (a) the precursor powder, (b) a magnetite model (c) conventional Al-based geopolymer (K_2O• Al_2O_3•$4SiO_2$•$11H_2O$) labeled as KGP and (d) iron-substituted leucite ($KFeSi_2O_6$). The PDF for the magnetite was calculated using PDFIT with parameters from Wechsler *et al.*[41] PDF data for KGP was from an earlier publication by the authors.[37] The parameters for $KFeSi_2O_6$ was taken from Bell *et al.*[42]

Table III – Calculated and Observed Fe-O Bonding Distances in Magnetite (from Fleet[40])

Coordination	Tetrahedral		Octahedral	
Iron State	Fe^{2+}-O	Fe^{3+}-O	(Fe^{2+}, Fe^{3+})-O	Fe^{3+}-O
Bond distance (Å)	2.03	1.89	2.113	2.045

The PDFs of KGP and GP50-361 have similar peaks at r = 1.6, 2.6, and 3.1 Å (Fig 5(c)). The first peak at r = 1.6Å was slightly broadened in KGP compared to GP50-361 as it is comprised of both Si-O and Al-O pairs which are centered near 1.6 and 1.75 Å, respectively. The sharpness of the first peak in GP50-361 indicates it is comprised of Si-O bonding pairs only. Initially, it was expected that the geopolymer would form a structure that was analogous to $KFeSi_2O_6$. As shown in Fig. 5(d), the first peak in in the $KFeSi_2O_6$ is much broader than that of GP50-361 and is comprised of both Si-O bonds as well as tetrahedrally coordinated Fe^{3+} which has an average bonding length of 1.86 Å.[42] The absence of this broad initial peak in GP50-361 suggests there is negligible quantity of tetraherally coordinated Fe^{3+}. This does not preclude the possibility of tetrahedrally coordinated Fe^{2+}, which has been also been found to occupy tetrahedral sites in some iron synthesized leucites.[43]

However, further analysis revealed that the presence of tetrahedrally coordinated Fe^{2+} is also unlikely. The peaks for GP50-361 centered at r = 2.6 and 3.1 Å are characteristic of O-O and T-T (T = Si, Al) bonding pairs in leucite. In the $KFeSi_2O_6$ structure, these peaks occur at a higher r value as the incorporation of tetrahedral Fe^{3+} expands the network and unit cell. In lieu of this trend, it was unlikely that a large portion of tetrahedral Fe^{2+} was present in GP50-361, as this would lead to an even larger unit cell than that shown for $KFeSi_2O_6$. The location of these peaks in GP50-361 at lower r values indicated that it was comprised of a tetrahedrally-coordinated Si network and contained very little, if any, terrahedrally-coordinated iron. The Fe was therefore mostly present in octahedral coordination. This is consistent with prior observations in which ferric nitrate solution or ferric hydroxide was added to metkaolin-based geopolymers resulted in octahedrally coordinated iron.[11]

Although the PDF results rule out an iron-leucite structure ($KFeSi_2O_6$), the versatility of iron permits a variety of other possibilities as shown in Table IV. In the phases shown in Table IV, silica is located in a tetrahedral network, while the iron is predominately in octahedral coordination. This would produce a structure that is consistent with the PDF results for GP50-361. Additionally, nontronite, glauconite, celadonite, and ferristilpnomelane all contain alkali. The formation of nontronite, minnesotaite, greenalite and ferrihydrite is favored in highly basic conditions,[44] which is similar to that used to produce GP50-361.

Table IV – Natural Iron-Silicates that have Predominately Octahedrally-Coordinated Iron.[45-47]

Phase	Empirical Formula
Nontronite	$Na0.3Fe^{3+}_2(Si,Al)_4O_{10}(OH)_2 \bullet n(H_2O)$
Glauconite	$(K,Na)(Fe^{3+},Al,Mg)_2(Si,Al)_4O_{10}(OH)_2$
Celadonite	$K(Mg,Fe^{2+})(Fe^{3+},Al)[Si_4O_{10}](OH)_2$
Hisingerite	$Fe^{3+}_2Si_2O_5(OH)_4 \bullet 2(H_2O)$
Neotocite	$(Mn,Fe^{2+})SiO_3 \bullet (H_2O)$
Ferristilpnomelane	$K(Fe3^{+},Mg,Fe^{2+})_8(Si,Al)_{12}(OH)_{27}$
Minnesotaite	$(Fe^{2+},Mg)_3Si_4O_{10}(OH)_2$
Berthierine	$(Fe^{2+},Fe^{3+},Al,Mg)_{2-3}(Si,Al)_2O_5(OH)_4$
Greenalite	$(Fe^{2+},Fe^{3+})_{2-3}Si_2O_5(OH)_4$

In addition, many of the phases shown in Table IV form a finely-grained or amorphous structure depending on formation conditions.[46] In the constrained environment of GP50-361, which contained little water and a highly viscous gel structure, the formation kinetics were severely hindered, making crystalline formation unlikely. EDS chemical analysis results for the insoluble part of GP50-361 suggest a chemistry of $0.48K_2O \bullet 1.38Fe_2O \bullet 4SiO_2$, which does not exactly match any of the phases shown in Table IV. However, the much lower potassium content compared to conventional geopolymer is a further reflection that a minimal amount, if any, of the iron is located in a tetrahedral bonding environment, as this would require potassium to play a charge balancing role.

CONCLUSIONS

An iron-based inorganic polymer was made by adding high surface area $Fe_2O_3•2SiO_2$ iron silicate powder to potassium silicate solution. The iron silicate power, which was fabricated synthetically, was found to be far less reactive than the equivalent aluminosilicate powder ($Al_2O_3•2SiO_2$) and required nearly one year to dissolve. Structural and compositional analysis suggested that a conventional iron-geopolymer analogue was not formed. Rather, the inorganic polymer obtained consisted of a tetrahedral silicate network and iron in predominately octahedral coordination. This sort of structure was consistent with a variety of naturally occurring iron-silicates.

The results of this work suggest that it is not feasible to produce an iron-based geopolymer analogue (i.e. $K_2O• Fe_2O_3•4SiO_2•13H_2O$ instead of $K_2O• Al_2O_3•4SiO_2•13H_2O$) via the conventional method of adding a $Fe_2O_3•2SiO_2$ powder to potassium silicate solution. However, the $Fe_2O_3•2SiO_2$ powder used in this study was found to be weakly crystalline, and have may have not been optimal for high reactivity. In prior zeolite work, the use of other source materials combined with hydrothermal processing was shown to be sufficient to obtain a ferrosilicate tetrahedral network. In this regard, more work needs to be done using different processing approaches to determine if it is possible to produce an iron geopolymer analogue.

ACKNOWLEDGEMENTS

This work was supported by Air Force Office of Scientific Research (AFOSR), USAF, under Nanoinitiative Grant No. FA9550-06-1-0221. Experimental work was conducted at the APS, Argonne National Laboratory, which is supported by the U. S. Department of Energy, Office of Science, and Office of Basic Energy Sciences, under Contract No. DE-AC02-06CH11357. The authors also acknowledge the use of facilities at the Center for Microanalysis of Materials, in the Frederick Seitz Research Laboratory, at the University of Illinois at Urbana-Champaign, which is partially supported by the U. S. Department of Energy under Grant No. DEFG02-91-ER45439. The authors would also like to thank Dr. Peter Chupas (Argonne National Laboratory) and Dr. Karena Chapman (Argonne National Laboratory) for their help with the PDF work, and Dr. Pankaj Sarin for many useful discussions.

REFERENCES

[1]J. Davidovits, "Geopolymers - Inorganic Polymeric New Materials," *J. Therm. Anal.*, **37** [8], 1633-56 (1991).
[2]K. Ikeda, K. Onikura, Y. Nakamura, and S. Vedanand, "Optical Spectra of Nickel-Bearing Silicate Gels Prepared by the Geopolymer Technique, with Special Reference to the Low-Temperature Formation of Liebenbergite (Ni_2SiO_4)," *J. Am. Ceram. Soc.*, **84** [8], 1717-20 (2001).
[3]R. Vallepu, A. Mikuni, R. Komatsu, and K. Ikeda, "Synthesis of Liebenbergite Nano-Crystallites from Silicate Precursor Gels Prepared by Geopolymerization," *J. Mineral. Petrol. Sci.*, **100** [4], 159-67 (2005).
[4]R. Vallepu, Y. Nakamura, R. Komatsu, K. Ikeda, and A. Mikuni, "Preparation of Forsterite by the Geopolymer Technique - Gel Compositions as a Function of pH and Crystalline Phases," *J. Sol-Gel Sci. Technol.*, **35** [2], 107-14 (2005).
[5]R. B. Borade, "Synthesis and Characterization of Ferrisilicate Zeolite of Pentasil Group," *Zeolites*, **7** [5], 398-403 (1987).
[6]C. V. A. Duke, K. Latham, and C. D. Williams, "Isomorphous Substitution of Fe^{3+} in LTL Framework Using Potassium Ferrate(VI)," *Zeolites*, **15** [3], 213-18 (1995).

[7]L. M. Kustov, V. B. Kazansky, and P. Ratnasamy, "Spectroscopic Investigation of Iron Ions in a Novel Ferrisilicate Pentasil Zeolite," *Zeolites*, **7** [1], 79-83 (1987).
[8]K. Latham, C. D. Williams, and C. V. A. Duke, "The Synthesis of Iron Cancrinite Using Tetrahedral Iron Species," *Zeolites*, **17** [5-6], 513-16 (1996).
[9]G. Zi, T. Dake, and Z. Ruiming, "Hydrothermal Crystallization of Zeolite Y from Na_2O-Fe_2O_3-Al_2O_3-SiO_2-H_2O System," *Zeolites*, **8** [6], 453-57 (1988).
[10]A. M. F. Jiminez, E. E. Lachowski, A. Palomo, and D. E. MacPhee, "Microstructural Characterisation of Alkali-activated PFA Matrices for Waste Immobilisation," *Cem. Concr. Compos.*, **26** [8], 1001-06 (2004).
[11]D. S. Perera, J. D. Cashion, M. G. Blackford, Z. M. Zhang, and E. R. Vance, "Fe Speciation in Geopolymers with Si/Al Molar Ratio of Similar to 2," *J. Eur. Ceram. Soc.*, **27** [7], 2697-703 (2007).
[12]J. Bell, M. Gordon, and W. M. Kriven, "Use of Geopolymeric Cements as a Refractory Adhesive for Metal and Ceramic Joins"; pp. 407-13 in Ceramic Engineering and Science Proceedings, Vol. 26, *The 29th International Conference on Advanced Ceramics and Composites*. Edited by D. Zhu, K. Plucknett, and W. M. Kriven. American Ceramic Society, Westerville, OH, 2005.
[13]B. A. Latella, D. S. Perera, T. R. Escott, and D. J. Cassidy, "Adhesion of Glass to Steel Using a Geopolymer," *J. Mater. Sci.*, **41** [4], 1261-64 (2006).
[14]M. A. Gülgün, M. H. Nguyen, and W. M. Kriven, "Polymerized Organic-Inorganic Synthesis of Mixed Oxides," *J. Am. Ceram. Soc.*, **82** [3], 556-60 (1999).
[15]S. Brunauer, P. H. Emmett, and E. Teller, "Adsorption of Gases in Multimolecular Layers," *J. Am. Chem. Soc.*, **60**, 309-19 (1938).
[16]P. J. Chupas, X. Y. Qiu, J. C. Hanson, P. L. Lee, C. P. Grey, and S. J. L. Billinge, "Rapid-Acquisition Pair Distribution Function (RA-PDF) Analysis," *J. Appl. Crystallogr.*, **36**, 1342-47 (2003).
[17]A. P. Hammersley, S. O. Svensson, M. Hanfland, A. N. Fitch, and D. Hausermann, "Two-Dimensional Detector Software: From Real Detector to Idealised Image or Two-Theta Scan," *High Pressure Res.*, **14** [4-6], 235-48 (1996).
[18]T. Ohsuna, B. Slater, F. F. Gao, J. H. Yu, Y. Sakamoto, G. S. Zhu, O. Terasaki, D. E. W. Vaughan, S. L. Qiu, and C. R. A. Catlow, "Fine Structures of Zeolite-Linde-L (LTL): Surface Structures, Growth Unit and Defects," *Chem-Eur. J.*, **10** [20], 5031-40 (2004).
[19]T. Proffen, and S. J. L. Billinge, "PDFFIT, A Program for Full Profile Structural Refinement of the Atomic Pair Distribution Function," *J. Appl. Crystallogr.*, **32**, 572-75 (1999).
[20]M. A. Gülgün, W. M. Kriven, and M. H. Nguyen, Processes for Preparing Mixed Metal Oxide Powders USA Patent 6,482,387, November 19th, 2002.
[21]G. M. Dacoata, E. Degrave, P. M. A. Debakker, and R. E. Vandenberghe, "Synthesis and Characterization of Some Iron-Oxides by Sol-Gel Method," *J. Solid State Chem.*, **113** [2], 405-12 (1994).
[22]K. Ishikawa, T. Yoshioka, T. Sato, and A. Okuwaki, "Solubility of Hematite in LiOH, NaOH and KOH Solutions," *Hydrometallurgy*, **45** [1-2], 129-35 (1997).
[23]P. N. Joshi, S. V. Awate, and V. P. Shiralkar, "Partial Isomorphous Substitution of Fe^{3+} in the LTL Framework," *J. Phys. Chem.*, **97** [38], 9749-53 (1993).
[24]M. Gordon, J. Bell, and W. M. Kriven, "Comparison of Naturally and Synthetically Derived, Potassium Based Geopolymers"; pp. 95-103, in Ceramic Transactions, Vol. 165, *Advances in Ceramic Matrix Composites X.*, Edited by N. P. Bansal, J. P. Singh, and W. M. Kriven, The American Ceramic Society, Westerville, OH, 2004.
[25]M. Gordon, J. Bell, and W. M. Kriven, "Geopolymers: Alkali Bonded Ceramics (ABCs) for High-Tech Applications"; pp. 215-24, in Ceramic Transactions, Vol. 175, *Advances in Ceramic Matrix Composites XI.*, Edited by N. P. Bansal, J. P. Singh, and W. M. Kriven, The American Ceramic Society, Westerville, OH, 2005.

[26]J. Davidovits, "Geopolymers: Inorganic Polymeric New Materials," *J. Mater. Educ*, **16**, 91-139 (1994).
[27]Y. S. Ko, and W. S. Ahn, "Isomorphous Substitution of Fe^{3+} in Zeolite LTL," *Microporous Mater.*, **9** [3-4], 131-40 (1997).
[28]M. Steveson, and K. Sagoe-Crentsil, "Relationships Between Composition, Structure and Strength of Inorganic Polymers - Part I - Metakaolin-Derived Inorganic Polymers," *J. Mater. Sci.*, **40** [8], 2023-36 (2005).
[29]E. G. Derouane, M. Mestdagh, and L. Vielvoye, "EPR Study of Nature and Removal of Iron(III) Impurities in Ammonium-Exchanged Nay-Zeolite," *J. Catal.*, **33** [2], 169-75 (1974).
[30]K. Lazar, L. Guczi, and B. M. Choudary, "State of Fe^{2+} Ions inside X-Zeolite Lattice During CO + H_2 Reaction," *Solid State Ionics*, **32-3**, 1000-05 (1989).
[31]S. R. Elliott, "Medium-Range Structural Order in Covalent Amorphous Solids," *Nature*, **354** [6353], 445-52 (1991).
[32]T. Egami, and S. J. L. Billinge, *Underneath the Bragg Peaks: Structural Analysis of Complex Materials*, 2003.
[33]P. J. Chupas, K. W. Chapman, and P. L. Lee, "Applications of an Amorphous Silicon-Based Area Detector for High-Resolution, High-Sensitivity and Fast Time-Resolved Pair Distribution Function Measurements," *J. Appl. Crystallogr.*, **40**, 463-70 (2007).
[34]V. Petkov, S. J. L. Billinge, J. Heising, and M. G. Kanatzidis, "Application of Atomic Pair Distribution Function Analysis to Materials with Intrinsic Disorder. Three-Dimensional Structure of Exfoliated-Restacked WS_2: Not Just a Random Turbostratic Assembly of Layers," *J. Am. Chem. Soc.*, **122** [47], 11571-76 (2000).
[35]B. E. Warren, *X-ray Diffraction*, Addision-Wesley, New York, 1969.
[36]V. Petkov, "Atomic-Scale Structure of Glasses Using High-Energy X-ray Diffraction," *J. Am. Ceram. Soc.*, **88** [9], 2528-31 (2005).
[37]J. L. Bell, P. Sarin, R. P. Haggerty, P. E. Driemeyer, P. J. Chupas, and W. M. Kriven, "X-ray Pair Distribution Function Analysis of a Metakaolin-Based, $KAlSi_2O_6 \bullet 5.5H_2O$ Inorganic Polymer (Geopolymer)," *J. Mater. Chem.*, **18** [48], 5974-81 (2008).
[38]B. H. Toby, and T. Egami, "Accuracy of Pair Distribution Function-Analysis Applied to Crystalline and Noncrystalline Materials," *Acta Crystallogr. A.*, **48**, 336-46 (1992).
[39]W. H. Bragg, "The Structure of the Spinel Group of Crystals," *Philos. Mag.*, **30**, 305-15 (1915).
[40]M. E. Fleet, "The Structure of Magnetite," *Acta Cryst.*, **B37**, 917-20 (1981).
[41]B. A. Wechsler, D. H. Lindsley, and C. T. Prewitt, "Crystal Structure and Cation Distribution in Titanomagnetites (Fe_3-xTi_xO_4)," *Am. Mineral.*, **69**, 754-70 (1984).
[42]A. M. T. Bell, and C. M. B. Henderson, "Rietveld Refinement of the Structures of Dry-Synthesized $MFe^{III}Si_2O_6$ Leucites (M = K, Rb, Cs) by Synchrotron X-ray Powder Diffraction," *Acta Crystallogr., Sect. C: Cryst. Struct. Commun.*, **50**, 1531-36 (1994).
[43]K. E. R. England, C. M. B. Henderson, J. Charnock, and D. J. Vaughan, "Investigation of Fe Structural Environments in Leucite Type Framework Silicates Using a Combination of Mössbauer and X-ray Absorption Spectroscopies," *Hyperfine Interact.*, **91**, 709-14 (1994).
[44]V. Chevrier, F. Poulet, and J. P. Bibring, "Early Geochemical Environment of Mars as Determined from Thermodynamics of Phyllosilicates," *Nature*, **448** [7149], 60-63 (2007).
[45]J. M. D. Coey, T. Bakas, and S. Guggenheim, "Mössbauer-Spectra of Minnesotaite and Ferroan Talc," *Am. Mineral.*, **76** [11-12], 1905-09 (1991).
[46]R. A. Eggleton, J. H. Pennington, R. S. Freeman, and I. M. Threadgold, "Structural Aspects of the Hisingerite Neotocite Series," *Clay Miner.*, **18** [1], 21-31 (1983).
[47]K. J. D. Mackenzie, C. M. Cardile, and I. W. M. Brown, "Thermal and Mössbauer Studies of Iron-Containing Hydrous Silicates. 7. Glauconite," *Thermochim. Acta*, **136**, 247-61 (1988).

CONSOLIDATED GEO-MATERIALS FROM SAND OR INDUSTRIAL WASTE

E. Prud'homme[1], P. Michaud[1], E. Joussein[2], C. Peyratout[1], A. Smith[1] and S. Rossignol[1]

[1] Groupe d'Etude des Matériaux Hétérogènes (GEMH-ENSCI) Ecole Nationale Supérieure de Céramique Industrielle, 47-73 Avenue Albert Thomas, 87065 Limoges
[2] GRESE, EA 3040, 123 avenue Albert Thomas, 87060 Limoges
■ corresponding author - sylvie.rossignol@unilim.fr – tel.: 33 5 55 45 22 24

Geo-materials *Dehydroxylated kaolinite* *Sand*
Silica fume *Foam* *Insulating material*

ABSTRACT

The synthesis of geopolymers based on potassium polysialate, was achieved at room or slightly elevated temperature, by alkaline activation of raw minerals, sand or industrial waste. The materials were prepared from a solution containing dehydrated kaolinite and KOH pellets dissolved in potassium silicate. Then the mixture was transferred to a polyethylene mould sealed with a top and placed in an oven at 65°C during 24 hours. For the reinforced geopolymer composites, sand or silica fumes were added to the previous solution. For all the composite geopolymer materials, following dissolution of the raw materials, a polycondensation reaction was used to form the amorphous solid. This was studied by attenuated total reflection Fourier transform infrared (ATR-FTIR) spectroscopy. The reinforcement of geopolymer by sand addition yielded a material with mechanical properties comparable to concrete. The in-situ inorganic foam based on silica fume has potential as an insulating material for an application in building materials due to its stability at medium acidity since the thermal measurement by "Flash laser" has given a value of 0.22 $W.m^{-1}.K^{-1}$ for the thermal conductivity.

INTRODUCTION

Geopolymers are amorphous, three-dimensional, alumino-silicate binder materials, which were first introduced to the inorganic cementitious world by Davidovits in 1978[1]. Geopolymers may be synthesized at room or slightly elevated temperature by alkaline activation of alumino-silicates obtained from industrial wastes, calcined clays, natural minerals or mixtures of two or more of these materials. In a strong alkaline solution, alumino-silicate reactive materials are rapidly dissolved into solution to form free SiO_4 and AlO_4 tetrahedral units[2, 3]. During the reaction of polycondensation, the tetrahedral units are linked in an alternate manner to yield amorphous geopolymers. Geopolymer concretes are based on compounds that are generally produced from one or more solid components (binders) and one or more liquid components (activators), which react together to form strong, durable materials. Some binders have been used, like PVA fibers or fly ash by technical extrusion to improve mechanical properties[4]. Furthermore little work has been devoted to the feasibility of making geopolymer synthesized foam without organic additives or PVA fibers[5, 6]. Amongst the different families of geopolymers, those based on potassium present modified thermal and mechanical properties due to the larger size of the potassium ion compared to sodium[7, 8, 9].

Currently, there is a political as well as a societal demand for products which require less energy for manufacturing and which are easy to recycle. The new materials have to display properties analogous or even improved with respect to those of existing materials. Traditional materials associating mineral binders and local raw materials can be found in all cultures and time periods.[10] However, the physico-chemistry of the exchanges between the various components of the composite are not well known. Furthermore, new cementitious materials known as geopolymers, including generally silicates or aluminosilicates, can be used as substitutes for conventional hydraulic binders.[11] However, the control of consolidation time, hydraulic behavior and the subsequent working properties of the materials remain difficult. Geopolymer materials could also be used to passivate industrial wastes and as an alternative to cements.

The aim of this work was to determine the composition of geopolymer based on potassium polysialate obtained from raw minerals. To obtain the mechanical properties similar to cement, sand was added to the mixture. The incorporation of silica fume was also studied to produce in-situ geopolymer foam. The understanding of the polycondensation reaction of the various samples was determined by in-situ ATR-FTIR spectroscopy, XRD and SEM measurements. The thermal and mechanical properties of the synthesized geomaterials are also examined.

EXPERIMENTAL PROCEDURES

Samples preparation

The initial geo-material was prepared from a solution containing dehydroxylated kaolinite and KOH pellets (85.7% of purity) dissolved in potassium silicate (Si/K=1.66, density 1.33 g/cc) described in Figure 1. The reactive mixture was then placed in a polystyrene sealed mould in an oven at 70°C for 4 hrs so that the polycondensation reaction was complete.

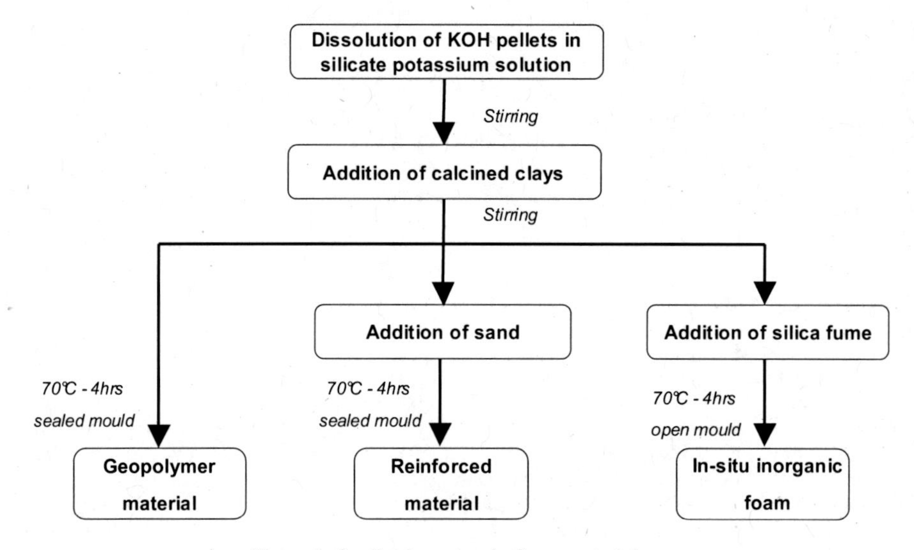

Figure 1: Synthesis protocol of geo-materials.

Then the material was removed from the mould and placed in an oven at 70°C for 24 hrs for drying. In order to obtain consolidated geo-material, two kinds of additives were tested, one to reinforce the mechanical properties by sand and the other to improve thermal properties by industrial waste. The sand was added to the previous reactive mixture, the paste was poured into rectangular sealed molds and placed in an oven at 70°C for 4 hrs, then removed from the mould and replaced in an oven at 70°C for 24 hrs. For an amount of 100 g of reactive mixture, 80 g of sand was added yielding a sample with the following dimensions: 55 mm x 10 mm x 6.5 mm. Silica fumes were added to the reactive mixture and placed in oven at 70°C in an open mold. Different materials with variation of molar ratio were synthesized according to Table I.

Characterization

FTIR spectra were carried out on a ThermoFischer Scientific 380 infrared spectrometer (Nicolet) using the attenuated total reflection (ATR) method. The IR spectra were scanned between

500 and 4000 cm^{-1} with a resolution of 4 cm^{-1}. The commercial software, OMNIC (Nicolet Instruments), was used for data acquisition and spectral analysis.

X-ray patterns were determined by X-ray diffraction (XRD) experiments from a Siemens D 5005 powder diffractometer using the $Cu_{K\alpha}$ radiation ($\lambda_{K\alpha}$ = 0.154186 nm) and a graphite black monochromator. XRD patterns were obtained using the following conditions: dwell time: 2 s; step: 0.04°. Crystalline phases were identified by comparison with PDF standards (Powder Diffraction Files) from ICDD.

The morphology of the final products was determined by using a Cambridge Stereoscan S260 scanning electron microscope (SEM). Prior to observation, a carbon layer was deposited on the samples.

The average pore size was calculated using the Image J program based on a counted number of 500 issues from several cut of samples.

Differential thermal analysis (DTA) and thermogravimetric analysis (TGA) were made in order to characterize nature of the solids. TDA–TGA experiments were carried out in a Pt crucible between 25 and 1200°C using a Cetaram Setsys evolution. The samples were heated at 10°C.min^{-1} in dry airflow.

Four point bending tests determined the mechanical strength. The bar-shaped samples with the rectangular cross-section of 55 mm x 10 mm x 6.5 mm were tested on a Lloyd EZ 20 machine with a maximum tension–compression effort of 5 kN. A constant crosshead displacement velocity of 0.5 mm/min was employed. Two strain gages had been bonded on the top and bottom faces in the middle part of the sample. The measures of strain and stress allowed observation of the behavior curve of materials. The material plastic strength theory is applied to calculate the Young's modulus on the tension sample face (bottom face). The rupture stress σ and the Young's modulus E are calculated according to the following equations 1, 2:

$$\sigma \equiv \frac{3F(L-l)}{2bh^2} \quad (1)$$

$$E = \frac{\sigma}{\varepsilon} \quad (2)$$

where ε is the stress measured by gages in the elastic domain (linear part of stress vs. strain curve).

Thermal properties were determined using the "Laser Flash" method as described by the work of Michot et al.[12] The thermal diffusivity was measured with the flash technique using a neodymium-glass laser operating at 1.053 μm as the flash source to heat the front face of the disc sample. The transient backface temperature was monitored using a liquid-nitrogen-cooled infrared detector (Hg–Cd–Te) connected to an amplifier and a storage oscilloscope. The temperature–time data were analysed using Degiovanni's method, which takes into account the effect of heat losses on the value of thermal diffusivity. The thermal conductivity (λ) was then calculated from the equation 3:

$$\lambda = \rho c \alpha \quad (3)$$

where α is the thermal diffusivity, c is the specific heat and ρ is the bulk density. Cylindrical samples of F_i were synthesized with a thickness of approximately 2 mm.

Table I: Composition of various samples

Ratio	$n_{Al_2O_3}/n_{SiO_2}$	$n_{Al_2O_3}/n_{K_2O}$	n_{SiO_2}/n_{K_2O}
Geopolymer	0.32	1.29	4.00
F 0 Si	0.22	0.66	2.97
F 25 Si	0.20	0.66	3.30
F 50 Si	0.17	0.66	3.84
F 75 Si	0.15	0.66	4.39
F_i	0.13	0.66	4.93
F 125 Si	0.12	0.66	5.48

RESULTS

1) Formulation of geo-material

A material corresponding to the formulation $K^{+}_{0.10}[(SiO_2)_{1.63}AlO_2]_{0.10}$, 0.67 H_2O was first synthesized by dehydrated kaolinite obtained from calcined (700°C) kaolin supplied by IMERYS France. The formation of this geomaterial was due to a geopolymerization reaction involving the consolidation of material. This phenomenon was followed by ATR spectroscopy. A drop of reactant mixture was put on the diamond crystal protected by a cover, which prevented water evaporation from the mixture at room temperature. These conditions were necessary to promote the polycondensation in a closed environment.[13] The spectra shown in the Figure 2(a) were recorded as a function of time.

The bands on the spectra at t = 0 h, respectively, at 3255 and 1620 cm^{-1}, were attributed to Si-O-, H bonds and to water. Their intensity gradually diminished with time.[14] The bands due to Si-O-M bonds[15, 16] (M= Si. Al. K) were located in the 1100-950 cm^{-1} range and their precise positions depended on the length and bending of the Si-O-M bond as given in Table II. The decrease of OH bands compared to the Si-O-Al bands was due to the increase of the polycondensation time and was characteristic of the formation of geo-material. Furthermore, the Si-O-M^{+} [17] shift from 979 to 946 cm^{-1} also in agreement with the literature data, gave evidence for dissolution of the metakaolin species by the basic environment.[18] This experiment as a function of time proved that, at room temperature, a time of 17 h was sufficient to achieve a consolidated geo-material.

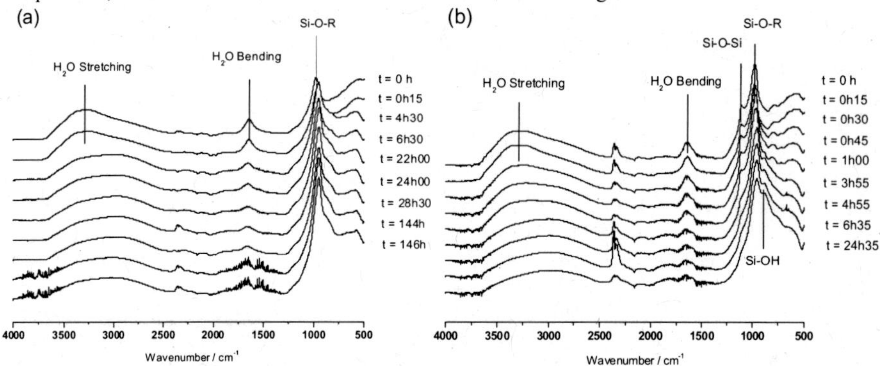

Figure 2: In-situ infrared spectra of the (a) geopolymer and (b) foam at 20°C as a function of the polycondensation reaction time.

Table II: Main vibration bands of Si-O species[15,19,17]

Wavenumber (cm^{-1})	Type	Bond	Reference
798 (m)	Symmetric stretching	Si - O - Si	17
727	Symmetric stretching	Si - O - Si	17
1115-1140	Asymmetric stretching	Si - O - Si	17
1165 (sh)	Asymmetric stretching	Si - O - Si	17
1080	Asymmetric stretching	Si - O - Si	15
466	Bending	Si - O - Si	17
560	Symmetric stretching	Al - O - Si	17
1077	Asymmetric stretching	Al - O - Si	19
840	Rocking	Si - OH	17

The resulting XRD pattern (Figure 3a), displayed a broad peak characteristic of an amorphous material. The calcination at 700°C of kaolin involved the transformation in dehydrated kaolinite, but some crystallized peaks were also detected and were due to traces of illite clay present as a contaminant. The shift in 2θ peak position observed in the geo-material XRD pattern by comparison with the dehydrated kaolinite XRD pattern provided evidence for dissolution of SiO_4 and AlO_4^- species which came from dehydrated kaolinite, into the alkaline environment, during the geopolymerization reaction as detected by ATR measurements.

The behavior at temperature was studied by thermal analysis (Figure 4). The endothermic peak below 400°C associated with a weight loss was attributed to free water and hydroxyl condensation. The weight loss of around 12% is in agreement with the work of Duxson et al.[20] Above this temperature, the geo-material was stable on temperature increase until 800°C.

2) Addition of silica fume

To increase the thermal resistance by creating porosity in the geo-material, addition of silica fumes has been investigated. The addition of silica fumes to the solution of SiO_2, K_2O and KOH in the same conditions as previously used, gave a very reactant mixture involving the formation of in-situ inorganic foam. This new compound can be considered as a geo-material since the XRD pattern is characteristic of an amorphous material (Figure 3(b)). Effectively, the maximum of diffracted intensity was positioned at similar 2θ values for both the original geo-material and the inorganic foam. Moreover, since the XRD peaks of the silica fume had disappeared in the foam XRD pattern, the formation of this compound involved dissolution or transformation of the small local order of silica fume.

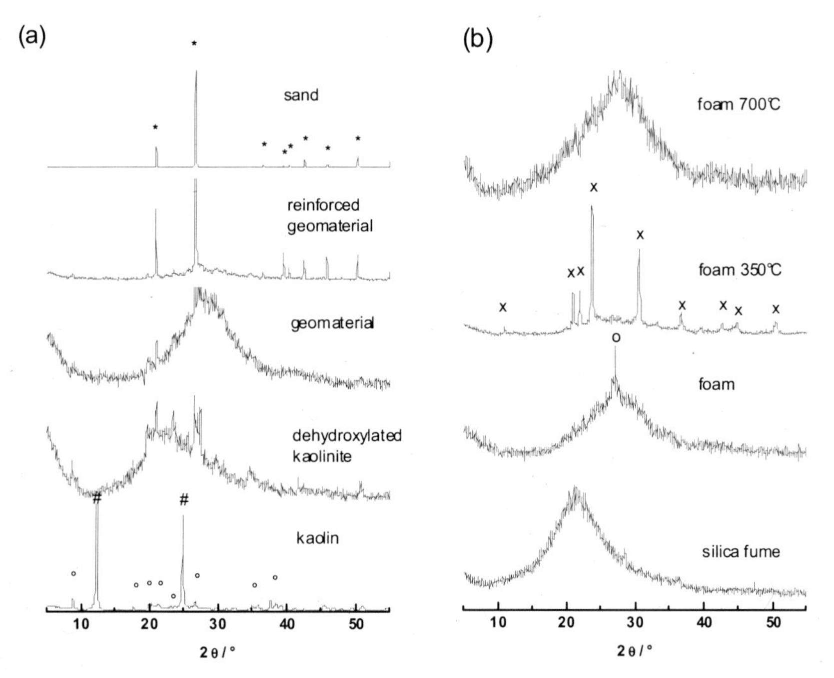

Figure 3: XRD patterns of (a) raw materials, geo-material and reinforced material and (b) foams at various temperatures (* sand, # Kaolinite ; x $KAl_4Si_2O_9(OH)_3$; ° Illite).

To find evidence of the disorder in the silica fume lattice, the synthesis of inorganic foam was studied by ATR spectroscopy (Figure 2(b)). The main bands detected for the geo-material were observed on the various foam spectra and in fact the decrease of the OH and Si-O-M bands was faster. The polycondensation reaction in the foam seemed to be finished after 4 h 30 min compared to 17 h for geo-material. However, new bands also appeared and were located respectively at 1110, 914, 880, 800 and 670 cm^{-1} corresponding respectively to Si-O-T[18], Al-OH, Si-O-Si[21], Si-O-Al[17], and O-Si-O[15]. First, the decrease of the band at 1110 cm^{-1} linked to amorphous silica, confirmed the loss of short range order of the silica fume, always seen in the XRD pattern. At the same time, the band at 880 cm^{-1} due to stretching vibration of Si-OH increased, revealing the consummation of water involving non-bridging oxygen atoms. Similarly, the bands located at 914 and 800 cm^{-1} showed the attack of Al-O species during the geopolymerisation, which was not clearly observed for the geo-material. Another band at 670 cm^{-1} could be attributed to zeolite species and must be verified. These experiments proved the almost complete dissolution of raw materials to create the in-situ inorganic foam.

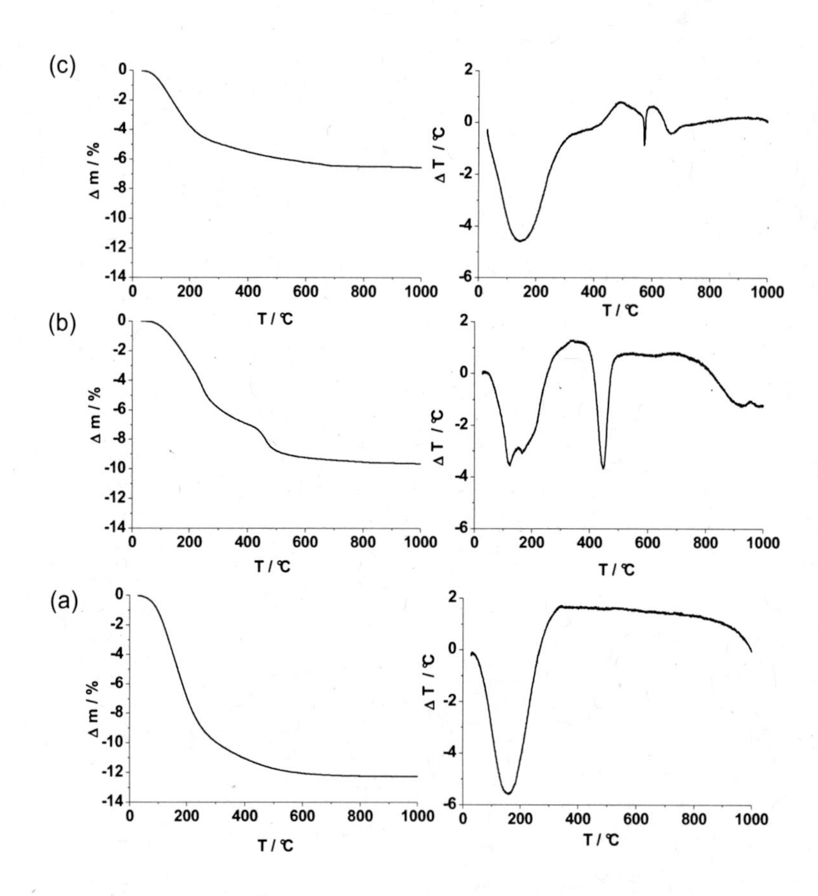

Figure 4: Thermal analysis curves of the (a) geopolymer (b) foam and (c) reinforced materials

To understand the inorganic foam formation as a function of silica fume constant, some experiments were performed (Figure 5). The F_i sample was characteristic of the first in-situ foam synthesized (Table I). The ratio of the final volume to initial volume was given to compare the influence of components depending on the molar ratio. The increase of the final volume in relation to the amount of silica fume revealed the important role of this compound for in-situ inorganic foam formation. Nevertheless, the amount of silica fume must be at least 50 wt. % to get a significantly porous material. For higher values, the volume enhancement was proportional to the silica amount and corresponded to an initial molar ratio of $n_{Si}/n_K \approx 2.5$.

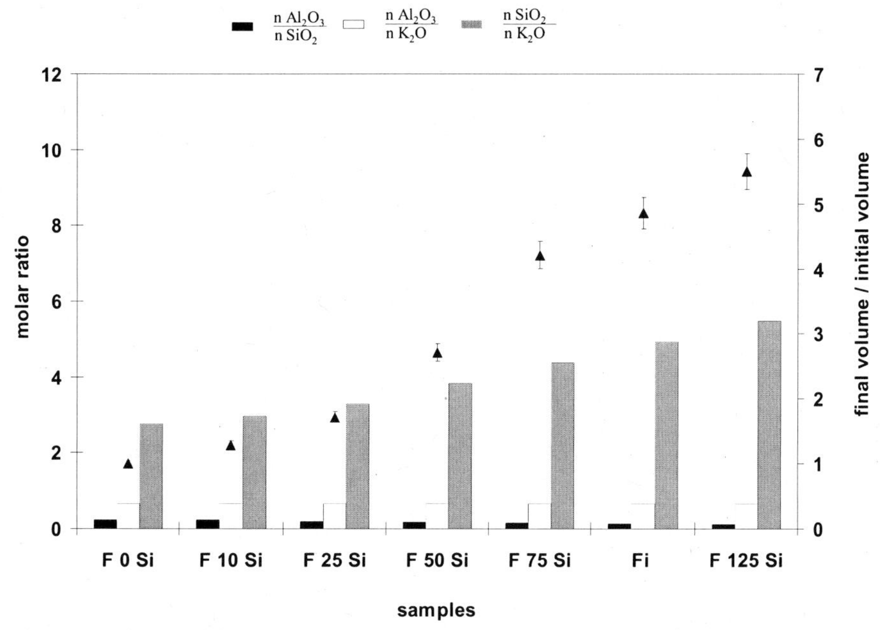

Figure 5: Ratio (▲) of final volume vs. initial volume obtained from each sample after a curing at (70°C, 6h) as a function of the various Si, Al and K molar ratios for silica fume constant amounts.

In order to understand how porosity might be controlled (Figure 6A (a, b)), investigations were performed during the drying of samples at 70°C through a period of 7 days. Only three experiments are given in Figure 7, representing the removal of sample from the mould after □ 6h, ▒ 7 days and ■ 2 days. The comparison of the values of pore size depended strongly on the time before removal of the mould. Without removal, the foam obtained displayed a large range of sizes with an average pore size at approximately 0.37 mm. Removal after 48 hrs involved a slight decrease in size, but the heterogeneous scale was displaced to the small pores. The smallest pore size of 0.13 mm was achieved by removal of the mould after 6 hrs. We deduced that choice of the drying conditions can be used to control, to some extent, the pore size.

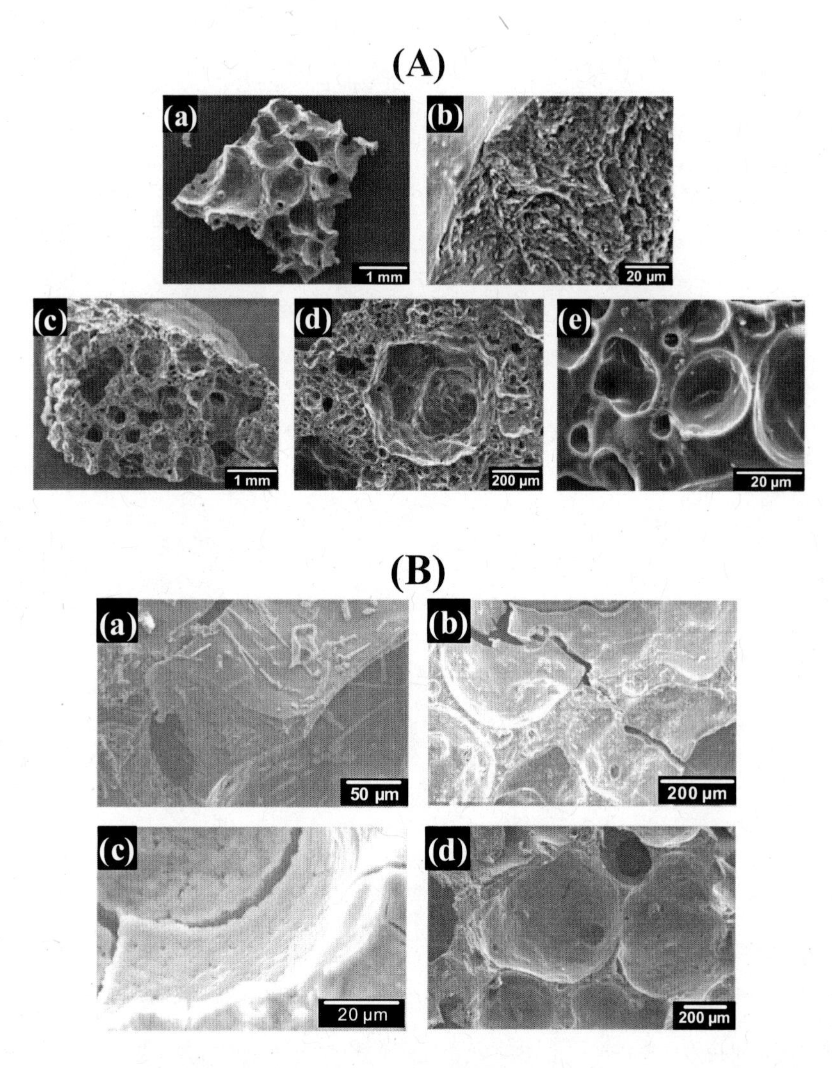

Figure 6: (A) Photos of geopolymer foam (a-b) before and (c-d-e) after curing at 750°C and (B) photos of geopolymer foam (a) fresh in contact with solution at pH = (b) 7.4. (c) 10.1 and (d) 4.5

SEM measurements confirmed the high pore volume fraction of the foam and the alveolar arrangement left behind after a thermal treatment at 750°C (Figure 6A (c, d, e)). This phenomenon could be attributed to the presence of potassium oxide. To understand more precisely the behavior of the foam under high thermal treatment, the thermal analysis was made, as presented in Figure 4(b). The two endothermic peaks, respectively at around 180 and 480°C, could be correlated to two weight losses, corresponding finally to a value close to that of the geo-material. The first peak matched to loss of free water, while the second corresponded to loss of structural water. This last point assumed the existence of a crystalline phase observed when the sample was heated at 350°C for 2 hours. This phase could be a hydrated potassium aluminum silicate such as the compound ($KAl_4Si_2O_9(OH)_3$). The most surprising was the disappearance of this phase at 750°C, from the

XRD pattern implying that it was a metastable phase. Some in-situ heating XRD experiments are in progress to clarify this phenomenon. The SEM photo at 750°C (Figure 6A (e)) also reveals the aspect of an amorphous compound.

In order to be used in building construction, these materials require stability in various aqueous environments. Foam samples were immersed in various solutions, respectively at pH = 4.6, 7.4 and 10.1 and dried before SEM study (Figure 6B). In an acid environment (Figure 6B (b, d)), the needle-shaped species observed on the surface, which could be attributed to potassium hydroxide, or carbonate, disappeared with drying time, meaning probably an excess of potassium during the synthesis. On the contrary, at basic pH levels, a deterioration of the compound was observed (Figure 6B (c)).

An explanation of inorganic in-situ foam formation can be proposed, based on (i) a production of a gas, (ii) an increase of the viscosity and (iii) a consolidation of material. The generation of porosity was probably due to the H_2 production produced by water reduction and by the oxidation of silicon in a basic medium ensuring the formation of $Si(OH)_4$ species. This fact can also be noted by the presence of the Si-OH band after 15 minutes (Figure 2 (b)) and in the XRD pattern (Figure 3 (b)) by the 2-theta shift, revealing small local order transformation in silica. Consequently, the mixture diminished in charged species involving the possibility of creating a chemical gel, which is concentrated in silica. As a result, the increase in viscosity arose from the lattice gel becoming more and more rigid where the diffusion of the gas was enhanced.

The thermal diffusivity was determined by the « laserflash » method with a value of 5.9×10^{-7} $m^2.s^{-1}$. The amorphous character of this foam implied that the nature of the bond in the sample was relatively complex and did not allow the establishment of an exact value of the heat capacity. Nevertheless the major constituents permitted an estimation of the value to be between 700 and 1000 $J.kg^{-1}.K^{-1}$. From the density of 534 $kg.m^{-3}$, a value for the thermal conductivity between 0.22 and 0.24 $W.m^{-1}.K^{-1}$ can be calculated at room temperature, classifying this material as an efficient thermal insulator material.

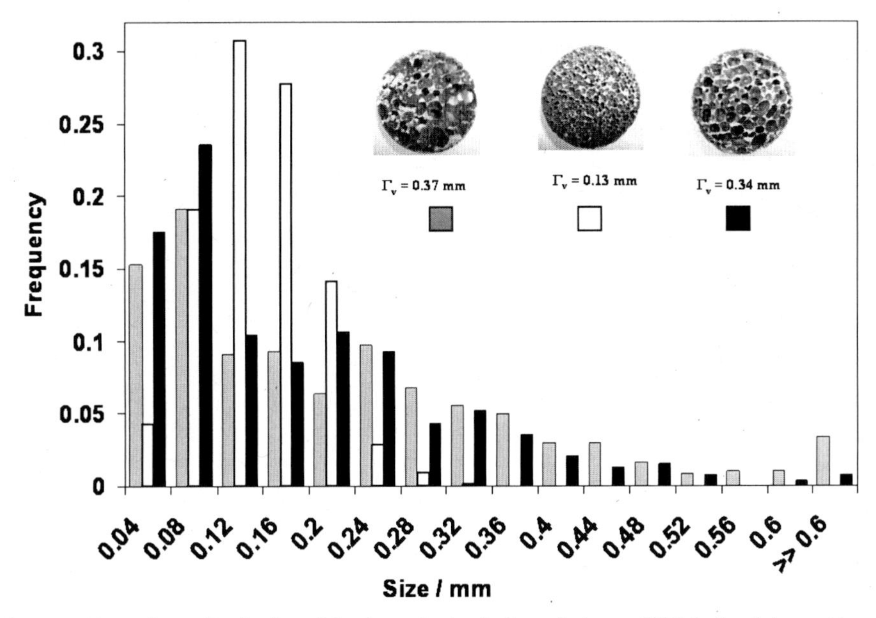

Figure 7: Pore volume distribution of the foam obtained after a drying at 70°C during 7 days with a removal of sample of his mould after □ 6h, ■ 7 days and ■ 2 days.

3) Addition of sand: reinforced material

As shown by the X-ray diffraction pattern (Figure 3 (a)), the addition of sand (80%) gave rise to a compound composed of two phases, a crystallized sand and an amorphous phase, respectively. This mixture suggested the existence of a composite material. X-ray mapping (Figure 8) revealed a homogeneous distribution of sand surrounded by a matrix phase coming from geopolymer composition, acting as a binder. The material was also studied by thermal analysis (Figure 4(c)) where the weight loss at around 6.5 % was proportional to the introduced geo-material mixture. Nevertheless, the weight loss process finished at a higher temperature (600°C) than from the geo-material due to the presence of the sand. In fact, this last compound was also responsible for the endothermic peak at around 500-600 °C in the thermal differential curve. This artifact could be due to some impurities present in the sand, which decomposed at this temperature, like clays. The thermal curves were in agreement with this hypothesis. The behavior of geopolymer and reinforced material determined by four point bending are given in Figure 9. Only, linear elastic behavior was observed, typical of the fragile material. The introduction of sand to the initial composition increased the Young's modulus value from 9.3 to 18 MPa and the rupture stress from 1.56 to 10.6 MPa. Consequently, the addition of sand improved the mechanical properties, in particular, a better rigidity of the material and a greater rupture stress. Further experiments with various amounts of sand are in progress to understand its effect on the geo-material mechanical properties.

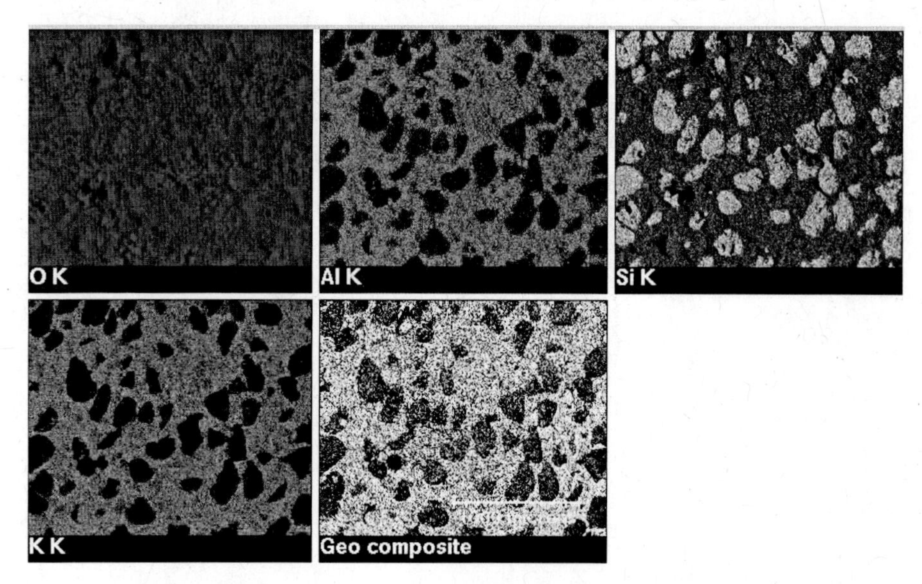

Figure 8: SEM X-ray mapping of consolidated geo-material containing sand.

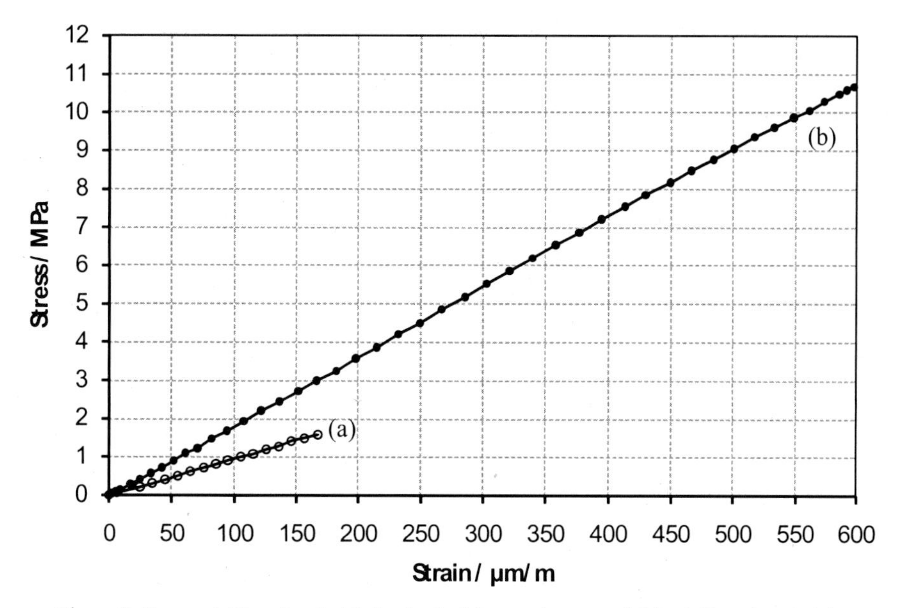

Figure 9: Four-point bending test behavior in (a) geopolymer and (b) reinforced material.

CONCLUSION

Geopolymers are materials with interesting chemical and physical properties in relation with their preparation feasibility. Their reinforcements at room temperature imply a potential application such as cements. We have achieved successfully the synthesis of various consolidated materials (geopolymer, in-situ inorganic foam, reinforced material) prepared from a mixture containing potassium silicate, potassium hydroxide and dehydroxylated kaolinite and additives such as sand or silica fume. The reaction of geopolymerization was studied by in-situ FTIR spectroscopy with particular attention to the presence of a band at 950 cm^{-1} which is characteristic of the end of the geopolymerization reaction. The amorphous nature of samples, revealed by XRD measurements, has the signature of the dissolution of raw materials leading to the presence of SiO_4 and AlO_4 species isolated on the geomaterial.

The addition of silica fume to this geo-material has involved modifications in terms of chemistry and in the porosity of the sample. The synthesized inorganic foam was characteristic of a porous material where the size can be controlled by drying and by the chemistry, notably the molar ratio of Si/Al and Si/K. This demonstrates the importance of the potassium element. Furthermore, this in-situ inorganic foam is characteristic of an insulating material.

Finally, the added sand reinforced the mechanical properties of the geo-material by a factor of 2. The compound exhibited the behavior of a composite material where the sand was coated by the geo-material acting as binder.

REFERENCES

[1] J. Davidovits, Chemistry and Applications, 19-36 (2008).
[2] H. Xu, Geopolymerisation of Aluminosilicate Minerals, *PhD Thesis*, Department of Chemical Engineering, University of Melbourne, Australia, (2001).

[3] M. W. Grutzeck and D. D. Siemer, Zeolithes Synthesised from Class F Fly Ash and Sodium Aluminate Slurry, *J. Am. Ceram. Soc.*, **80 (9)**, 2449-2458, (1997).
[4] Z. Li, Y. Zhang and X. Zhou, Short Fiber Reinforced Geopolymer Composites Manufactured by Extrusion, *Journal Materials in Civil Engineering*, vol **17(6)**, 624-631, (2005).
[5] J. P. Wu, A. R. Boccaccini, P. D. Lee and R. D. Rawlings, Thermal and Mechanical Properties of a Foamed Glass Ceramic Material Produced from Silicate Waste, *Eur. J. Glass Sci. Technol. A*, **48 (3)**, 133-141, (2007).
[6] V. Barbosa and K. Mackensie, Synthesis and Thermal Behaviour of Potassium Sialate, *Mater. Letters*, **57**, 1477-1482, (2003)
[7] P. Duxson, G. C. Lukey, and J. S. J. Van Deventer, Thermal Conductivity of Metakaolin Geopolymers Used as a First Approximation for Determining Gel Interconnectivity, *Ind. Eng. Chem. Res.*, **45**, 7781-7788, (2006).
[8] B. G. Nair, Q. Zhao and R. F. Cooper, Geopolymer Matrices with Improved Hydrothermal Corrosion Resistance for High Temperature Applications, *J. Mater. Sci*, **42**, 3083-3091, (2007).
[9] J. W. Phair, and J. S. J. Van Deventer, Effect of the Silicate Activator pH on the Microstructural Characteristics of Waste Based Geopolymers, *International Journal of Mineral Processing,* **66 (1-4)**, 121-143, (2002).
[10] A. V. Kirshner and H. Harmuth, Investigation of Geopolymer Binders with Respect to their Application for Building Materials, *Ceramics – Silikáty*, **48**, 117-120, (2004).
[11] Z. Yunsheng, S. Wei, L. Zongjin, Z. Xiangmimg and C. Chungkong, Impact Properties of Geoplymer Based Extrudated Incorporated with Fly Ash and PVA Short Fiber, *Construction and building Materials*, **22**, 370-383, (2008).
[12] A. Michot, D. S. Smith , S. Degot and C. Gault, Thermal Conductivity and Specific Heat of Kaolinite : Evolution with Thermal Treatment, *J. Euro. Ceram. Soc.*, **28**, 2639-2644, (2008).
[13] J. Davidovits, Synthetic Mineral Polymer Compound of the Silicoaluminates Family and Preparation Process, *US Patent, 4, 472, 199*, (1984).
[14] P. Innocenzi Infrared Spectroscopy of Sol-Gel Derived Silica-Based Films : a Spectra-Microstructure Overview, *Journal of Non Crystaline Solids*, **316**, 309-319, (2003).
[15] M. Criado, A. Polomo and A. Fernandez-Jiménez, Alkali Activation of Fly Ashes, Part 1 : Effect of Curing Conditions on the Carbonation of the Reaction Products, *Fuel*, **84**, 2048-2054, (2005).
[16] J. Davidovits, Scientific Tools, X-rays, FTIR, NMR, *Geopolymer : Chemistry and Applications*, 61-76, (2008).
[17] W. K. W. Lee and J. S. J. Van Deventer, Use of Infrared Spectroscopy to Study Geopolymerization of Heterogeneous Amorphous Aluminosilicate, *Langmuir*, **19**, 8726-8734, (2003).
[18] C. A. Rees, J. L. Provis, G. C. Lukey and J. S. J. Van Deventer, Attenuated Total Reflectance Fourier Transform Infrared Analysis of Fly Ash Geopolymer Gel Aging, *Langmuir*, **23**, 8170-8179, (2007).
[19] J. Davidovits, Scientific Tools, X-rays, FTIR, NMR, *Geopolymer: Chemistry and Applications*, 61-76, (2008)
[20] P. Duxson, G. C. Lukey and and J. S. J. Van Deventer, Thermal Evolution of Metakaolin Geopolymer, Part 1: Physical Evolution, *J. of Non-Crystalline Solids*, **352**, 5541-5555, (2006).
[21] T. Uchino, T. Sakka, K. Hotta and M. Iwasaki, Attenuated Total Reflectance Fourier-Transform Infrared Spectra of a Hydrated Sodium Silicate Glass, *J. Am. Ceram. Soc.* **72 (11)**, 2173-2175, (1989).

ALKALI ACTIVATED AEROGELS

Forrest Svingala, Benjamin Varela
Department of Mechanical Engineering
Rochester Institute of Technology
76 Lomb Memorial Drive
Rochester, New York 14623-5604
United States of America

ABSTRACT

Clay aerogels are unique materials formed through the sublimation drying of alumino-silicate clay hydrogels. Aerogels have been an area of increased research interest in the past decade due to their very low density, high surface area/porosity, and very low thermal conductivity. Significant efforts have been made to increase the mechanical strength and moisture resistance of these materials through the incorporation of both organic polymers and fiber reinforcement. Alumino-silicates can also be alkali activated, producing a highly-crosslinked 3d network polymer with generally excellent mechanical strength and chemical resistance, but with high density. Some efforts have been made to reduce density through the use of blowing agents, like H_2O_2 and aluminum powder. This work presents a preliminary investigation into the combination of aerogel production techniques with alkali activation, with the goal of producing a high specific strength, alkali activated aerogel. Both metakaolin and S-Type furnace slag were investigated as alumino-silicate sources. It was found that it is possible to create a stable solid material with density of approximately 1g/cc, and porosity on the order of 10-20 microns.

INTRODUCTION

Aerogels are formed from conventional liquid-solid gels by replacing the liquid component with a gas, without allowing shrinkage to occur. In clay aerogels, this is typically accomplished by a freeze drying process. Clay aerogels have recently enjoyed increased research interest due to their unique properties. Their low density and low thermal conductivity lends these clay aerogels great potential as a petroleum-free alternative to many organic polymer foams, especially in packaging and insulation applications. Pure clay aerogels have bulk densities near 0.1 g/cc, with upwards of 90% porosity [1]. These aerogels typically have extremely low strength (less than 2 KPa compressive strength), and readily reconvert to a liquid-solid gel upon contact with bulk liquids [2]. To improve the properties of these aerogels, most past research has centered on the addition of organic polymers [1] [3], or the addition of organic fibers [2]. With the addition of both PVOH and silk fibers, D. Schiraldi [2] was able to obtain a composite aerogel with compressive strength of 108 KPa, and a compressive modulus of 8.8 MPa.

Geopolymeric materials have long been shown to have high early compressive strength and excellent physical/chemical environmental resistance. However, most geopolymers have relatively high density making them less than ideal in many situations. Past work by E. Liefke [4] produced a low density (.1-.8 g/cc), low thermal conductivity geopolymer foam through the addition of blowing agents like hydrogen peroxide and sodium perborate. These blowing agents reduce the density of the foam by releasing oxygen gas during the gel phase of geopolymerization, ultimately creating large pores with 0.5 mm-3 mm diameter. This TROLIT foam has a very high application temperature (1000°C) and retains good compressive strength (0.5 MPa to 2 MPa). However, pore size can be difficult to control, and careful control of the mix is required.

J. L. Bell and J. W. Kriven [5] have demonstrated the creation of geopolymer foams through the use of hydrogen peroxide or metallic Al powder. In alkaline environments, Al powder reacts with hydroxide ions and releases hydrogen gas, producing pores and reducing bulk density. The goal of these authors was the creation of a percolating network of interconnected porosities, allowing conversion of the hardened geopolymer to a ceramic without cracking due to buildup of internal steam pressure. Through the use of a constant-volume mold, these authors were able to obtain much smaller and more consistently sized pores than are produced by the TROLIT process. Using between 0.5%wt and 1.5%wt hydrogen peroxide, bulk densities between 1.23 g/cc and 1.09 g/cc were produced, with compressive strength greater than 49 MPa. Use of 60 %wt Al powder produced a foam with bulk density of 1.56 g/cc, with an interconnected network of pores and unreported compressive strength.

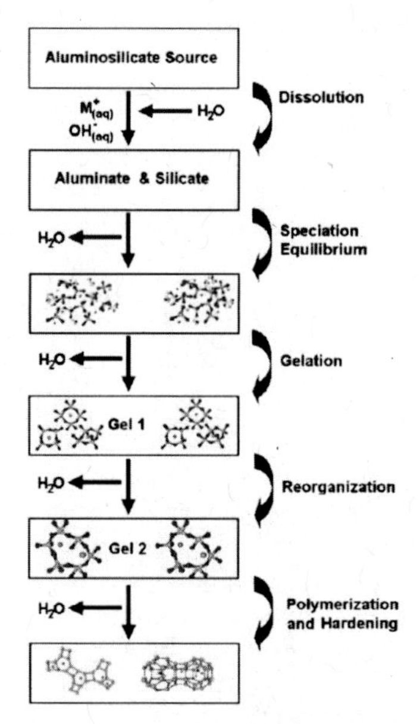

Figure 1: Simplified steps in the geopolymerization reaction (from J. L. Provis, et. al. [7])

The basic steps in geopolymerization are shown in Figure 1. Initially, a source of alumino-silicate is dissolved in a highly alkaline environment. As dissolution proceeds, a complex mixture of aluminate and silicate species is produced. These dissolved alumino-silicate species begin to condense into a loose network, rejecting water and organizing into the first of two gel states. As the reaction proceeds, the degree of polycondensation increases, forming a second gel state consisting of a highly interconnected, 3D, alumino-silicate, polymer network, surrounding pores of water rejected by the condensation process. Systems with the appropriate composition can continue to react and cure, producing the final hardened geopolymer.

The production of the clay aerogel and the foamed geopolymer both rely on alumino-silicate materials. The present work is a preliminary investigation into a combination of these two production techniques, with the goal of creating an alkali activated aerogel. A composition is selected that is observed to gel, but lacks sufficient alumino-silicate concentration to harden under ambient conditions. After a maturation time, this gel is then processed in the manner of an aerogel: frozen in an alcohol-dry ice bath, and freeze dried to remove excess water. This creates a material with similar bulk density to the TROLIT foams, but with a much smaller and more consistent pore size of 10-20 microns.

SAMPLE PREPERATION

A commercially available metakaolin and S-type ground granulated blast furnace slag were used as aluminosilicate sources. The metakaolin was supplied by Engelhard Corporation (Iselin, New jersey) while the slag was provided by St.Marys Cement (Ontario, Canada).

The activating solution was prepared by mixing 66.6 g of STARSO sodium silicate solution (PQ Corporation) and 33.4 g of a 15 molal NaOH solution. The NaOH solution was prepared by dissolving reagent grade NaOH pellets (Fisher Scientific) in water. Sodium aluminate (Fisher Scientific) was used as a source to increase the amount of alumina in samples where slag was used.The mass composition in terms of oxides according to the supplier's information is presented in Table 1.

Table 1. Chemical composition of raw materials in mass percentages.

	SiO_2	Al_2O_3	Na_2O	K_2O	CaO	H_2O
Metakaolin	53	43.8	0.23	0.19	0.02	--
Slag	39	7.7	0.28	0.45	39	--
Sodium Silicate	24	--	13.5	--	--	62.5
Sodium Aluminate	--	53	43	--	--	--

The activating solution was prepared immediately prior to use. 5 g of aluminosilicate were added to 100 g of the activating solution and stirred in a high shear impeller type mixer, until an increase in viscosity was observed. In this work, samples based on metakaolin are referred to as M series while those based on slag are referred to as S series. In cases where sodium aluminate was added, the sample name bears the suffix "+ Al". Table 2 presents the theoretical molar ratios for the mixtures used in this study.

Table 2. Molar ratios of samples

Sample	**SiO_2/Al_2O**	**$(Na_2O$ +**	**$H_2O/(Na_2O$ +**	**$(Na_2O + CaO)/Al_2O_3$**
M-Series	14.4	0.97	12.03	13.98
S-Series	79.18	1.12	10.77	89.18
S+Al	10.04	1.24	9.77	12.47

Once a viscous slurry was obtained, it was cast into plastic molds and cured for 48 hours under conditions generally reported in the geopolymer literature The curing conditions studied were room temperature, oven cure at 65°C and moist cure at 65°C.

After curing, the samples were deep frozen in a dry ice/alcohol bath for 2 hours. The samples that were cured at elevated temperature were left to cool to room temperature in order to avoid thermal shock. The last step consisted in placing the samples in a vacuum chamber for 72 hours.

Initially, the chamber was held at -20°C while the samples were loaded and the chamber evacuated. Once the vacuum reached a 1 Torr of pressure, the chamber was allowed to warm to room temperature for the rest of the cycle.

Sublimation Drying Apparatus

Sublimation drying was performed in a custom-built dryer consisting of an eight liter vacuum chamber mounted in a commercial chest freezer, a dry ice cold trap, solvent filter, and rotary vane vacuum pump (Fig 2). This system has an ultimate vacuum pressure of approximately $2x10^{-2}$ Torr, and the ability to control chamber temperatures between 0°C and -25°C. This dryer allows for greater possibilities in sample size and geometry, at a lower cost than commercially available freeze dryers.

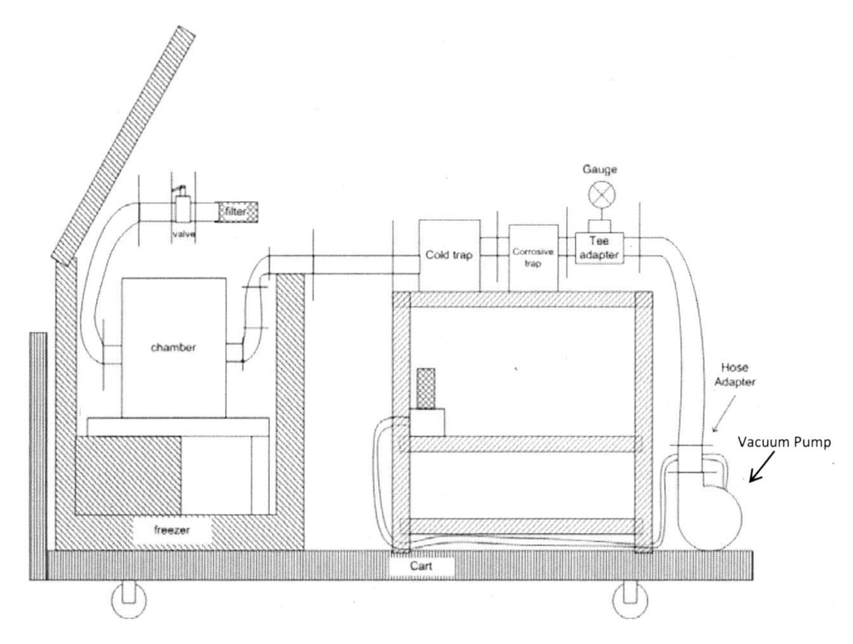

Figure 2: Sublimation dryer schematic diagram

RESULTS

In processing the samples, it was found that the M-series and S-series samples required different mix times to produce a viscous fluid. S-series samples were found to form a viscous fluid after a mix time of approximately 5 minutes, and to completely gel in approximately ten minutes. M-series samples failed to form a viscous liquid after a 5 minute mix, and attempts to cure at this point resulted in separation of the metakaolin and activator solution, rather than a stable gel. As a result, mix time for the M-series was increased until the sample was observed to form a viscous fluid, resulting in a final mix time of 30 minutes. The decreased reaction time of the S-series was consistent with the high CaO content of the furnace slag. CaO has been found to accelerate the setting, and increase the compressive strength of geopolymer systems [8].

Sample condition after cure

The condition of samples prior to drying was strongly influenced by the curing method used. Furnace cured samples formed a glassy, semi-rigid gel, with negligible mechanical strength. Precipitation of a solid phase was also observed. These samples also experienced significant shrinkage due to evaporation. Moist cured samples maintained their initial volume after curing, and otherwise resembled the furnace cured samples. Samples cured at room conditions formed a dull semi-solid, with appreciable rigidity, and slight shrinkage due to evaporation.

Sample condition after sublimation dry

After the drying stage all samples were found to have lost 15%-20% of their mass due to water loss. Final density was typically between 1 g/cc and 1.3 g/cc. Samples cured at room temperature formed homogeneous solids, with pore size observed by SEM to be approximately 10-20 microns. Furnace cured samples were sticky on the surface with unreacted silicates, with generally higher densities due to partial evaporation collapse of the gel during curing. On the surface, moist cured samples appeared hard and dry, often exhibiting significant efflorescence. However, the interior of the samples often remained somewhat moist, retaining properties similar to the pre-dry state. Extended drying times did not appear to completely dry the interior of these samples. It is possible that after a critical drying time, efflorescence or other factors create a layer of material that is difficult for water vapor to penetrate, preventing further removal of water.

Compressive Strength

Compressive testing was performed on quartered cylindrical samples with nominal size of 1.125" in diameter and 0.350" thick. All samples tested were matured under ambient conditions for 12-24 hours prior to freeze drying. An Instron compressive tester with 1000lb load cell was used, with a strain rate of 0.02 in/min. The average results over 10 samples are presented below in Table 3.

Table 3: Average compressive strength, density, and specific strength for N=10

Sample	Average Density, g/cm^3	Average Compressive Strength, MPa	Average Specific Strength, m^2/s^2
S-series	1.25	4.67	3.74E+03
M-series	1.03	2.93	2.85E+03
S+Al series	1.03	9.55	9.25E+03

SEM characterization

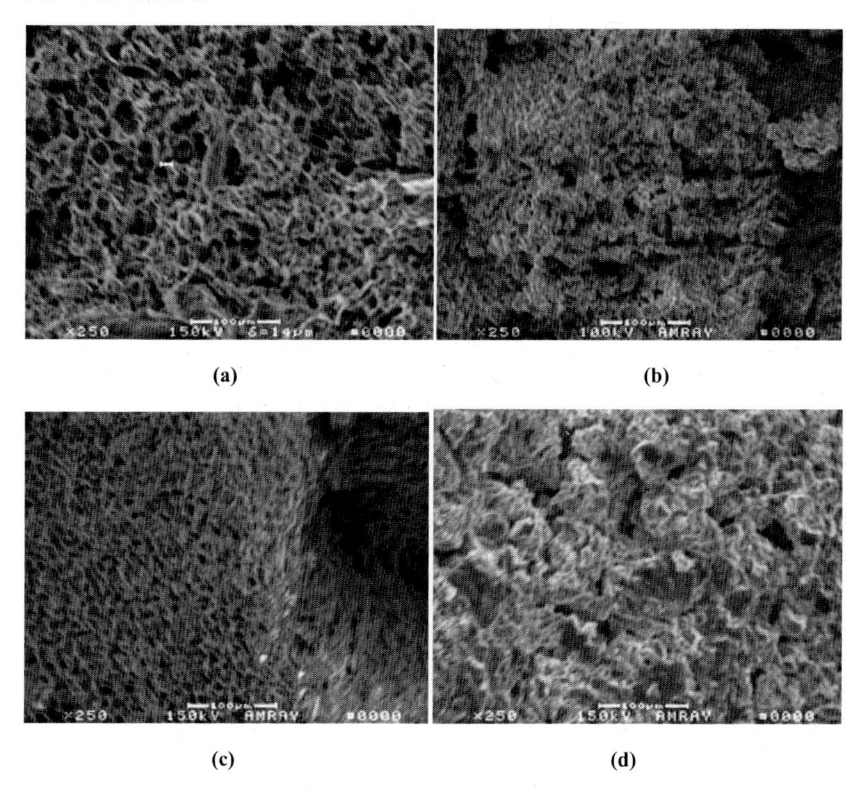

Figure 3: Scanning electron micrographs of samples; (a) MetaMax, ambient cure; (b) MetaMax, moist cure;(c) S-type slag and sodium aluminate , moist cure; (d) S-type slag, moist cure

All samples showed a similar fibrous microstructure, with a high number of pores. These pores ranged from approximately 10 to 20 microns in diameter. The curing method did not appear to have a significant influence on the final microstructure. The extremely high Si:Al ratio of the S-type slag sample has created a more disordered microstructure, with fewer and larger pores (Fig 3(d)). As the Si:Al ratio is reduced, the microstructure became more ordered, with a greater number of smaller pores, regularly distributed throughout the sample (Fig 3C). In clay/water aerogels, the clay particles are aligned into parallel plates by growth of ice crystals during the freezing process, as seen in Figure 4. This effect was not observed in our alkali activated clay samples, as the metakaolin particles are completely dissolved before the gel precursor is formed, eliminating their plate-like morphology.

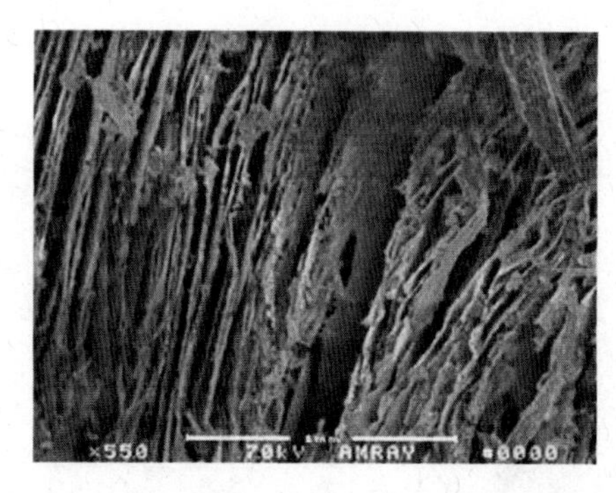

Figure 4: A clay aerogel, formed from a clay-water gel

CONCLUSIONS

As discussed in the geopolymer literature, the molar ratios of the mixtures used for this study are beyond the recommended ranges to synthesize a solid material with high compressive strength. The ratio of liquid to solid components was too high, the reaction is diffuse and no condensation occurs. It was observed that when these mixtures were left to mature at room temperature and cured at 65°C no condensation reaction occurs and no solid material was obtained. When these mixtures were vigorously mixed in a mechanical mixer an increased in viscosity followed by the formation of a gel was observed. This gel became stiffer with time until it became rigid. It was observed that this gel was stable at room temperature but undergoes phase separation when heated.

When the gel was freeze dried and the water was removed by sublimation, as in aerogel processing, the gel changed into a stable solid material with densities between 1 and 1.3 g/cc. SEM micrographs showed a fibrous structure with porosities in the range of 10 - 20 microns.

Compressive testing results show greater strength and specific strength in the furnace slag system. This is likely due to the greater reactivity of the furnace slag, relative to the metakaolin[8]. The addition of sodium aluminate is shown to greatly increase compressive strength of the slag samples, owing to the increase in available reaction sites. Manipulation of the Si:Al ratio may allow production of samples with even greater strength.

This work is a first attempt to combine alkali activation and aerogel synthesis to produce a low density foamed material with mechanical strength. Further work must be done to more extensively evaluate this material, especially mechanical and thermal stability, and to optimize the formulation and processing.

ACKNOWLEDGMENTS

The authors would like to thank the Center for Layered Polymeric Structures at Case Western Reserve University for funding and support of this project. This material is based upon work supported by the National Science Foundation under Award No. DMR-0423914

REFERENCES

[1] Arndt, E.M.; Gawryla, M.D.; and Schiraldi, D.A. "Elastic, low density epoxy/clay aerogel composites" *Journal of Materials Chemistry* Vol. 17 pp. 3525-2529, 2007

[2] Finlay, K.; Gawryla, M.D.; Schiraldi, D.A. "Biologically based fiber-reinforced/clay aerogel composites" *Ind. Eng. Chem. Res.* Vol. 47 pp. 615-619, 2008

[3] Bandi, S.; Schiraldi, D. A. "Glass Transition Behavior of Clay Aerogel/Poly(vinyl alcohol) Composites" *Macromolecules* Vol. 39 pp 6537-6545, 2006

[4] Liefke, E.; "Industrial Applications of Foamed Inorganic Polymers" Geopolymer '99 Proceedings, pp 189-199, 1999

[5] Bell, J. L.; Kriven, W.M.; "Preparation of Ceramic Foams From Metakaolin-Based Geopolymer Gels" In Developments in Strategic Materials. Edited by Hua-Tay Lin, Kunihito Koumoto, Cer. Eng. Sci Proc. Vol 29 [10] pp. 97-112, 2008

[6] Provis, J. L.; van Deventer, J. S. Js; "Geopolymerisation Kinetics 1. In-situ Energy Dispersive X-ray Diffractometry" *Chemical Engineeing Science* Vol. 62, pp 2309-2317, 2007

[7] Duxsun, P.; Fernandez-Jimenez, A.; Provis, J.L.; Lukey, G.C.; Palomo, A.; van Deventer, J.S.J. "Geopolymer Technology: The current state of the art" *Journal of Material Science* Vol. 42, pp. 2917-2933, 2007

[8]] Davidovits, J. Geopolymer Chemistry and Applications 2nd Ed., Institut Geopolymere, St. Quentin France, 2008 [7] Hofmann, U. *Nature* Vol. 171 pp. 682, 1953

Author Index